Emerging Trends in Open Source Geographic Information Systems

Naveenchandra N. Srivastava
Bhaskaracharya Institute for Space Applications and Geo-Informatics, India

A volume in the Advances in Geospatial Technologies (AGT) Book Series

IGI Global
DISSEMINATOR OF KNOWLEDGE

Published in the United States of America by
IGI Global
Engineering Science Reference (an imprint of IGI Global)
701 E. Chocolate Avenue
Hershey PA, USA 17033
Tel: 717-533-8845
Fax: 717-533-8661
E-mail: cust@igi-global.com
Web site: http://www.igi-global.com

Library of Congress Cataloging-in-Publication Data

Names: Srivastava, Naveenchandra N., 1971- editor.
Title: Emerging trends in open source geographic information systems / Naveenchandra N. Srivastava, editor.
Description: Hershey PA : Engineering Science Reference, 2018. | Includes bibliographical references.
Identifiers: LCCN 2017034169| ISBN 9781522550396 (hardcover) | ISBN 9781522550402 (ebook)
Subjects: LCSH: Geographic information systems. | Open source software.
Classification: LCC G70.212 .E45 2018 | DDC 910.285--dc23 LC record available at https://lccn.loc.gov/2017034169

This book is published in the IGI Global book series Advances in Geospatial Technologies (AGT) (ISSN: 2327-5715; eISSN: 2327-5723)

British Cataloguing in Publication Data
A Cataloguing in Publication record for this book is available from the British Library.

For electronic access to this publication, please contact: eresources@igi-global.com.

Advances in Geospatial Technologies (AGT) Book Series

ISSN:2327-5715
EISSN:2327-5723

Editor-in-Chief: Nilanjan Dey, Techno India College of Technology, India

MISSION

The geospatial technology field is a fast-paced, high growth industry that is involved in a variety of fields, including military planning, public health, land use, environmental protection, and Google Earth mapping. With such a diverse body of applications, the research in geospatial technologies is always evolving and new theories, methodologies, tools, and applications are being developed. **Advances in Geospatial Technologies (AGT) Book Series** is a reference source and outlet for research that discusses all aspects of geographic information, including areas such as geomatics, geodesy, GIS, cartography, remote sensing, and other areas. Because geospatial technologies are so pervasive in such a variety of areas, AGT also includes books that address interdisciplinary applications of the technologies.

COVERAGE

- Global Positioning Systems
- Public Health Applications of Geospatial Technologies
- Geographical information systems
- Environmental Monitoring
- Geovisualization
- Digital Terrain Modeling
- Surveying Technologies
- Disaster Informatics
- Military Security and Geospatial Technologies
- Spatial Reference Systems

The Advances in Geospatial Technologies (AGT) Book Series (ISSN 2327-5715) is published by IGI Global, 701 E. Chocolate Avenue, Hershey, PA 17033-1240, USA, www.igi-global.com. This series is composed of titles available for purchase individually; each title is edited to be contextually exclusive from any other title within the series. For pricing and ordering information please visit http://www.igi-global.com/book-series/advances-geospatial-technologies/73686. Postmaster: Send all address changes to above address.

Titles in this Series

For a list of additional titles in this series, please visit:
https://www.igi-global.com/book-series/advances-geospatial-technologies/73686

Geospatial Technologies in Urban System Development Emerging Research and...
Alok Bhushan Mukherjee (Birla Institute of Technology, India) Akhouri Pramod Krishna (Birla Institute of Technology, India) and Nilanchal Patel (Birla Institute of Technoloy, India)
Engineering Science Reference • ©2018 • 179pp • H/C (ISBN: 9781522536833) • US $145.00

Spatial Analysis Techniques Using MyGeoffice®
João Garrott Marques Negreiros (University of Saint Joseph, China & Macau Polytechnic Institue, China)
Engineering Science Reference • ©2018 • 336pp • H/C (ISBN: 9781522532705) • US $225.00

Handbook of Research on Geospatial Science and Technologies
Joyce Gosata Maphanyane (University of Botswana, Botswana) Read Brown Mthanganyika Mapeo (University of Botswana, Botswana) and Modupe O. Akinola (University of Lagos, Nigeria)
Engineering Science Reference • ©2018 • 457pp • H/C (ISBN: 9781522534402) • US $345.00

Emerging GIS Applications for Emergency and Disaster Management
Shruti Mantri (Kirti College, India) and Seema Purohit (Kirti College, India)
Information Science Reference • ©2017 • 117pp • H/C (ISBN: 9781683180098) • US $165.00

Volunteered Geographic Information and the Future of Geospatial Data
Cláudio Elízio Calazans Campelo (Federal University of Campina Grande, Brazil) Michela Bertolotto (University College Dublin, Ireland) and Padraig Corcoran (Cardiff University, UK)
Information Science Reference • ©2017 • 357pp • H/C (ISBN: 9781522524465) • US $205.00

Remote Sensing Techniques and GIS Applications in Earth and Environmental Studies
Abhisek Santra (Haldia Institute of Technology, India) and Shreyashi Santra Mitra (Haldia Institute of Technology, India)
Information Science Reference • ©2017 • 335pp • H/C (ISBN: 9781522518143) • US $185.00

For an entire list of titles in this series, please visit:
https://www.igi-global.com/book-series/advances-geospatial-technologies/73686

IGI Global
DISSEMINATOR OF KNOWLEDGE

701 East Chocolate Avenue, Hershey, PA 17033, USA
Tel: 717-533-8845 x100 • Fax: 717-533-8661
E-Mail: cust@igi-global.com • www.igi-global.com

Table of Contents

Detailed Table of Contents

The availability of new mobile devices (tablets and smartphones) equipped with many sensors is changing or, better, enriching the way we monitor and sense the world that surrounds us. The internet has permeated completely not only our scientific and technological development, but also our life. Only some years ago, we used geospatial data and GIS software installed within our computers. Nowadays, data and operators are provided via the net by means of distributed and shared geo-services and a simple and powerless mobile device is enough to connect them. The possibility of interaction has become not only faster and more user friendly but also active, being individuals and communities free of adding, deleting, and changing contents in real time in the new GeoWeb2.0. This chapter explores GeoWeb2.0.

There has been a significant development in the area of free and open source geospatial software. Research has flourished over the decades from vendor-dependent software to open source software where researchers are paying increasing attention to maximize the value of their data. It is often a difficult task to choose particular open source GIS (OGIS) software among a number of emerging OGIS software. It is important to characterise the projects according to some unified criteria. Each software has certain advantages and disadvantages and it is always time consuming to identify exactly which software to select for a specific purpose. This chapter focuses on the assessment criteria enabling developers, researchers, and GIS users to select suitable OGIS software to meet their requirements for analysis and design of geospatial application in multidisciplinary fields. This chapter highlights the importance of assessment criteria, followed by an explanation of each criteria and their significance with examples from existing OGIS software.

For new generations of citizens in all countries, a level of proficiency in geospatial concepts and skills will be required to realize the potential of professional and developmental opportunities. The teaching of geospatial skills links into traditional science, technology, engineering, and mathematics (STEM) curriculum objectives, community-wide concerns and initiatives, and global citizenship. Therefore, by the pre-university and undergraduate level, it is desirable for each student to have acquired such competencies. Free and open-source tools that are accessible and affordable in most areas of the world, along with data availability, offer an opportunity to support teaching such a curriculum. Here, core geospatial concepts are introduced, along with available data and tools. Then, using three scenarios, it is shown how the core concepts can be applied to different settings for educational purposes.

Everyone involved in geospatial information systems has heard of Environmental Systems Research Institute (Esri)company. Most people have tried ArcGIS software at least for one time. Esri has been the pioneer in this industry for a long time; it also defines the future of geospatial trends. In this chapter, the author adopts a neutral position to discuss the relation of Esri and open-source community. They cover almost every aspect where Esri and open source co-occur. Even the latest achievement of Esri, which is R-bridge, is discussed here. Going into the details of everything is not the goal of the chapter; however, a minimum description is provided for each section. Proper references are given to the reader for further study.

Data mining is one of the current vital fields for all types of data including spatial data. An example of useful extracted patterns from spatial data is to describe changes in metropolitan poverty rates based on city distances from major highways. Geospatial is a term that has recently been gaining in popularity; moreover, many applications on geospatial have different uses in different fields such as geo-tagging, geospatial technology, and geo-fencing. Analyzing spatial data is considered a complex task due to its details as it is related to locations with a special representation such as longitude and latitude. Other attributes are involved in the description of objects which can be analyzed using different data mining techniques. In this chapter, a demonstration of the basic information is performed considering spatial data and spatial data mining including all aspects such as the different type of data, different methods of analysis, different mining techniques, and other related topics.

Water resources management is dependent on knowledge and understanding of water quantity and quality information with the latest developments in information technology such as geographic information system (GIS) to develop effective hydrological modeling within the water-based systems. The efficiency of such hydrological modeling relies on the accuracy of applied data. In this way, the application of low-quality data in developing models for integrated management of water resources can impose irreparable financial and human resources and environmental costs in the catchment area. Thus, in regions that shortage of data is the issue, semi-distributed modeling is a useful tool. In this chapter, three aims are followed: (1) effect of digital elevation model (DEM) type and resolution on extracted hydro-geomorphologic parameters, (2) effect of wavelet-based de-noising method on extracted hydro-geomorphologic parameters, (3) determination of the optimal cell size to extract topographic attributes with good agreement to the real features.

Most web and mobile applications are based on searching the location-based objects called spatial objects. In spatial database systems, searching such objects is a challenging task since it deals with geo-spatial capabilities. Sometimes, the spatial queries are associated with text information in order to obtain the most relevant answers nearest to the given location. Such queries are called spatial textual query. Conventional spatial indexes and text indexes are not suitable for resolving such queries. Since these indexes use various approaches to perform searching, they can cause performance degradation. Effective processing of the query mainly depends on the index structure, searching algorithms, and location-based ranking. This chapter reviews the different hybrid index structures and search mechanisms to extract the spatial objects, the different ranking model it supports, and the performance characteristics.

Rivers represent one of the sources of freshwater. The existence and survival of humanity is intricately connected with the availability of freshwater resources. It is valuable to provide comprehensive information of all water resources through online geo-visual interfaces. Specific web slices can be of use, such as the illustration of headwater sources along with details of streams of various orders and tributaries up to the point of merging with the sea downstream. Since the local self-government departments, Grama Panchayat, can play a crucial role, they too are eager to know the details of the drainage network flowing within each administration unit. The possibilities emerging out of integration of free and open source GIS is enormous and can eventually succeed in bringing forth a ray of hope to the forlorn riverine ecosystem.

Foreword

The author and I belong to the same community (i.e., GIS). The frequently upgrading itself science, giving a bird's eye synoptic view of the entire land.

Analysing data has been rendered easy via Open Source Geographic Information Systems which is prevalent on the servers that provide online services on the Internet.

Free software—open source software—has become the backbone of modern GIS. It runs on your phone, on your laptop and desktop computers, and is embedded in microcontrollers for appliances, automobiles, industrial machinery and countless other devices. This renders the previously complicated and expensive GIS Processing easy and cost effective.

How it is rendered easy and cost effective-is explained by the various contributors/ authors of *Emerging Trends in Open Source Geographic Information Systems.*

Madhulika Singh
Amity University, India

Foreword

From five decades of satellite technology to prolific use of smartphones, we are intentionally and unintentionally amassing geospatial data that continue to influence our lifestyles in more ways than we recognize. Geographic Information Systems (GIS), that provide the framework to acquire, search, query, extract, analyze these geospatial data and to extract meaningful relationships, patterns, and trends for decision making, are constantly improving with enhanced user-defined functionalities. Whereas the traditional analysis of satellite data in conjunct with field data continues to guide major decisions related to natural resource exploration an management, real-time analysis of crowd-sourced data using open-source analytical tools that operate in the cloud are driving our everyday decisions. What we buy, how we travel, how we plan our day, and even how we react is manipulated by the current data-driven technology. Therefore, this book on 'Emerging Trends in Open Source Geographic Information Systems' is both timely and exciting.

In a world of global citizenship, embracing an open-source trend is arguably the only logical way forward. Web-enabled open-source software tools harness the collective wisdom of many individuals, and in turn make analytical tools globally accessible. New trends in cloud-computing are also changing ways we approach geospatial data access, management, analysis, and visualization. There is growing evidence that for a majority of users the new technologies are more affordable, accessible, and usable, and we will continue to see a rise in their development and use.

Editor Naveenchandra N. Srivastava has a strong academic background in the GIS science. He has first-hand experience in applying the technology in industry setting as well as in teaching the topic in a university environment. This broad experience reflects well in the topics that he chooses to include in this book. The chapters present an excellent mix of finding free or low-cost geospatial data; selecting open-source software to extract and manipulate data; querying large volumes of data; adding value to archived maps; case studies especially for hydrological applications; as well as resources of training the next generation of a geospatially savvy workforce.

Whether you are a GIS educator or practitioner, you will find something new to learn from this book.

Anupma Prakash
University of Alaska – Fairbanks, USA

Preface

This book aims to showcase new concepts, thoughts, ideas, and methods in the field. It presents collections of informative and relevant chapters concerning open source GIS which is a continuously evolving area. The idea is to encourage the use of open source GIS software as a backbone in various industrial and research projects with the aim of providing better solutions to manifold challenges that the open source geospatial community is facing today. The ten chapters within this book intend to summarize recent progress and identify key research issues concerning applications and development of open source GIS tools. The scope of this book is to illustrate the key issues and challenges concerning open source GIS.

In Chapter 1, the authors give an overview of geospatial data and operators provided via the net by means of distributed and shared geo-services. They explore the possibility of active interaction through geographic information systems and geo-services. Current GIS software provides faster and easier communication as well as promotes collaborative communities within the realm of GeoWeb 2.0.

In Chapter 2, the author examines assessment criteria enabling developers, researchers, and GIS users to select suitable OGIS software to meet their requirements for analysis and design of geospatial applications in multidisciplinary fields. This chapter highlights the importance of assessment criteria, followed by an explanation of each criteria and their significance with examples from existing OGIS software.

In Chapter 3, the authors introduce core geospatial concepts along with accessible and affordable data and tools that support teaching geospatial development and skills. Additionally, using three scenarios, the chapter explores how the core aspects can be applied to different settings for educational purposes. Finally, the authors examine the importance of geospatial concepts and skills in professional development as well as in different communities and in various fields of study.

In Chapter 4, the author adopts a neutral position to discuss the relation of Esri and the open source community. The chapter covers multiple aspects of Esri and open source software, including R-bridge. Additionally, the author touches on further topics of open source software, including geoportal servers and Geopackage, providing descriptions as well as references for further study.

In Chapter 5, the authors focus on spatial data and spatial data mining, including aspects such as the different types of data, different methods of analysis, different mining techniques, and other related topics. The chapter explores various approaches to analyzing spatial data in relation to longitude and latitude, as well as other attributes involved in the description of objects through different data mining techniques.

In Chapter 6, the authors study three areas of geographic information systems in water resources management, including the effect of digital elevation model and resolution on extracted hydro-geomorphologic parameters, the effect of the wavelet-based de-noising method on extracted hydro-geomorphologic parameters, and the determination of the optimal cell size to extract topographic attributes with agreement to features. The chapter explores the role of semi-distributed modeling in providing quantity and quality information within water-based systems.

In Chapter 7, the authors review the different hybrid index structures and search mechanisms to extract spatial objects, the different ranking models they support, and the performance characteristics. The chapter highlights techniques in spatial database systems and emerging approaches to using geospatial capabilities to improve textual query.

In Chapter 8, the authors examine the possibilities emerging out of the integration of free and open source geographic information systems and their role in improving riverine ecosystems. The chapter highlights the use of geographic information systems in providing comprehensive information of all water resources through online geo-visual interfaces.

In Chapter 9, the authors review the methods of how the value of printed or hard copy maps can be enhanced by adding QR codes with multimedia content placed on the cloud attached to it. The chapter examines the use of QR codes in the creation of inventories, product tracking, and document management as they can provide a larger storage capacity and faster readability.

I hope that readers will find the chapters of this book interesting. This book should serve as a helpful reference for the open source GIS community.

Naveenchandra N. Srivastava
Bhaskaracharya Institute for Space Applications and Geo-Informatics, India

Chapter 1
Free and Open Source Tools for Volunteer Geographic Information and Geo-Crowdsourcing

Maria Antonia Brovelli
Politecnico di Milano, Italy

Blagoj Delipetrev
University Goce Delcev, Macedonia

Giorgio Zamboni
Politecnico di Milano, Italy

ABSTRACT

The availability of new mobile devices (tablets and smartphones) equipped with many sensors is changing or, better, enriching the way we monitor and sense the world that surrounds us. The internet has permeated completely not only our scientific and technological development, but also our life. Only some years ago, we used geospatial data and GIS software installed within our computers. Nowadays, data and operators are provided via the net by means of distributed and shared geo-services and a simple and powerless mobile device is enough to connect them. The possibility of interaction has become not only faster and more user friendly but also active, being individuals and communities free of adding, deleting, and changing contents in real time in the new GeoWeb2.0. This chapter explores GeoWeb2.0.

DOI: 10.4018/978-1-5225-5039-6.ch001

INTRODUCTION

Geoweb2.0, Citizen Science, and VGI

Web 2.0 has been rapidly changing the way in which geographic data are served and consumed: information sharing, participation and collaboration among users are concepts which can describe the Web 2.0. In this frame people are given the possibility of mashing-up geographic contents and tools, i.e., combining data and functionalities from different sources, in ways original designers had not considered before (Nascimento & Raghavan 2008). And at the same time applications also are becoming distributed, geo-computations will move from local computers to distributed facilities operated by computational utilities and we live in the era of cloud computing, as we will see in the third paragraph.

A simple consideration of the success in using geospatial facilities, like for instance those provided by Google (e.g.: Google Maps and Earth), demonstrates the dramatic curiosity and active interest of people in the space where they live. Some authors emphasize this interest of people in geographic information, calling it the "GIS wikification" (Sui, 2008). Lowering barriers to contributing, making easier the interaction and the possibility of participation will result in more photos, maps and other pieces of information uploaded and therefore plays an important role in the success of usage of geography. Several million people are contributing to the growing of this process, called the Geospatial Web (or Geoaware Web) or simply Geo Web (Herring, 1994; McKee, 2004; Scharl & Tochtermann 2007). Some researchers describe the phenomenon saying that citizens act as volunteer sensors creating a global and collaborative patchwork of geographic information (Goodchild 2007a).

Many lay people have been approaching environmental and scientific researches with noteworthy contributions. This gave birth to a new discipline, called Citizen Science. To be more precise, Citizen Science, is not really a new discipline, because people have been contributing for many years to scientific researches. It is more correct to say that it is a new name given to an already real practice. As a term, it was added only in July 2014 to the Oxford English Dictionary with the following definition: "scientific work undertaken by members of the general public, often in collaboration with or under the direction of professional scientists and scientific institutions". The phenomenon of Volunteered Geographic Information (VGI) is one face of this medal, where citizens have become important players by voluntarily contributing georeferenced information about the Earth's surface and near-surface (Goodchild, 2007). Besides, geo-crowdsourcing has been emerging. This concept, coined by Howe (2006), refers to a work or activity performed by an undefined public or a large network of potential laborers. The merging of the two terms "outsourcing" and "crowd" brought to the crowdsourcing name, to which the prefix geo- is added every time the location is a relevant attribute. Compared to VGI, in general, geo-crowdsourcing refers to people who are less skilled and who are sharing and contributing multimedia data more unconsciously (e.g. by sending twitter messages or flick imagery). This is the main difference between these two approaches even if, often, authors tend to use the two terms as synonymous.

Examples of collaborative sharing of locative information are for instance the Flickr site), enjoying an increase of more than 2500 geo-tagged photos every second and the OpenStreetMap (OSM) project (http://www.openstreetmap.org/): "a free editable map of the whole world, made by people like you". Maps are created by using low cost GPS receivers and free aerial imagery and are available for downloading under a Creative Commons Attribution-ShareAlike 2.0 license. OSM was founded in July 2004 by Steve Coast and it has been a fast-growing movement, which at the end of September 2015 reached more than two million registered users (http://www.openstreetmap.org/stats/data_stats.html). Contributors upload their individual maps, but also mapping parties are organized to gather mappers for intensively surveying a particular area. Parties, besides specific open mapping courses given in those occasions and annual conferences, increase the spirit of community. People gain awareness of working on something useful and are almost immediately able to visualize their data into the general map. From the technological point of view, there are many free and open source (FOS) tools used in the OpenStreetMap project: part of them are available from other projects (like for instance OpenLayers and PostgreSQL; part, like the core which manages the databases, the user interaction and the APIs, were expressly developed for the project and are GPL (General Public License) licensed.

From the richness of the information provided, there is there is now a shared opinion that these data, whether they cannot be put in place of, they certainly can be put in support of authoritative data collected by local, regional and national agencies. Not only cooperation among citizens, but also with governments and commercial data providers are part of the collaborative model pursued for instance by the OSM Foundation.

The awareness about the necessity of integrating these potential tools for environmental and territorial analyses has been growing at every administrative and governmental level. Spatial data infrastructures (SDI) are the framework for spatial data maintenance, sharing, access and usage. Data generally come in this context from agencies. The SDI implementation improves globally the systems efficiency avoiding data duplication and reducing time and resources needed to search for the existing spatial data or to establish whether they may be used for a specific purpose. The SDI data providers are at the moment GIS experts because a high skill is required to create a new data server and this approach, at the moment, can be considered as a top-down one. In fact the emphasis is focused on providing reliable and updated data necessary for supporting the decision-making policies and activities. Nevertheless the spectrum of possible actors has been spreading in recent years and the two approaches (GeoSpatial Web and SDI) will find a common point, enriching each other with the best aspects they present nowadays.

FOSS4G

We shall start with a quick remark about where we have decided to delimit the FOSS4G (Free and Open Source Software for Geo-information). In fact, generally speaking, the number of FOSS4G applications has been increasing enormously in the recent years. But the simple availability of a piece of code is not enough for ensuring the real usability of the code itself and its sustainability in time.

Having in mind this weakness, in February 2006 some leading teams in the FOSS4G world joined their efforts to create OSGeo, a not-for-profit organization whose mission is to support the collaborative development of open geospatial technologies and data and to promote their widespread use.

The formal birth of such a kind of organization highlights on one hand the maturity reached by the different "tribes" aware of the necessity of synergizing the tool developments and requiring the nature of geospatial software to have a high level of interaction. On the other hand the Internet challenge, bringing to new exigencies and subsequent products, and the exit of geospatial data and applications from the niche status, were decisive in that choice.

From the beginning the main benchmarks were interoperability and the choice of Open Source Initiative (OSI) certified licenses allowing different technologies not only to work together but also to integrate and exchange pieces of code among them.

At the same time OSGeo guarantees the quality of the software following a procedure in which each software project before its release and licensing goes through an incubation process within a specific committee of the organization.

The principles of the OSGeo projects (globally called the "OSGeo Way") can be summarized as follows:

1. **Consensus/Inclusiveness**: The participation from all people - from beginning users to advanced developers - is welcomed;
2. **Fostering**: As most contributions are donated, projects encourage and gratify the participation of its volunteers;
3. **Openness**: Projects adopt open standards and collaborate with other osgeo projects; and
4. **Responsibility**: Projects are responsible for checking their code integrity with respect to the open source basics.

Different kinds of supports are made available by the global community for people embracing the usage or development of FOSS (Free and Open Source Software) tools: the OSGeo portal and specific websites for the different projects, forums, wikis, mailing lists, blogs, Web tutorials and open courses offered by the Education Committee.

An international FOSS4G conference (plus two regional in the continent where the international one does not take place), that goes beyond OSGeo, being represented by a larger array of projects, is held every year. This events are good occasions to be updated on the new projects, applications and tools. Besides standard presentations, there are workshops held in computer rooms, "Installfests" in which Community members help participants in installing FOSS software on their laptops and in taking first steps, "Birds of a Feather" meetings for likeminded people and Hackers' Code Sprints.

A relevant element of OSGeo is that an incubation committee exists, which oversees the quality of the new projects entering the foundation. All packages, before becoming OSGeo products, have to pass an incubation stage at which their compliance with OSGeo principles and requirements is evaluated and that could be considered a sort of OSGeo guarantee of quality. Stability, reliability and maintenance of the code, which lacked in some free and open source products, are characteristics of OSGeo tools.

Therefore in the following paragraphs when we speak about FOSS4G, it means that we have in mind OSGeo products or applications with characteristics similar to those required to become OSGeo projects.

Being charter members or supporters of the OSGeo Foundation, the authors in developing their applications, tend to use as much as possible OSGeo certified projects and, when they have to use (for various reasons) some application not belonging to the OSGeo family, to verify that it conforms ideally to the generally principle of OSGeo:

1. *Projects should manage themselves, striving for consensus and encouraging participation from all contributors - from beginning users to advanced developers.*
2. *Contributors are the scarce resource and successful projects court and encourage them.*
3. *Projects are encouraged to adopt open standards and collaborate with other OSGeo projects.*
4. *Projects are responsible for reviewing and controlling their code bases to insure the integrity of the open source baselines.*

These general principles apply to ODK (see paragraph 2), which is not a geospatial software but a more general set of tools helping organizations to manage mobile data collection solutions. Or for NASA World Wind Java SDK, which is a geospatial Software Development Kit whose quality can be proved (if it needs this) to have been awarded in 2009 the Software of the Year by NASA's Invention and Contributions Board (ICB).

After these two pieces of introduction, that can be together considered the cultural framework our work refers to, in next paragraphs we will present, as examples, two relevant applications we had developed.

A COLLABORATIVE MULTIDIMENSIONAL PLATFORM

VGI and Virtual Globes

The first Web technologies (nowadays called Web 1.0, to distinguish them from the new Web 2.0) were characterized by poor interaction with users, making their experience not specially pleasurable (Haklay et al., 2008). The geographical applications based on this paradigm were inevitably featured by the same limitation. In this first stage, geospatial information was supplied by public mapping agencies according to a top-down, authoritarian paradigm (Goodchild, 2007b). The extraordinary shift brought by Web 2.0 (O'Reilly, 2005) allowed the development of rich Web applications, which have become more and more similar to the traditional desktop ones (Graham, 2005), by enhancing user interaction.

The new, bi-directional approach gave people the chance to also generate and share contents, e.g. within forums, wikis, social networks, etc. The corresponding change of the Internet mapping tools caused by GeoWeb 2.0 (Maguire, 2007) was indeed introduced at the end of last century by the work of the Open Geospatial Consortium (OGC), which began to establish standards for Web mapping interoperability (Peng & Tsou, 2003).

The term VGI (Goodchild, 2007a), which expresses the geospatial application of the more general phenomenon of crowdsourcing (Howe, 2006), has been one of the major result of GeoWeb 2.0 and neogeography (Turner, 2006). It fully represents the concept of humans as sensors able to register geographic information from the ground. This possibility has exceptionally grown thanks to the huge variety of sensors installed on current mobile devices. As a consequence of all the factors discussed up to now, VGI systems have been evolving into Web-based, shared platforms allowing users to dynamically add and edit their own contents and create customized mash-ups.

The Open Data Kit Suite

In order to implement a VGI system and allow users to take advantage of their mobile devices for reporting significant Point of Interests (POIs), a flexible and enough simple suite, the Open Data Kit (ODK http://opendatakit.org), can be used.

ODK is composed of three modules, providing complementary functions: ODK Build, ODK Aggregate and ODK Collect. ODK Build allows developers to design the form (i.e. the questionnaire) that users will then fill in on the field. It is is a HTML5 web application and it works well for designing simple forms with a drag-and-drop user interface.

For more powerful form designers, XLSForm (formerly XLS2Xform) can be used. It is a tool which makes possible the creation of a form template using Excel spreadsheets and to convert them to a standard format (XForms) that can be used with the ODK tools.

The form template is subsequently loaded into ODK Aggregate, that is the server-side component of the ODK suite and that is also responsible for users' administration.

ODK Aggregate makes available a ready-to-deploy server and a data repository to provide blank forms to ODK Collect, to accept finalized forms (submissions) from it and to manage the collected data, i.e. visualizing them on maps and simple graphs, exporting them (e.g., as CSV files for spreadsheets, or as KML files for Google Earth) and publishing them to external systems (e.g., Google Spreadsheets or Google Fusion Tables).

ODK Aggregate can be deployed on Google's App Engine, enabling users to quickly get running without facing the complexities of setting up their own scalable web service. ODK Aggregate can also be deployed locally on a Tomcat server (http://tomcat.apache.org) or any servlet 2.5-compatible web container, backed with a MySQL (http://www.mysql.com) or PostgreSQL (http://www.postgresql.org) database server.

With ODK Collect, users are asked to provide basic information about the POI (i.e. typology, name and description), to take pictures of it and to register its location using e.g. the GPS device.ODK Collect renders forms into a sequence of input prompts that apply form logic, entry constraints, and repeating sub-structures. Users work through the prompts and they can save the submission at any step. The finalized submissions can then be sent to the server. Currently, ODK Collect has the limitation of working only with the Android platform, but it supports a wide variety of prompts (text, number, location, multimedia, barcodes), and works well without network connectivity. In Figure 1, just as an example, the form for collecting cultural Points of Interest can be seen.

Figure 1. Example of form compilation from a smartphone using the ODK application

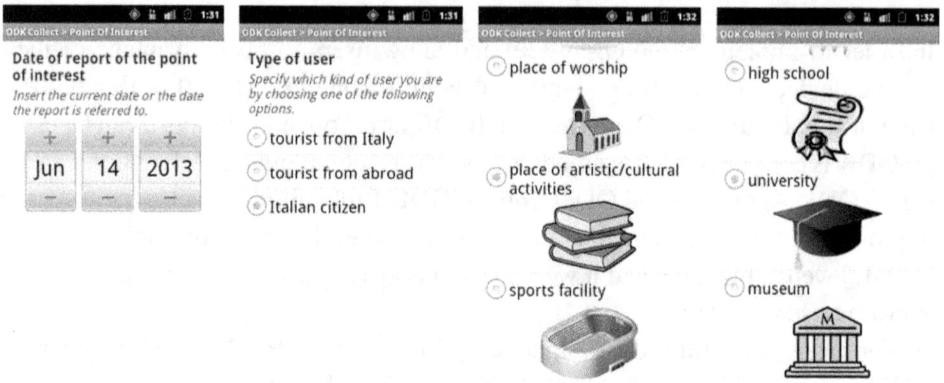

Virtual Globes and NASA World Wind

Thanks to virtual globes, the three-dimensional access to geospatial information has gained interest in recent years, allowing users to explore the Earth in 3D while streaming satellite imagery, elevation and other data from the Internet (Schultz et al., 2008).

Virtual globes can be considered as multi-purpose and multi-dimensional geographic environments, not only 3D but also 4D if the temporal dimension is considered (Brovelli et al., 2012a; Brovelli et al., 2012b). They can represent any geo-localizable data in a nearly infinite range of scales, allowing users to freely move around the globe by dynamically changing the point of view and to interact with the information represented on the surface (Elvidge & Tuttle, 2008). Using a virtual globe it is possible for instance to show a spatial distribution of an environmental variable (e.g. temperature, pressure and concentration of a pollutant) around the entire Earth, and zoom into the city streets to show the detailed structure of an urban setting (e.g. position of environmental monitoring stations, 3D representation of polluting factories and related technological networks).

Virtual globes are excellent tools that extend the expressive power through a conceptual representation of GIS data in a more realistic and intuitive medium. At the same time, ease of access and use of these tools allows us to share and communicate information, which was formerly "restricted" only to users of highly specialized software tools, and often accessible only in highly specific scientific environments.

Since many 3D virtual globes are available, their properties must be evaluated in order to select the most appropriate for the given specific needs. The first main feature to take in account is the platform required to run on. Some of them requires a specific operating system; others can be installed on diverse platforms due to the availability of releases for each operating system; and still others are platform-independent, due technology used for their development.

The second element to evaluate is the license type. Some of them are available as free and open source software, while others are "closed" (code not available for the public) and released with proprietary licenses. Some of the latter require the purchase of a license while others are available both with a freeware license or a paid license depending on the featured capabilities. The third feature to be considered is the type of application: some are desktop applications and require just a local installation, others are pure Web applications, and others again can be executed either by installing the application or directly as a Web application thanks to a specific browser plugins.

The first one is to use high-level programming interfaces made available from the virtual globes through external Application Programming Interfaces (APIs). This approach does not require very high programming skills and it is the only usable method with proprietary virtual globes, where the engine-core is visible only as a "black box". Some external APIs enable, with a scripting approach, the ability to programmatically interact with the features shown on the globe, i.e. draw and move geometries, drape images over the terrain, add 3D models, allowing one to build custom 3D map applications.

The second approach is to take advantage of the availability of the source code when the virtual globes are released with free and open source license. Having a full control of the application makes it possible to both improve the basic functionalities of the engine-core and to build new plugins. This approach allows to implement complex analytical functionalities (not achievable using an external API) and of combining the power of 3D visualization with the computational power of the typical GIS functionalities inside a unique multi-dimensional geographic platform.

In order to implement a collaborative realistic 3D scenario, the NASA World Wind virtual globe (http://goworldwind.org) was chosen. World Wind is NASA's open source virtual globe technology, providing the infrastructure for spatial data as a Software Development Kit (SDK). It is based on open standards and can be configured to accommodate any desired data format. The characteristic of being written in Java makes it platform-independent (software directly executable on different operating systems) and usable like desktop application or accessible by simple Web browsers (as an applet or via Java Web Start technology).

The openness of the code allows the complete control of both the customization and the extensions required for creating the 3D model and all the functionalities needed for interacting with the model itself.

NASA World Wind was conceived to work with large quantities of data and information, all of which can be stored on local disk or on remote data servers. The platform makes available a collection of pre-configured classes which project spatial data on a 3D virtual globe of the Earth. This virtual Earth is composed of many satellite images: BlueMarble (BMNG 500m), i3-Landsat (15m), USGS Orthophoto (1m US only), USGS Urban Area Orthophoto (0.5 - 0.25m selected US city), MS Virtual Earth Aerial Imagery, etc. Some available digital elevation models are SRTM30Plus (30 arc-sec, ~900m), SRTM3 v2 v4.1 (3 arc-sec, ~90m), ASTER (~30m), USGS NED (~30m, ~10m US only), which are dynamically served by NASA and USGS WMSs (Web Mapping Services). Moreover, it is possible to access any OGC-compliant WMS server providing geo-referenced images and maps to be projected on the globe (code implementation is also made available by NASA) and the digital elevation models to be superimposed onto the geoid model which is implemented within the platform. Finally, besides imagery and elevation data, it is possible to accurately locate on the NASA World Wind virtual globe and in its surrounding space both 2D objects such as lines, polygons, markers, callouts and multimedia viewers, and 3D objects built up from geometric primitives such as parallelepipeds, spheres and extruded polygons. The complete freedom of customizing the presented data makes the platform suitable for any spatial data application, thanks to the possibility of controlling the quality and accuracy of both the horizontal component (by varying the texture) and the vertical component (by varying the digital elevation model).

The 3D Social Platform

System Architecture

This section offers a technical overview of the 3D social platform developed through the integration of a VGI system based on ODK servers, OGC compliant WMS servers and a Virtual Globe. The architecture is shown in Figure 2. It consists of two principal components: the client and the server. The former, available as a Java Web Start (JWS) Application, is based on the NASA World Wind Java core and developed taking advantage of the NASA World Wind SDK. It provides a 3D visualization of the available geospatial contents and it interacts with the server through a Restful Web Service Protocol. The server is deployed on a J2EE application server (i.e. GlassFish, https://glassfish.java.net) backed with a relational database (i.e. PostgreSQL, http://www.postgresql.org) to store all the data relative to the platform (i.e. the system data and data uploaded by the users). The client can search and retrieve data from any WMS and ODK server, thus allowing to visualize and interact on the globe even with contents created through ODK Collect mobile clients.

Figure 2. Architecture of the 3D social platform

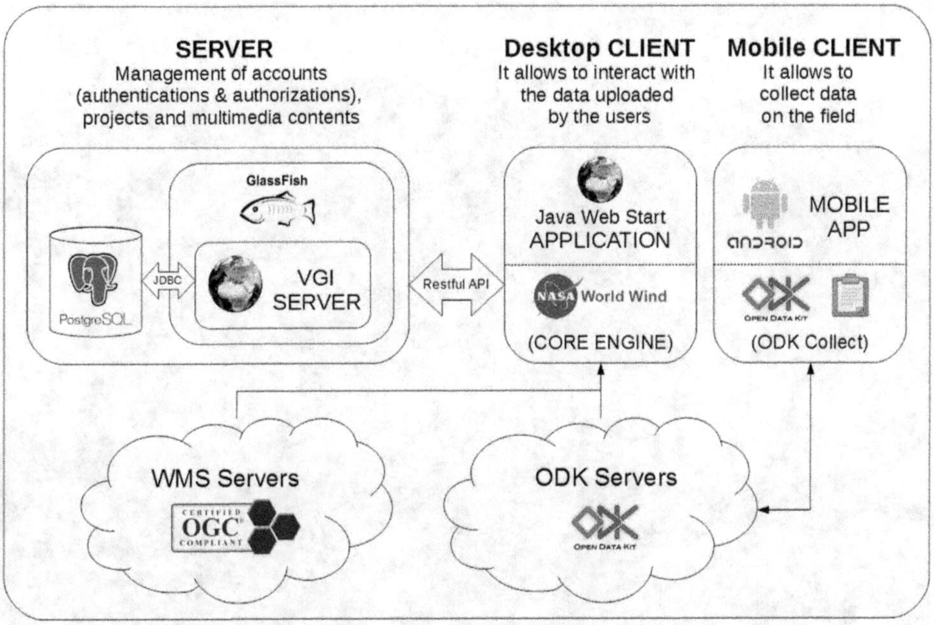

Main Functionalities

The 3D viewer allows users to interact with two different kinds of geodata. A first level of information is provided by several base layers such as OpenStreetMap, the aerial orthophotos and the satellite imagery made available, as standard geoservices, by the geoportals of institutional administrations (e.g. INSPIRE Geoportal, Italian National Geoportal, Lombardy Region Geoportal, etc.). In order to show these data, the application is able to connect to all the institutional WMS compliant servers available on the network and to any dedicated WMS server specifically built-up and populated for the specific customization of the platform.

A second level of information is provided by the crowdsourced data collected by users through the ODK Collect application. For these specific kinds of information, the platform has an innovative capability to connect with any ODK servers and publish the related data through a marker-based representation (see Figure 3).

Taking advantage of this technology and using a dedicated ODK server specifically configured for the project, users can report significant Points Of Interest (POIs) through their mobile devices (smartphone and tablets), upload them to the server and each content will be automatically geo-localized on the virtual globe as markers. All the markers are open to collaborative contribution: everyone can upload and

Figure 3. Markers related to the POIs collected by users

link to them comments and multimedia contents (image, video, and audio files) in order to enrich and share the information related to each POI.

By clicking on a marker symbolizing a POI, the related details collected through the ODK mobile app are displayed (see Figure 4) and a dedicated section gives also access to all the textual and multimedia contents (comments, images, audios and videos). These elements are entirely managed by the client: images are displayed in a picture viewer, while, for videos and audios, the application relies on a customizable external player.

The POIs visualization is fully customizable thanks to a suitable layer management interface. The fields to be displayed can be selected, and the marker icons can be customized by picking them from a default collection or providing them manually.

The application can manage different projects populated with different WMS and ODK layers depending on the emphasized topics (cultural, environmental, touristic, historic, etc, theme). All the projects created by the users are publicly accessible by design, in order to promote the community participation.

Finally, the introduction of the fourth dimension (time) provides a more in-depth navigation through the content. A time bar enables temporal filtering of all the POIs on the globe, just by picking a given date or setting a range. The same capabilities

Figure 4. Visualization of the detailed data for the selected POI

are also made available for multimedia data, which can be independently filtered according to the date of the content upload or (if provided) to the actual date of the element, e.g. useful to filter historical data.

Data Customization

The VGI platform is able to display a wide variety of data sources having different purpose (educational, environmental, tourism, landscape, naturalistic, etc.) since the platform can be connected to any WMS and ODK Aggregate server available on the Internet.

More in detail, the ODK Application Programming Interfaces (APIs), specially integrated into the virtual globe Java code, enable the client to retrieve all the information required to connect to any ODK Aggregate server, i.e. the list of available forms and the corresponding data submitted, which can then be visualized on the 3d platform. From a logical perspective, the connection to an ODK Aggregate server is treated in the same way as to a standard WMS server: the user has to select the type of server (see Figure 5), enter its URL and then pick up the layer(s) of interest.

Figure 5. Form to connect transparently to any WMS or ODK server available on the network

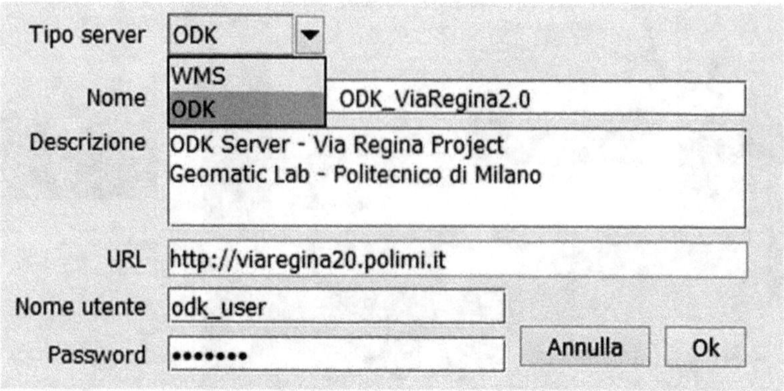

The only difference is that, while WMS servers provide raster map layers, ODK Aggregate servers provide marker layers.

Any server (both WMS and ODK Aggregate server) to which a user connects is saved into a persistent catalog which is available to all the users in order to facilitate its reuse. Once users have created 3D customized mash-ups, they can save their projects in a catalog (i.e. are available for reuse) for the whole community. The user's credentials, the list of available servers and projects, and the projects themselves (i.e. the available layers and their on/off status, the location and the camera orientation of the point of view over the globe) are stored in the PostgreSQL database on the server.

View Customization

Consistently with its multi-thematic nature, the platform allows the dynamic customization of the icons of the markers on the client-side according to the specific project the user is working on. Unlike the WMS layers, where the graphic layout is predefined on server side (WMS layer style), the client provides a graphical interface (GUI) to manage the styles of markers of the layers derived from ODK Aggregate servers. When adding an ODK layer to a project, users can select the attribute of the point to be used for styling layer markers among those available, e.g. the type of point. Automatically, as it happens for the typical styling tools in GIS packages, all the values assumed by the markers for the selected attribute are displayed. For each of them, users can pick up an icon from a rich collection of icon libraries available within the platform, or they can manually provide an icon by uploading a file (see Figure 6).

Figure 6. Marker icons thematization by attributes

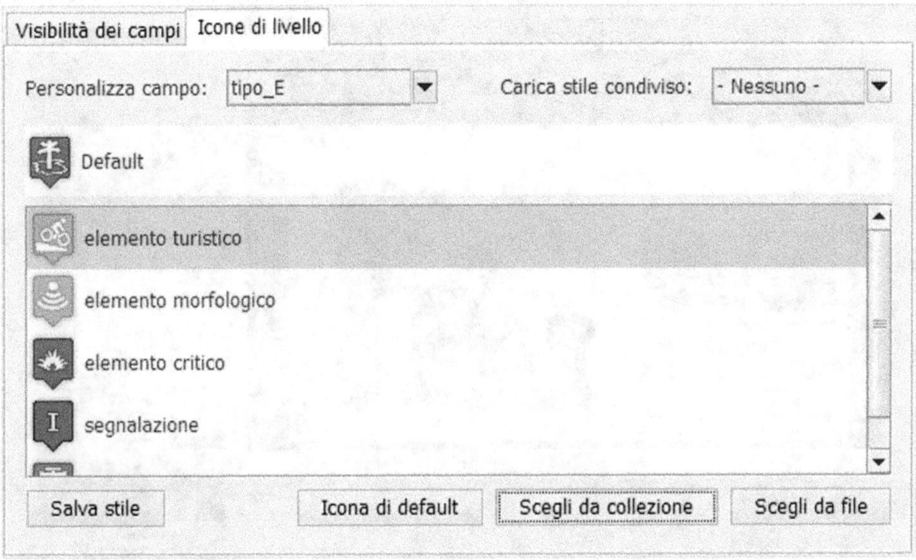

In the same way of servers and projects, styles are also shareable, so that users adding the same layers in their projects can save time by opting for a ready-made icon set defined by another user.

Management of Multimedia Contents

The social nature of the VGI platform allows any user to add comments and multimedia contents (images, audios and videos) to any ODK layer. This is possible from a built-in interface on the client-side available when clicking on a marker. Besides showing all the information related to the specific selected marker, the interface displays the available multimedia contents and allows users to add new ones (see Figure 7). The client integrates all these functionality in the same client-side graphics environment, thus making visualization and data upload easy and intuitive. Moreover, a dedicated web-application based on the Java Server Faces technology was designed and implemented to facilitate the remote administration of the platform contents, allowing all the CRUD (create, read, update and delete) basic functions on the database.

Figure 7. Marker icons thematization by attributes

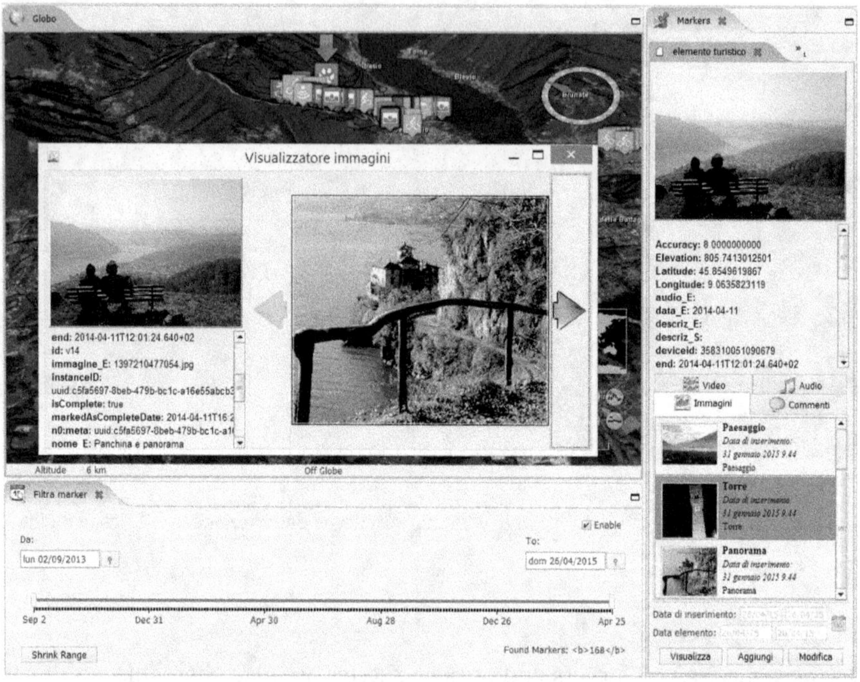

Time Filtering of Data

A temporal slider available on the bottom side of the interface (see Figure 7) allows the time filtering of both the ODK marker layers and their multimedia contents introducing the 4th dimension in the data visualization. While markers are filtered according to their submission date (i.e. the date they were created from ODK Collect), multimedia contents can be filtered either by the date they were uploaded on the platform or by the actual date the content was created (e.g the picture was taken). The date of creation (together with a description of the data) can be added by users as a metadata during the upload of multimedia contents. Filtering capability makes the platform particularly useful for historical projects and, more in general, projects where temporal information plays a key role.

Authentication and Authorization

Every user (authenticated or anonymous) can access, view and explore all the projects created with the platform and automatically shared with the community.

The authenticated users can create new projects for specific topics selecting WMS layers (from any WMS servers available on internet), ODK layers (from any

ODK Aggregate servers available on internet) and uploading new data (markers and related multimedia contents). These projects are maintained by the creator (owner). This means that metadata (project name, project description, etc) layers list (both WMS and ODK) and markers styles can be changed only by the owner.

For each project all the authenticated users can add and share new data with the community in order to enrich the information about the specific context. In detail every users can upload new POIs and add new multimedia data (images, videos, sounds, comments, …) both on the own POIs and on the pre-existing POIs uploaded by other users. The authorization policies ensure that only the owners of the context (projects, POIs, and multimedia data) can perform editing operations on it.

COLLABORATIVE CLOUD APPLICATION FOR WATER RESOURCES

Overview

Presently, most of the existing software for water resources is desktop-based, designed to work on a single computer. This represents a major limitation in many ways, starting from limited computer processing and storage power. Often scientists are complaining about running model time, that can take days, even weeks, making research quite difficult. Another limitation is a multiple user support. The classical desktop application do not support multiple users to work together, sharing data, models, and results. Data and model portability is an issue, and depending on the software type or version, sometimes even the same software on another computer can make problems with data and model portability. The software availability and accessibility are often restricted by licenses that must be used inside the organization building, or the user must be physically there to use the desktop software. These are all major limitations of any desktop software. Which brings us the main research question of this article: Is it possible to develop water resources application that alleviates most of the issues mentioned before that provides seamlessly endless storage and computing power, is available everywhere and anytime, provides multi-user platform, diminish data portability issue, and it is flexible for upgrading the existing and connecting/adding other software components.

The only feasible solution lies in the web and cloud. There are various examples of web applications (Choi et al. 2005; Horak et al. 2008; Delipetrev et al. 2010; Reed et al. 2010; Delipetrev et al. 2014; Arias Munoz et al. 2015), cloud web services (Bürger et al. 2012; Quiroga et al. 2013) and mobile applications (Jonoski et al. 2012; Jonoski et al. 2013) in the water domain. The CUAHSI (Consortium of Universities for the Advancement of Hydrologic Science, Inc.) group from the

United States has developed the hydrologic information system used for publishing and accessing water data and a collaborative environment for sharing hydrologic data and models respectively (Peckham & Goodall 2013). Previously mentioned systems often need "only a web browser" to be used, hiding all implementation details from the end-users. These advantages can significantly increase the number and the diversity of users.

Hiding the implementation details from end-users and providing services using the network (web) as an extension of our devices is the essence of cloud computing. Cloud computing (Armbrust et al. 2010) is a technology through which everything – from computing power, applications, infrastructure, systems, models are delivered to users wherever and whenever they are needed, without end-user knowledge of the physical location of system components. Similar to this idea is the electric grid, where users utilize electrical power without understanding system components (Carr 2008). This technology, in fact, makes computation and information always available. Of course, the drawback of cloud computing is network dependence e.g. if there is no network connection the systems will not work.

The cloud application for water resources is a continuation of previous research (Delipetrev et al. 2014) that presents the development of a web application for water resources based on open source software. The web application for water resources has three web services. The cloud application has an additional web service and in total has four web services for

1. Data infrastructure (DI),
2. Support for water resources modelling (WRM),
3. User management, and
4. Water resources optimization (WRO).

The cloud application enhancement are the following:

1. The web application is distributed / deployed on two VMs (Virtual Machines). The VM_1 is running as a micro instance of Amazon Web Services (AWS) public cloud, and the VM_2 is running on a Xen cloud platform at the University Goce Delcev in the Republic of Macedonia.
2. The web service for supporting WRM that runs on VM_1, and the DI web service that runs on VM_2, are communicating with WFS-T (Web Feature Service Transactional), XML (eXtensible Markup Language) messages over the internet, demonstrating distributed computer environment.
3. An additional simple web service for user management is developed, to comply with the NIST (National Institute of Standard and Technology) definition (Mell & Grance 2009) for cloud computing.

4. Additional optimization algorithms are included into the web service WRO, improving its capabilities.
5. Clear demonstration of hybrid cloud is presented, where VM_1 is part of the AWS public cloud, and VM_2 is running in the private cloud. The advantage of this distributed computer environment is that the data security and protection can reside in the private cloud (VM_2), while the web services can work in the public one (VM_1).
6. The main enhancement compared to the web application is the scalability. The presented design and implementation demonstrate that the cloud application can support seamlessly unlimited number of users.

The cloud computing application was tested using data from the Zletovica river basin located in the North-Eastern part of the Republic of Macedonia. The test is a proof of concept that this application can be a foundation for a modern cloud based solution. The application url_www.delipetrov.com/his/ provides video presentation and explanation of the system components, guides how to use the services, etc.

Design and Implementation

The cloud application has four web services:

1. DI.
2. Support of WRM.
3. User management.
4. WRO.

Figure 8 shows the web services and the data communication links represented by arrows. The communication between the web service for supporting of WRM and the DI web service is asynchronous, or on demand.

Important milestone is the deployment of the cloud application between the two VMs running on separate physical servers. The first VM_1 is a micro instance on the AWS, and the second VM_2 is running on the Xen cloud platform. The VM_1 has 8 GB HDD, 1 GB RAM and Ubuntu 13 as an operating system. The VM_2 has 30 GB HDD, 1 GB RAM and Fedora 16 as an operating system. The VM_2 is running on a physical server IBM x3400 M3 with four-core Intel Xeon E5620 2.40 GHz with 12 MB of cache per processor. The AWS management console, and the Citrix XenCenter, respectively manage the VM1 and VM2. Both management consoles provide tools to control the cloud application and server environment i.e. control the CPU usage, the memory, the disks and the network connections on all of the virtual machines.

Figure 8. Design of the cloud computing application for water resources

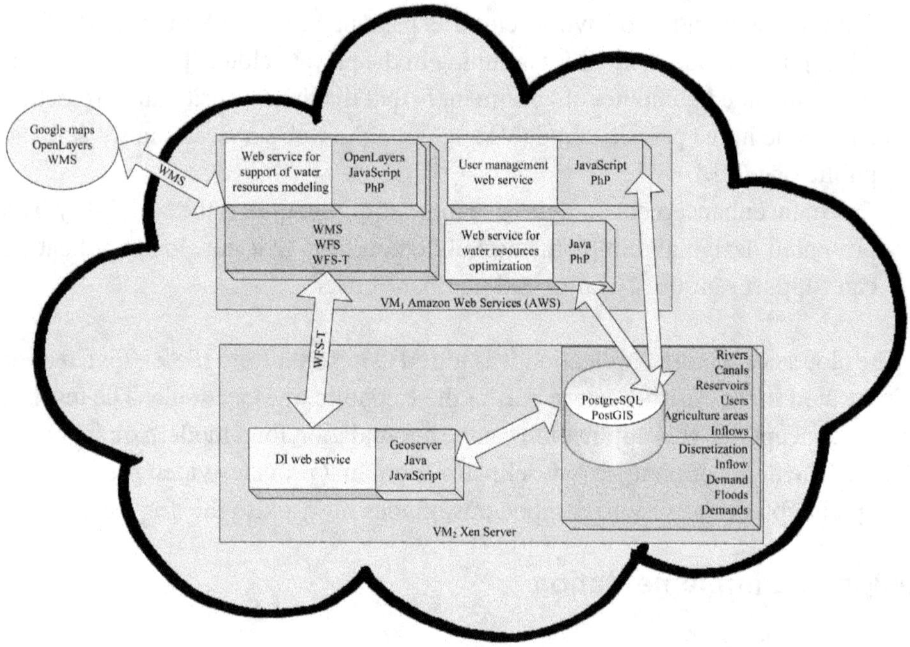

DI Web Service

The DI web service is composed of two components

1. HMak database created in PostgreSQL and PostGIS and
2. GeoServer application.

The HMak database is the central database in this cloud application storing all data. The DI web service is running on VM_2 that resides in the private cloud created by the Xen cloud platform. The HMak stores geospatial data, including topographic, hydro-geological, rivers, roads, municipal, etc. The GeoServer shares the HMak geospatial data on the internet. The HMak stores six geospatial vector layers: rivers, canals, reservoirs, users, inflows, and agriculture land, and their attribute tables, that are used by the web service for supporting of WRM. Additionally, the HMak stores around 40 time series data tables used by the WRO web service.

GeoServer is an OSGeo server for sharing geospatial data. It is interoperable and can publish data from any major geospatial data sources using open standards. The combination of these two components is exceedingly used for creating spatial data infrastructure (SDI) (Infrastructures 2004), where PostgreSQL and PostGIS are geo-database repository, and GeoServer manages geospatial data. The current cloud application configuration is a prototype SDI solution that permits storage and management of geospatial data. More importantly, GeoServer is as a middle tier application that connects the HMak database on one side and provides OGC web services (WFS-T) on the other side. In our cloud application, GeoServer provides WFS-T interface for the web service for supporting WRM.

Web Service for Support of WRM

The web service for supporting of WRM is intended to provide a web interface for creating and editing geospatial water resources elements just like in classical desktop applications (see, for example, RIBASIM of Deltares, Mike Basin of DHI, and other applications). This web service provides tools to work with the geospatial objects from the six layers described before, that represent water resources components and its infrastructure, shown in Figure 9. These geospatial elements are WRM building blocks, corresponding to the name of the service "for support of WRM." The main difference between the web service for support of WRM and desktop interfaces is that the web service for WRM is accessible by multiple users simultaneously over the internet using simply a web browser.

The six geospatial layers are designed to allow only a specific type of geospatial data for each layer, e.g. points for reservoirs, inflows, users; polylines for rivers and

Figure 9. Cloud application interface

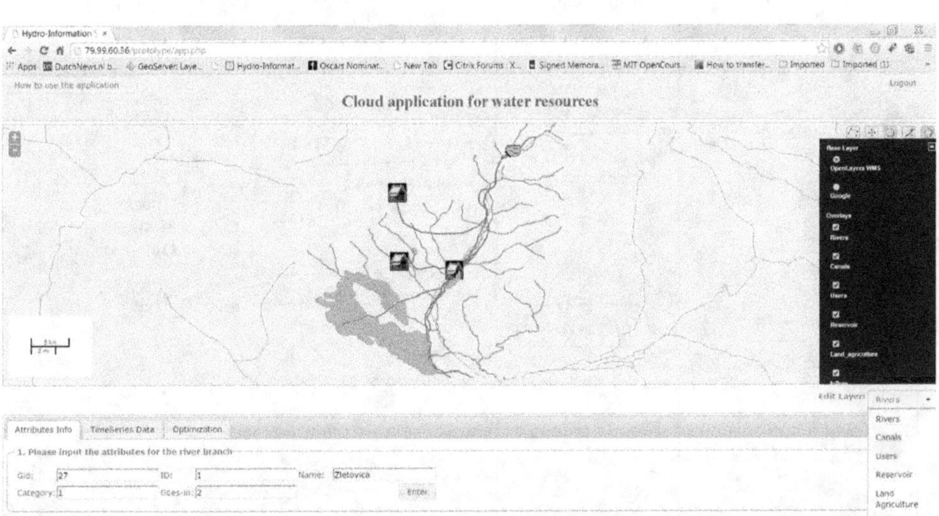

canals; and polygons for agricultural areas. Each layer has a simple attribute table, only used in demonstrational purposes. Generally, the web service for support of WRM is a customized web GIS service designed for water resources.

The web service for supporting of WRM is developed using PHP, Ajax, JavaScript and the OpenLayers library, which supports the OGC standards Web Map Services (WMS) and Web Feature Services - Transactional (WFS-T). Taking advantage of the former protocol, the system is connected with Google Maps and OpenLayers WMS. This allows users to select the background map from the menu as shown in Figure 9, where OpenLayers WMS is used. Important is that other users can select different background maps e.g. Google Maps without interfering between each other.

The OpenLayer library creates WFS-T communication between the web service for support of WRM user interface running on VM_1, and the geospatial data stored in HMak where the GeoServer acts like a middle tier running on VM_2. The WFS-T communication provides a framework to create, update, and delete geospatial data over the internet and the two VMs. The WFS-T communication is based on XML messages. The browser request also generates a WMS request that fetches the data from WMS providers with HTTP "get" and "post" methods. Figure 10 shows the activity diagram of a web browser refresh. As the WFS-T does not support attribute data that accompany every geospatial object, additional PHP and Ajax scripts were developed to work with the geospatial data attributes. When a geospatial object is

Figure 10. Activity diagram of a browser refresh

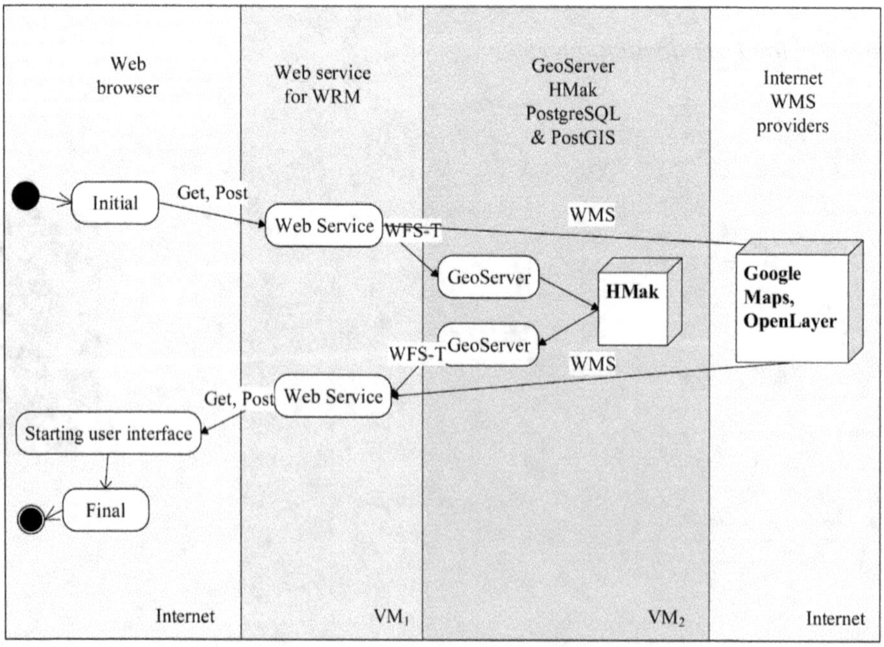

selected from the user interface, scripts are executed that read the selected geospatial object attribute data from HMak, and present it to the tab "Attribute info". The attributes can be updated and, afterwards, saved back in HMak. The "Time series data" provides a possibility to attach time series data for each geospatial object that is stored in HMak.

Web Service for Water Resources Optimization

The WRO web service implements three algorithms for the optimal reservoir operation named

1. Nested dynamic programming (ndp)
2. Nested stochastic dynamic programming (nsdp) and
3. Nested reinforcement learning (nrl).

The nDP and nSDP cores are created using examples provided by Loucks (2007), while nRL is developed on the basis of several books and articles (Kaelbling 1996; Sutton & Barto 1998; Bhattacharya et al. 2003; Comité & Delepoulle 2005). Further discussion about the algorithms is not explained, because it goes beyond the scope of this article. The three algorithms are coded in Java and developed as prototype applications.

The WRO web service has several components: PHP web form for data input and upload into the HMak database running in VM_1, the three prototype Java applications for each algorithm residing in VM_2, and a separate page for results presentations again in VM_1. The web application directory on VM_2 stores the three algorithms executable JAR files. Each of the algorithms has its own list of tables concerning the water users' demands, reservoir storage discretization, reservoir inflow, etc., making over 40 data input tables for the three algorithms in HMak. The "Optimization" tab provides user interface for the WRO web service. There are three buttons, and when a button is pressed, the appropriate prototype algorithm application is selected which connects to HMak database, reads the data, calculates the optimal reservoir operation and stores back the optimization results into HMak. The JavaScript library "highcharts.js" presents the results (A screenshot of an example result plot is presented in the following section).

Web Service for Users' Management

The web service for users' management is simple with a main purpose to control the cloud computing application access. Each user receives its own login and password to access the cloud application monitoring the time spent on the system. The user profiles stores the cloud application usage time. Further development of this service will include users' computer power and storage usage. Using these information's, the administrator can effectively manage the cloud application users.

Results and Test

Figure 11 shows the hydro system Zletovica model created by the web service for WRM. The hydro system contains the reservoir Knezevo, river network, canal network, towns as users and agricultural areas. The towns and reservoir titles are added additionally and are not part from the web service.

The WFS-T support of GeoServer, allows other desktop applications to connect in the same manner as the web service for support of WRM does. Two desktop software MapInfo and uDig successfully connected to the GeoServer and the DI web service. The desktop applications share the same working environment as the web service for support of WRM and can insert, edit the six geospatial data layers. This demonstrates that it is possible to connect a desktop software to the cloud application

Figure 11. Water resource model of the Zletovica river basin

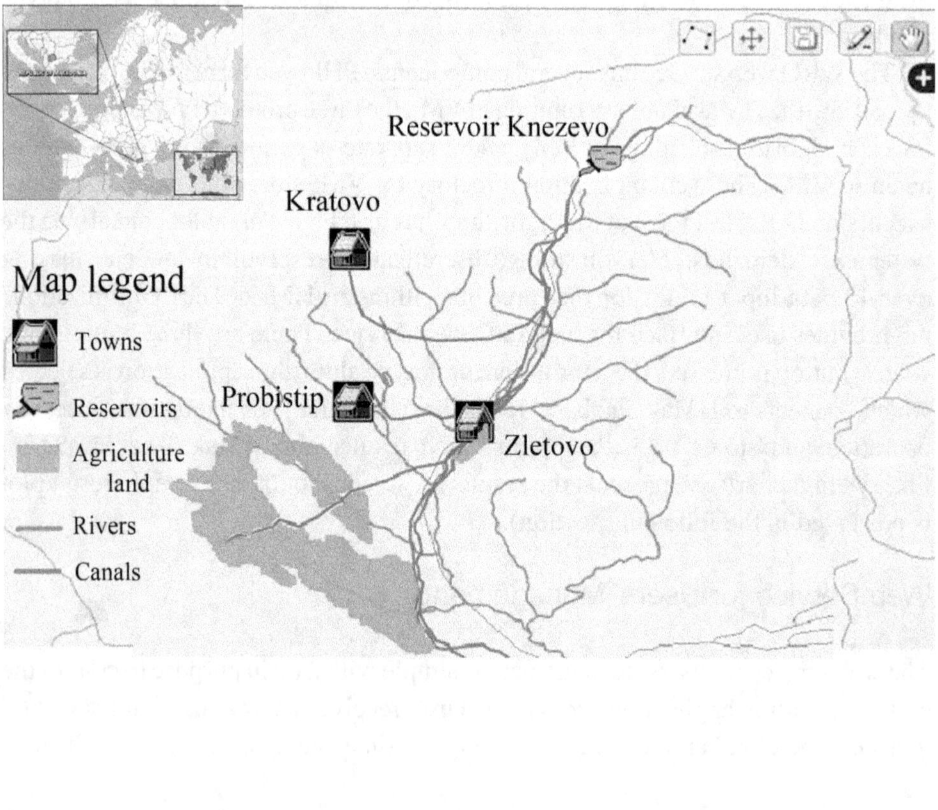

web services. This option demonstrates flexibility in using a combination of already existing desktop software together with the cloud application.

The WRO web service was tested using data from an existing study (GIM, 2009) of the Zletovica river basin. Three nDP, nSDP and nRL datasets were uploaded as CSV files into the HMak using the WRO web service interface. Figure 12 presents the optimal reservoir storage results of the three algorithms.

Six students who worked with the web services for several hours carried out the test of the cloud application. The students were accessing the cloud application from their homes and communicated thanks to instant messaging services (Skype and Facebook) between each other. The idea was to demonstrate that the cloud application supports multiple geographically dispersed users that can collaborate in the same working environment.

The main conclusions are:

1. The students collaborated among each other modelling water resources (often from Macedonia river basins) with the web service for support of WRM just as using a desktop application on their own computer, with a major difference: all modelling was done online, in real time and everyone participated in the joint model. With only one web browser refresh the student can view the last version of the water resources model.

Figure 12. Optimal reservoir operation graph from the three algorithms using the WRO web service

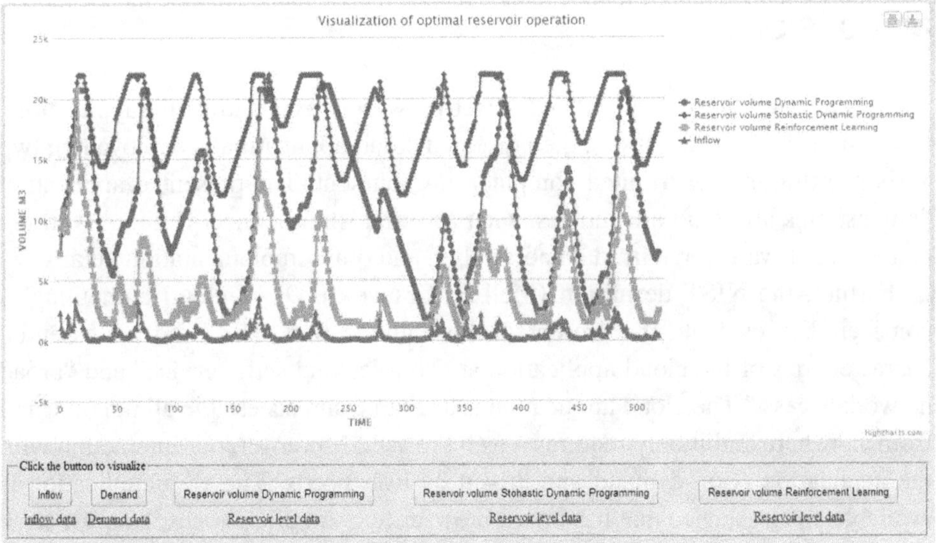

**For a more accurate representation see the electronic version.*

2. At the beginning, there is a need for task separation, because it is possible that several users can work on the same geospatial data, therefore constant communication between users is needed that was provided by the instant messaging services. The miscommunication happened in our first trials, and was corrected afterwards.
3. The students used the WRO web service executing optimizations and viewing the results. The WRO web service does not support parallel optimizations, and they need to be executed one by one. The teams can jointly upload data and view optimization results.
4. Important aspect is the data flow between VM_1 and VM_2 or the web service for support of WRM and the DI web service and the users' computers. With each browser refresh, the DI web service generates six WFS-T XML files that are sent to the service for support of WRM and from there to the user's computer. Even a small model can generate substantial network traffics. Our simple example of the Zletovica river basin model at each refresh generated around 3 MB. This can be an important issue if a complex model with many geospatial objects is considered.

The general conclusion is that the cloud application is stable and functional, supporting multiple users with increased workload. The system performance measures, the workload and other characteristics were not performed, although both AWS management console and Citrix Xen control provides that information. The main test was to demonstrate that the cloud computing application works as designed.

DISCUSSION

The cloud computing paradigm "Only a web browser is needed to use the application" is accomplished. Furthermore, the presented cloud application is deployed on two VMs working in a distributed computer environment. The presented application demonstrates all cloud advantages, such as diminishing concerns about working platform, software versions, data portability, and other implementation details.

Further, the NIST definition (Mell & Grance 2009) of cloud computing is considered to evaluate the presented cloud application. The first two essential characteristics of the cloud application are "on-demand self- service" and "broad network access." The cloud application is available and accessible all the time and from anywhere and it only requires a web browser. Moreover, our interaction with the application is on-demand and driven by user needs. The web application is available on any device that has a web browser (e.g. mobile phones, laptops, etc.).

The third and fourth essential cloud application characteristics are the capability for "resources pooling" and "rapid elasticity." The basic adjustment concerning the workload can be performed by increasing the current VMs computational power. The VMs workload can be monitored over AWS console and XenCenter and adjust appropriately. The cloud application components, standards and programming languages are interoperable and can be deployed on an unlimited number of servers and connect appropriately. The issues concerning scalability and resource pooling can be resolved by creating many data repositories similar to the HMak deployed on a number of VMs, storing large quantities of geospatial and other types of data. Furthermore, several GeoServer instances can connect to multiple data repositories, creating a giant DI system. The web service for support of WRM can connect to the multiple GeoServer instances, and finally the web services can be replicated on several VMs. Depending on the number of users, the workload, the storage capacities, the processing power, the number of servers available, etc., the optimal cloud application environment can be adapted.

The last essential characteristic of cloud computing is "measured service" which is rudimentary supported by measuring the time of each user's usage of the system. This satisfies cloud computing criteria, but needs to be vastly improved (e.g. with measuring processing power consumption, storage capacity utilization, etc.).

Concerning service models, the presented cloud computing application belong to software as a service (SaaS). Users with a web browser access the cloud application and do not care about underlying cloud infrastructure. The current deployment model is hybrid of public - private cloud because the VM_1 is running in public cloud AWS, and VM_2 is on private Xen cloud platform.

The most valuable feature of the cloud application is its real time collaboration platform capabilities. Multiple users using only a web browser can work jointly with the web services and collaborate in the same working environment in real time. An example is the web service for supporting WRM is when a user saves the current work. After that moment all other distributed users with just refreshing the web browser window can see the change (new/modified rivers, users etc.). All of the data and models are stored in HMak and users do not have to be concerned about hardware and software support infrastructure.

Another important concern about implementing cloud solutions is the data protection and safety. Often many companies and organization dismiss implementation of cloud computing solutions just because their data will be stored somewhere on the internet raising major concerns about its safety and protection. This prototype cloud application makes an elegant solution where services are residing in the public cloud, while the data is stored in the private cloud. If for instance an attack happens on the cloud application, to protect the data the private cloud VM_2 can be disconnected from the public cloud AWS VM_1. Another key point is that data resides inside the institution, and only the web services are "outsourced." This concept can be applied in many organizations where the data needs to be stored internally.

The presented cloud application is developed using open source software and custom prototype code. This increases the application value, because many software companies could do the same. This cloud application is a prototype solution and development to a final stage would need significant human and material resources.

This research demonstrates that there are available open source software and technologies to create complex and robust cloud application for water resources. The application is a SaaS hybrid cloud solution running on two virtual machines VMs, from which VM_1 runs on AWS and VM_2 on Xen cloud platform. The cloud application was tested in the Zletovica case study and with real-time multiple users that collaborated and jointly modelled and optimized water resources.

The cloud application can be further improved by connecting to new data repositories. Additional modelling, optimization and other decision support services have already been envisioned and can be added to the existing platform, so that it can evolve into a fully cloud based water resources modelling system.

CONCLUSION

The chapter wants to be a practical and effective contribution in the frame of collaborative geospatial applications.

We presented two solutions developed by the authors: the former is a multi-dimensional platform able to connect any OGC WMS and any ODK server and to publish the related data. Taking advantage of the ODK components, users are able to report significant Points Of Interest (POIs) through their mobile devices (smartphones and tablets) and to upload them to the server. A multidimensional webGIS is able to concurrently publish on the virtual globe both the WMS layers (used generally for contextualizing the project) and the POIs stored on the ODK server(s). Each content is automatically geo-localized on the virtual globe as a marker. All markers are open to collaborative contribution: everyone can upload and can link to them comments and further multimedia contents (image, video, and audio files). The platform allows users to navigate in four dimensions thanks to the time bar, which enables the temporal filtering of all the POIs on the globe.

The second application presented is a collaborative cloud application for water resources. It was deployed and tested in a distributed computer environment running on three virtual machines, two commercial and one open source cloud. The cloud application has four web services, all of them are interconnected and communicate between each other.

The collaborative cloud application for water resources provides a real time collaboration platform where multiple users, only with a web browser can jointly collaborate in real time on the same working environment. All of the data and models are stored in the cloud and users do not have to be concerned about hardware and software support infrastructure. The presented collaboration platform demonstrates all cloud advantages, such as diminishing concerns about working platform, software versions, data portability, and other implementation details.

The proposed innovative applications were successfully built thanks to two ingredients that authors believe fundamental for the development of new geospatial systems: flexible, extensible and reliable FOS libraries and software components (OpenLayers, JQuery, GeoServer, PostgreSQL, PostGIS, GlassFish, PrimeFaces, NASA World Wind Java SDK, Open Data Kit) and well defined, open protocols and standards (XForms, WMS, WFS and WFS-T). For those who want to play with them, they are respectively available at the following URLs: http://geomobile.como.polimi.it/policrowd2.0 and http://delipetrov.com/his/.

REFERENCES

Arias Muñoz, C., Brovelli, M. A., Corti, S., Micotti, M., Weber, E., & Soncini Sessa, R. (2015). A FOSS approach to Integrated Water Resource Management: The case study of Red-Thai Binh rivers system in Vietnam. *Geomatics Workbooks*, *12*, 471–476.

Armbrust, M., Fox, A., Griffith, R., Joseph, A. D., Katz, R., Konwinski, A., ... Stoica, I. (2010). A view of cloud computing. *Communications of the ACM*, *53*(4), 50–58. doi:10.1145/1721654.1721672

Bhattacharya, B., Lobbrecht, A. H., & Solomatine, D. P. (2003). Neural networks and reinforcement learning in control of water systems. *J Water Res Pl-Asce*, *129*(6), 458–465. doi:10.1061/(ASCE)0733-9496(2003)129:6(458)

Brovelli, M. A., Valentini, L., Zamboni, G. 2012b Multi-frame and multi-dimensional historical digital cities: The Como example. *International Journal of Digital Earth*. DOI:10.1080/17538947.2012.739662

Brovelli, M. A., & Zamboni, G. (2012). Virtual globes for 4D environmental analysis. *Applied Geomatics*, *4*(3), 163–172. doi:10.100712518-012-0091-3

Bürger, C. M., Kollet, S., Schumacher, J., & Bösel, D. (2012). Introduction of a web service for cloud computing with the integrated hydrologic simulation platform ParFlow. *Computers & Geosciences*, *48*, 334–336. doi:10.1016/j.cageo.2012.01.007

Carr, N. G. (2008). *The big switch: Rewiring the world, from Edison to Google*. WW Norton & Company.

Choi, J., Engel, B., & Farnsworth, R. (2005). Web-based GIS and spatial decision support system for watershed management. *Journal of Hydroinformatics*, *7*, 165–174.

Cohn, J. P. (2008). Citizen science: Can volunteers do real research? *Bioscience*, *58*(3), 192–197. doi:10.1641/B580303

Comité, F., & Delepoulle, S. (2005). PIQLE: a platform for implementation of Q-learning experiments. *NIPS workshop: reinforcement learning benchmarks and bake-offs II*.

Delipetrev, B., Jonoski, A., & Solomatine, D. P. (2014). Development of a web application for water resources based on open source software. *Comput Geosci-Uk*, *62*, 35–42. doi:10.1016/j.cageo.2013.09.012

Delipetrev, B., Jonoski, A., & Solomatine, D. P. (2014). A novel nested dynamic programming algorithm for optimal reservoir operation. Hydroinformatic conference.

Delipetrev, B., Mihajlov, D., Delipetrev, M., & Delipetrov, T. (2010). Model of the Hydro-Information System of the Republic of Macedonia. *CIT. Journal of Computing and Information Technology*, *18*(2), 201–204. doi:10.2498/cit.1001804

Elvidge, C. D., & Tuttle, B. T. 2008 How virtual globes are revolutionizing earth observation data access and integration. The International Archives of the Photogrammetry, Remote Sensing and Spatial Information Sciences, 37, 137-140.

Goodchild, M. F. (2007a). Citizens as Voluntary Sensors: Spatial Data Infrastructure in the World of Web 2.0. *International Journal of Spatial Data Infrastructures Research*, *2*, 24–32.

Goodchild, M. F. (2007b). Citizens as Sensors: The world of volunteered geography. *GeoJournal*, *69*(4), 211–221. doi:10.100710708-007-9111-y

Haklay, M., Singleton, A., & Parker, C. (2008). Web Mapping 2.0: The Neogeography of the GeoWeb. *Geography Compass*, *2*(6), 2011–2039. doi:10.1111/j.1749-8198.2008.00167.x

Herring, C. (1994). *An Architecture of Cyberspace: Spatialization of the Internet*. U.S. Army Construction Engineering Research Laboratory. Retrieved from http://citeseerx.ist.psu.edu/viewdoc/download?doi=10.1.1.37.4604&rep=rep1&type=pdf

Horak, J., Orlik, A., & Stromsky, J. (2008). Web services for distributed and interoperable hydro-information systems. *Hydrology and Earth System Sciences*, *12*(2), 635–644. doi:10.5194/hess-12-635-2008

Howe, J. (2006). The rise of crowdsourcing. *Wired Magazine*, *14*(6), 176–183. doi:10.1016/j.isprsjprs.2010.06.005

Infrastructures, D.S.D. (2004). *The SDI Cookbook*. GSDI/Nebert.

Jonoski, A., Alfonso, L., Almoradie, A., Popescu, I., van Andel, S. J., & Vojinovic, Z. (2012). Mobile phone applications in the water domain. *Environmental Engineering and Management Journal*, 11.

Jonoski, A., Almoradie, A., Khan, K., Popescu, I., & Van Andel, S. (2013). Google Android mobile phone applications for water quality information management. *Journal of Hydroinformatics*, *15*(4), 1137–1149. doi:10.2166/hydro.2012.147

Kaelbling, L. P. (1996). Special issue on reinforcement learning - Introduction. *Machine Learning*, *22*(1-3), 7–9. doi:10.1007/BF00114721

Loucks, E. B. (2007). *Water Resources Systems Planning and Management - an introduction to methods, models and applications*. UNESCO Publishing.

Maguire, D. J. (2007). GeoWeb 2.0 and Volunteered GI. *Workshop on Volunteered Geographic Information*, 104-106.

McKnee. (2004). *The Spatial Web. OGC White Papers*. Retrieved from portal.opengeospatial.org/files/?artifact_id=3859

Mell, P., & Grance, T. (2009). The NIST definition of cloud computing. *National Institute of Standards and Technology*, *53*, 50.

Nascimento, D. M., & Raghavan, V. (2008). Participatory Free and Open Source GIS in the Web 2.0 - Exploring trends in GIS in times of Collaborative Creation. *Proceedings of the FOSS4G 2008*. Retrieved from http://wgrass.media.osaka-cu.ac.jp/gisideas10/papers/c89a8e86e7c71486cbea29a3ac62.pdf

O'Reilly, T. (2005). *What is Web 2.0: Design patterns and business models for the next generation of software*. Retrieved from http://oreilly.com/pub/a/web2/archive/what-is-web-20.html

Peckham, S. D., & Goodall, J. L. (2013). Driving plug-and-play models with data from web services: A demonstration of interoperability between CSDMS and CUAHSI-HIS. *Comput Geosci-Uk*, *53*, 154–161. doi:10.1016/j.cageo.2012.04.019

Peng, Z. R., & Tsou, M. H. (2003). *Internet GIS*. Hoboken, NJ: John Wiley &Sons.

Quiroga, V. M., Popescu, I., Solomatine, D., & Bociort, L. (2013). Cloud and cluster computing in uncertainty analysis of integrated flood models. *Journal of Hydroinformatics*, *15*(1), 55–70. doi:10.2166/hydro.2012.017

Reed, P. M., Bills, B., Anderson, M., Ketchum, B., & Piasecki, M. (2010). The Mid-Atlantic Watershed Atlas (MAWA): Open access data search & watershed-based community building. *Environmental Modelling & Software*, *25*(6), 808–812. doi:10.1016/j.envsoft.2009.12.004

Scharl, A., & Tochtermann, K. (2007). *The Geospatial Web, Advanced Information and Knowledge Processing Series*. London: Springer.

Schultz, R. B., Kerski, J. J., & Patterson, T. C. (2008). The use of virtual globes as a spatial teaching tool with suggestions for metadata standards. *The Journal of Geography*, *107*(1), 27–34. doi:10.1080/00221340802049844

Silvertown, J. (2009). A new dawn for citizen science. *Trends in Ecology & Evolution*, *24*(9), 467–471. doi:10.1016/j.tree.2009.03.017 PMID:19586682

Sui, D. Z. (2008). 2008 The wikification of GIS and its consequences: Or Angelina Jolie's new tattoo and the future of GIS. *Computers, Environment and Urban Systems*, *32*(1), 1–5. doi:10.1016/j.compenvurbsys.2007.12.001

Sutton, R., & Barto, A. (1998). *Reinforcement learning: An introduction*. The MIT press.

Turner, A. 2007 Introduction to Neogeography, Sebastopol, CA: O'Reilly Media Short Cuts.

Chapter 2
Development of Assessment Criteria for Various Open Sources GIS Software Packages

Shahriar Shams
Institut Teknologi Brunei, Brunei

ABSTRACT

There has been a significant development in the area of free and open source geospatial software. Research has flourished over the decades from vendor-dependent software to open source software where researchers are paying increasing attention to maximize the value of their data. It is often a difficult task to choose particular open source GIS (OGIS) software among a number of emerging OGIS software. It is important to characterise the projects according to some unified criteria. Each software has certain advantages and disadvantages and it is always time consuming to identify exactly which software to select for a specific purpose. This chapter focuses on the assessment criteria enabling developers, researchers, and GIS users to select suitable OGIS software to meet their requirements for analysis and design of geospatial application in multidisciplinary fields. This chapter highlights the importance of assessment criteria, followed by an explanation of each criteria and their significance with examples from existing OGIS software.

INTRODUCTION

There has been a noteworthy progress in the area of free and open source geospatial Software. In the past proprietary software such as ArcView, ArcGIS, Arc info, have been extensively used for teaching, learning and research purposes due to wider functionalities, the user friendliness of their interfaces, and also demand

DOI: 10.4018/978-1-5225-5039-6.ch002

from industry for graduates having knowledge and skills of using a particular (or widely used) GIS software product. However, over the past decade there has been an intensive attempt to design and expand free, i.e. non-proprietary, desktop GIS software products (Sherman, 2008; Steiniger & Bocher, 2009; Donnelly, 2010).

Most of the industries are involved with solving problems related to basic needs of capture and visualization of spatial data and storage capability. Further advances in technology have increased interoperability across data repositories as well as user friendliness. These advancements have failed to address current development in Geographical Information Science (GIScience), which include research areas such as spatio-temporal data models, geographical ontologies, spatial statistics and spatial econometrics, dynamic modeling and cellular automata, environmental modeling, and neural networks for spatial data (Câmara, 2003). Developing a faster, cheaper and spatially enabled database is a major challenge for researchers and software developers than the software developed by the commercial vendors. A second important reason for developing open-source spatial analysis tools is the need to resolve the "knowledge gap" in the process of deriving information from images and digital maps. Difficulties in extracting information from images and digital maps are another important reason for developing open-source GIS tools. The complexity further arises due to sophisticated and sensitive data collecting process through remote sensing satellites, digital cameras, and GPS. In fact, very few techniques for image data mining in remote sensing archives are currently available in our large earth observation data archives (Câmara, 2003).

Studies on vendor dependent software to open source software has been increasing rapidly in the last 15 years due to growing consideration to maximize the value of spatial data so that resources can be utilised more efficiently (Shekar et al., 2001; Hartnett & Bertolotto, 2003; Morrison & Purves, 2002; Caˆmara & Onsrud, 2004; Purves et al., 2005). The XML and Java languages being the major languages for open source GIS software, has contributed to the advancement of the GIS software (Chen et al., 2010). Government, Academics and Developers are the major users of open source software such as web-based GIS and mapping applications as it is free of cost (Confino & Laplante, 2010). Researchers, academics and students can learn and collaborate more easily because of the open source concepts, coding, and algorithm. Robustness as well as user friendliness have made open source GIS widely popular and accepted in the industry. The continuous progress and development of free and open source software has extensively influenced and assisted in dissemination of information technology (Tiemann, 2009). Open source software will emerge as the largest distributer of software due to it continuous contribution to development of information technology leading to cheaper software and new business model (Peter

Galli, 2007). Overall, the open source GIS is a great platform to share and manage the geographical related information in a ubiquity, visualization and interactive way.

It is often a difficult task to choose particular open source GIS (OGIS) software among a number of emerging OGIS software. It is essential to categorise the projects based on specific criteria. Each software has certain advantages and disadvantages and it is always time consuming to identify exactly which software to be selected for a specific purpose. Therefore, it is important to develop certain assessment criteria based on which a user can select OGIS software. This chapter focuses on the assessment criteria enabling developers, researchers and GIS users to select suitable OGIS software to meet their requirements for analysis and design of geospatial application in multidisciplinary fields. This chapter highlights the importance of assessment criteria, followed by an explanation of each criteria and their significance with examples from existing OGIS software.

CONCEPT OF OPEN SOURCE AND FREE SOFTWARE

Open source software is a freely distributed software consists of source code hidden in the software as a programming language that the developers, researchers or users can easily modify or add features for customization or any other purposes. The software is also known as "libre software" or "open software". One important features of open source software is "free software licenses", which states the legal rights to users with freedoms to amend, run, and redistribute source codes of the licensed software (http://www.fsf.org/licensing/essays/free-sw.html, Free Software Foundation). There is often misunderstanding about what 'free' software indicates. The term 'free', does not indicate that it is free-of-cost, rather it grants freedoms to the developers, users by the software license. These freedoms include the right to (i) run the program for any purposes (e.g. educational or business use); (ii) customize and selling the program; (iii) freely copy and distribute the program; and (iv) modify the program, and distribute the modified version. Typical free software licenses such as the GPL, LGPL, Mozilla, Apache, MIT-style and BSD-style licenses grant these four freedoms. The figure 1 illustrates that free downloadable software is not necessarily equivalent to open source software. For example, Google Earth is one of the most popular free Virtual Globe GIS applications that everyone can download and use for free. But Google Earth is not an open source software rather it is a proprietary software. A red box was added to represent the new boundary of public domain software as shown in Figure 1. Because some public domain software may have no copyright restrictions or licenses, the public domain software has been altered to

Figure 1. Categories of free and non-free software
(Adopted from http://www.gnu.org/philosophy/categories.html, by Chao-Kuei Hung).

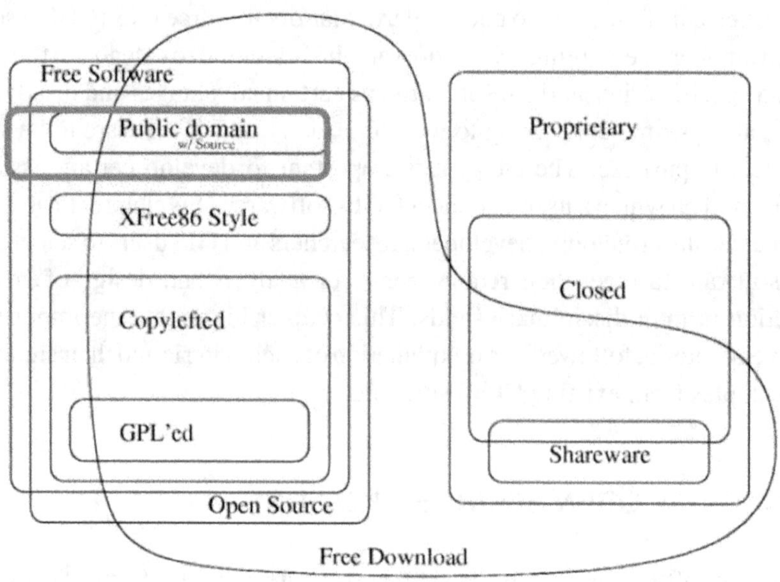

both include and exclude the open source domain. In GIS software, GRASS GIS is a famous example of public domain software. Public domain software can be re-structured and re-packaged to sell as proprietary software or licensed as open source software (Valdes, 2008).

Requirements for Open Source GIS

The three major prerequisites for Open source GIS system architecture are interoperability, sharing of data, and accessibility of resource browser of heterogeneous types. Although each may be seen as a distinct set of capabilities, they all must coexist in a common framework that defines how system components interact (http://www.regis.berkeley.edu/gardels/envmodel.html).

The Open GIS Consortium has following guiding principles for geospatial information to be available across any network, application or platform:

- Geospatial information should discover with easy access to data, irrespective to its physical location.
- Data should be easy to integrate, merge, superimpose and render for display or use in spatial analyses from different geospatial sources, even when sources contain dissimilar types of data.

- Special displays and visualizations, for specific audiences and purposes, should be easy to generate, even when multiple sources and types of data are involved. (http://www.eweek.com/article2/0,4149,1184832,00.asp).

Assessment Criteria

It has been found based on various software developed for GIS utilities that there are broadly two types of open source GIS Softwares. One is developed for catering the needs of specific users which are based on Desktop Applications primarily for their modelling work; analysis of data etc. and others are WebGIS based users who want to consume the information in a readymade mode as available for making a decision support system. As the GIS development became much simpler than before, it had promoted the development of open source GIS and a number of projects are still undergoing for utilising available resources to cater the needs of a large base of users as shown in Figure 2.

Figure 2 Open source GIS tools

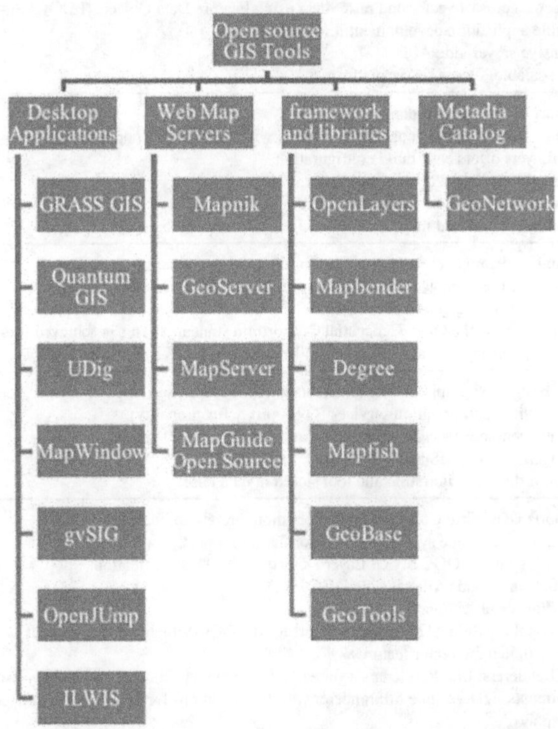

It is important to identify the key features of various existing GIS software so that important criteria can be selected based on these key features. A number of GIS software with key features is documented in Table 1.

Table 1. Key features of various existing open source GIS Software

GIS Software	Key Features
Map Server	• Rendering spatial data. • Supports industry standard data formats and spatial databases. • On-the-fly projection for both raster and vector data. • Provides a wide variety of spatial and attribute-based queries. • Supports popular Open Geospatial Consortium (OGC) standards including WMS,WFS and WCS. • Integrates with popular front-end environments such as ka-Map, Chameleon, Mapbender, MapBuilder and Cartoweb.
GeoServer	• Rendering images, serving and editing geospatial data. • Outputs are available in: WMS, WFS, KML, GeoJSON, GeoRSS format. • Open standards allow publishing from any major spatial data source including: shape files, SQL Server, PostGIS, DB2, Oracle, WFS, TIFF Images, MySQL; • Integrates with existing API's like Google, yahoo, etc. • Connects to Any GeoServer- WMS services can be easily integrated with any available server for publication of web service over the web.
Map Guide	• Windows and Linux, Apache and IIS, and multiple browser support • Interactive map viewing • Quality cartographic output • Built-in storage of XML resource documents • Uniform access of vector and raster data using Feature Data Objects (FDO) API • Flexible application development: PHP, .NET, Java • Extensive server-side APIs • Fast, scalable, secure server platform
Open Layers	• Support for a variety of data sources. • Support for displaying geographic features, with markers and popups. • OpenLayers offers easy build configuration, • designed to help into other applications. • Javascript API to allow full control over • OpenLayers-powered map from within Javascript on a web page.
MapFish	• Flexible and complete framework for building rich web-mapping applications. • Extends Pylons with geospatial-specific functionality such as for creating web services that allows querying and editing geographic objects. • Compliant with the Open Geospatial Consortium standards. This is achieved through OpenLayers or GeoExt supporting several OGC norms, like WMS, WFS, WMC, KML, GML etc.
Mapbender	• Displaying, navigating, editing and querying spatial data and maps. • Map services authorization services (OWS proxy functionality). • Management interfaces for user, group and service administration. • Management of WMS and WFS. • User interface configuration and tool stored in data base.
GeoTools	• Supports OGC Grid Coverage implementation, graphs and networks. • Coordinate reference system and transformation support. • Symbology using OGC Styled Layer Descriptor (SLD) specification. • Attribute and spatial filters using OGC Filter • Encoding specification. • Java Topology Suite (JTS) - with support for the OGC Simple Features Specification used as the geometry model for vector features. • Two Renderers: Lite Renderer - a stateless, low memory renderer, particularly useful in server side environments. J2D - a state full renderer with optimizations for update rate and good for interactive client-side displays.

continued on following page

Table 1. Continued

GIS Software	Key Features
Geospatial Data Abstraction Library (GDAL/ OGR)	• Library access from Python, Java, C#, Ruby, VB6 and Perl • Vector data model closely aligned with OGC Simple Features • Coordinate system engine built on PROJ.4 and OGC Well Known Text coordinate system descriptions • Utilities for data translation, image warping, subsetting, and various other common tasks • Highly efficiency raster data access, taking advantage of tiling and overviews • Support for large files - larger than 4GB
QGIS	• Create, edit, visualize, analyses and publish geospatial information for Windows, Mac, Linux, BSD and Android operating systems. With the use of QGIS • Browser, user can browse and preview data, metadata, drag and drop his data from one data store into the other one. • OGC compatible WMS and WFS services. QGIS Web Client allows publishing QGIS projects on the web with ease. • Powerful symbology, labeling and blending features helps in serving impressive maps. *(Source-* http://www.osgeo.org*)*
GRASS GIS	• Used for data management, image processing, graphics production, spatial modeling, and visualization of many types of data.
Open JUMP	• GIS Data interoperability via Open GIS and Cross platform • Highly efficient and reliabile
gvSIG	• user-friendly interface, being able to access the most common formats, both vector and raster ones. • features a wide range of tools (query tools, layout creation, geoprocessing, networks, etc.). • desktop application designed for capturing, storing, handling, analyzing and deploying any kind of referenced geographic information • GIS Data interoperability via Open GIS and Cross platform • Highly efficient and reliabile • Multilingual
ILWIS	• Vector and Raster capacity integrated into one package. • user-friendliness, it takes less time to learn. • Free open source software available for downloading by anyone • Some of the greatest selections of import and export modules of widely used data formats • On-screen and tablet digitizing • Comprehensive set of image processing tools, Orthophoto, image georeferencing, transformation and mosaicing • Advanced modeling and spatial data analysis, 3D visualization with interactive editing for optimal view findings • Geo-statistical analyses, with Kriging for improved interpolation, Production and visualization of stereo image pairs; Spatial Multiple Criteria Evaluation.
GeoNetwork	• Immediate search access to local and distributed geospatial catalogues • Up- and downloading of data, graphics, documents, pdf files and any other content type • An interactive Web Map Viewer to combine Web Map Services from distributed servers around the world • Online editing of metadata with a powerful template system • Scheduled harvesting and synchronization of metadata between distributed catalogues • Support for OGC-CSW 2.0.2 ISO Profile, OAI-PMH, Z39.50 protocols • Fine-grained access control with group and user management • Multi-lingual user interface

Cross Platform Operation: Cross-platform refers to software that can be made to work on multi-platforms, such as, the UNIX systems including Linux, Windows and Mac OS. Cross-platform has been achieved in various ways. In recent years, cross-platform has been adopted by more developers as a basic requirement for software development.

Installation: The users always prefer GIS software that is easy to install. The installation procedure should be straight forward with package being unbundled and installed with minimum efforts. Some GIS software requires a number other supporting packages to be downloaded and installed simultaneously making it inconvenient for the users. Potential candidates for open sources GIS software were subsequently downloaded and installed firstly on a PC (Pentium (R), Dual-Core CPU, E5200@ 2.50 GHZ, 1.98 Gb of RAM, Microsoft Windows, XP Professional, Version 2002, Service Pack2). A relatively new Sony laptop computer with MS Windows Vista has also been used to test the Vista compliance (Chen et al., 2010). The prerequisites and outcome of the installations are demonstrated in Table 2.

Open source Operation System: As we are looking for Open source GIS, the issue of open source operation system (OS) would naturally be brought up. Linux is a well-known open source OS and is getting more popular; however MS windows are still the dominant.

Programming languages: There are programming languages used in the development of open source GIS, like Python and Eclipse in Java programming and .NET framework in the Windows environment. A number of Open source GIS softwares with operating system and programming languages are shown in Table 3.

Maturity: Some packages may have great potential in the future but are currently still in the early or middle part of their developmental stages. Therefore each type of OGIS software need to carefully analysed in terms of maturity.

Popularity of the Software: For various reasons, there are some software packages which have drawn more attention to the user community, and they may be used more widely and developed further in the future. For example, some popular softwares mentioned in the OSGeo homepage for DeskTop GIS are GRASS GIS, OSSIM, Quantum GIS, gvSIG.

LEVEL OF COMPLEXITY: SOME SOFTWARE IS MORE COMPLEX IN TERMS OF ANALYTICAL FUNCTIONS

User Friendliness: It is an important criterion as ease of using various functions for GIS analysis can even encourage novice user to use the software. It also increases popularity among various users staring from beginners to expert users.

Data format, database and interoperability: Some software has limited capability to recognise and read wide varies of data formats. There is a need to have compliance with the openGIS standards. Also access to the database should be less time consuming. There are software packages which require less start up time to reading data while some may be faster and download large size of data as compared to other Open GIS software. The lack of an interoperable and object oriented database system are

Table 2. Prerequisites and Download file size for installations of potential Open source GIS software

Name	Prerequisites	Download File Size (Mb)	Installation
DIVA GIS V6.03.0	Java and Eclipse	120	Suitable for both XP and Vista
FWTools	Java, Python	20.3	Suitable for both XP and Vista
GeoServer	Need JAVA JDK1.4 or newer	36.7	Easy to install but not run on both Vista and XP
GeOxygene	Java	10.6	Problems in installation
GRASS GIS	Python	44.3	Suitable for XP
gvSIG	VJM1.05.12, JAI 1.1, JAI image I/O 1.0 (16 + 6 + 10 Mb)	69.4	Easy, need to install 3 Java programmes. Suitable for XP,
HidroGIS	Jre1.4.2	155	but cannot read image files in Vista Installation instruction in Spanish, hard to guess
ILWIS	No	13.5	Run directly
MapWindow GIS	No need for JAVA; Net framework 2.0, is needed for XP (can be downloaded during the installation)	34.1	Easy and suitable for XP
OpenJUMP	JRE1.5.0 (happy with the jre1.05.12 for gvSIG)	11	Easy
QGIS V0.11.0	Python 2.5.2 (10.7 Mb)	73	Cannot be installed in Vista; Suitable for XP
SAGA GIS	No	7.4	Run directly
SavGIS	No	10.5	Easy
uDig	jre1.05.12 + a special uDig SDK kit (not Java SDK) including Java and Eclipse, 79.7 Mb	85.4	Easy

(Source: Chen et al, 2010)

the major problems regarding storage of enormous environmental and hydrological data produced over the years. Therefore a system capable of storing data based on specific characteristics or defined schema will enable easy storage and sorting of data. Database compatibility and data format for various GIS softwares are shown in Table 4.

Table 3. Operation system and programming language of various open source GIS

Name	Operating System	Programing Language
Apache Batik	Windows, Linux	Java 1.3
DIVA GIS	Windows only	Java with Eclipse
Deegree	Windows only	Java 1.5, tomcat 5.5
Fmaps	Linux and Gnome	C
FWTools	Windows, Linux	Java, python
GeOxygene	Independent	Java
GeoServer	Mac, Unix, and Windows	Java 1.4
Generic mapping tools GRASS GIS	Windows, Linux, mac Unix (Linux,) Window	C/C++ C
gvSIG	Windows, Linux, Mac OS	Java
HidroSIG	Windows and Linux	Java
ILWIS	Windows only	MS Visual 6
KOSMO	Windows, Mac OS X, Linux	Java
JTS Topology Suite	Windows, Linux	Java
Mapnik MapWindow GIS	Windows, Linux, Mac OS Windows only	C++, python, NET(VB,C++,C#),. Net framework 2.0
mezoGIS	Linux and Windows	Python
monoGIS	Linux and Windows	OGR/GDAL (C++), Shapelib (C), Net Topology Suite, (.NET, C#), Geotools.
NRDB OpenJUMP	Windows only Windows, Mac OS X, Linux	C++, Java
OpenMap	Windows, Linux, Mac OS	Java
OSSIM PostGIS	Windows, Linux, Mac OS Windows, Linux, Mac OS	C++, C
Quantum GIS	Linux, MS Windows Mac OS X, POSIX	C++
SAGA GIS SAMT	Windows and Linux Unix	C++ ?
SavGIS	Windows only	?
SharpMap	Windows only	C#,.NET, meno, Ermaper ECW SDK
Thuban	Linux, Windows and Mac OS	Python
uDig	Windows (not window2000), Mac OS X, Linux	Java with Eclipse

(Source: Adopted from Chen et al., 2010).

Functionalities: Some packages only served as tools or libraries so their functionality is limited. For example read and write access (Vector and Raster Data), database link, OGC compatibility, thematic mapping, GPS support, scripting

Table 4. Database compatibility and data format for various GIS softwares

Name	OpenGIS	Database	Data Format
Diva GIS	WMS, WFS	Yes	Shp, grd, tif, ipg, sid, arc
gvSIG	WMS, WCS,WFS (ArcIMS)	Yes	shp, gml, dxf, dwg, dgn, geoBD, WFS, WMS, WCS, ArcIMS
MapWindow GIS	WMS, WFS	Access, ArcXML	shp, bgd, bil, asc, ESRI grid, img,ESRI FLT, ddf, aux, dhm, bt, bmp, ecw, map, sid, LF2, kap, wmf
OpenJUMP	WMS	Yes	ESRI (shp), ecw, gml, xml, fme, jml, wkt, txt, WMS, database query
QGIS	WMS	PostGreSQL/PostGIS	ESRI (shp), mapinfo (mif), cadd, ddf, gml, tif, img, dem, asc, dt0
SAGA GIS	No	Via ODBC	ESRI E00, GPX, GDAL, DXF, SBF, ODBC
FWTools	WMS	MySQL	VPF (i.e. VMAP, VITD), RPF (i.e. CADRG, CIB), and ADRG
GeoServer	WMS, WFS	PostGreSQL/PostGIS, Oracle	VPF, MapInfo, and Cascading WFS, MrSID, ECW, JPEG2000, DTED, Erdas Imagine, and NITF
HidroSIG	WMS, WFS	MySQL	ESRI (shp), dxf, jpeg, tiff, gif, bmp
GeOxygene	WMS, WFS	Oracle and PostGreSQL/ PostGIS	ESRI (shp), dxf, jpeg, tiff, gif, bmp
SavGIS	WMS, WFS	SavGIS databases (called SavBase, proprietary format)	ESRI (shp), dxf, tiff, gif, bmp
ILWIS	WMS, WFS	PostGreSQL/PostGIS	ESRI (shp), dxf, jpeg, tiff, gif, bmp
GRASS GIS	WMS, WFS	MySQL, PostGreSQL/ PostGIS, SQLite	ESRI (shp), dxf, TINs, jpeg, tiff, gif, bmp
uDig	WMS, WFS	ArcSDE,DB2,Oracle, ArcSDE, PostGIS	ArcSDE, DB2, Map Graphic, Oracle Spatial, PostGIS, WFS, WMS

functionality, coordinate projection are all important features based on which Open GIS software can be differentiated.

Performance, Geo-spatial simulation and visualization: The geo-spatial simulation including lines, polylines, areas, multi-areas, ellipses, circles, and rectangles as 2D graphics and visualizing them in triangulated irregular networks (TINs) and digital terrain models (DTMs), relief shading and surface draping of raster images in 3D is an added functionality for Open GIS. Following open source GIS software were tested for reading and panning an image size of 125 MB using a PC (Pentium (R), Dual-Core CPU, E5200@ 2.50 GHZ, 1.98 Gb of RAM, Microsoft Windows, XP Professional, Version 2002, Service Pack2) as shown in Table 5.

Security: Security is an important issue as some Open GIS may have connectivity to database allowing user to update and modify data based on particular schema. It is important to protect data, such as a database, from destructive forces and from the unwanted actions of unauthorized users.

Table 5. Performance of open source GIS

Name	Reading 125 Mb tif Image (Second)	Panning a 125 Mb Image	Time for Zooming (Second)	Comments
Diva GIS	V6.03 is very slow as uDig	Took a long time	Long	Not acceptable for large size images
gvSIG	2	Easy	<1	
MapWindow GIS	9	Easy	1	
OpenJUMP	2	Easy	<1	
ILIWS	4	Easy	<1	
HidroSIG	5	Easy	<1	
SavGIS	3	Easy	<1	
GeOxygene	11	Took longer time	Long	
GeoServer	13	Took Longer time	Long	
GRASS GIS	4	Easy	<1	
FWTools	3	Easy	<1	
QGIS SAGA GIS	2 Cannot read.tif	Easy	<1	Too few image formats
uDig	125 Mb is too big to handle, read a 47.7 Mb tif image in 317s	Took more than 350 s for a simple navigation function or zooming on a 47.7 Mb image	359s	Not acceptable for large size images

(Source: Chen et al., 2010)

Statistical and Geostatistical Analysis: Statistical analysis extract additional information from GIS data such as how attribute values are distributed, whether there are spatial trends in the data, or whether the features form spatial patterns. It is often used to examine the distribution of values for a particular attribute or to spot extreme high or low values. It is useful when defining classes and ranges on a map, when reclassifying data, or when looking for data errors. Geostatistical Analysis uses sample points taken at different locations in a landscape and creates a continuous surface. The sample points are measurements of some phenomenon, such as radiation leaking from a nuclear power plant, an oil spill, or elevation heights. Therefore, Statistical and Geostatistical Analysis is an important criteria for assessing the Open GIS software.

It is difficult to make a comparison among various criteria. However, a matrix and multi-criteria analysis method can be used to select specific OGIS software. This will enable users to quickly identify specific OGIS software that will satisfy the user's requirement.

Matrix Analysis

Matrix Analysis for decision making is an important tool for making a decision. It is mostly useful when many features need to be taken into consideration to select an alternative from number of potential alternatives. This makes it easy to come up with a decision when the option is not clearly understood. Matrix Analysis enables one to take decisions smartly and logically when confusion may arise due to complexity of having too many criteria to be taken into consideration. It is very handy at a time when one might be struggling to take a decision. For example Table 6 shows the matrix anlaysis for various Open source GIS softwares.

Multi-Criterion Analysis

The multi-criterion analysis also known as weighted scoring method includes identification of all the attributes of criteria that are to be evaluated. The distribution of weights to each of them indicates their relative importance; and the obtained scores to each option represent how it performs with respect to each feature. The result is a single weighted score for each option, which may be used to indicate and compare the overall performance of the options.

This process necessarily assigns numeric values to judgements. These judgements should not be arbitrary or subjective, but should reflect expert views, and should be supported by objective information. To achieve meaningful results which decision-makers can rely on, it is important that:

- The exercise is not left to the "experts", but is undertaken by a group of people who represent all of the interested parties;
- The group possesses the relevant knowledge and expertise required to make credible measurements and judgments of how the options will impact upon the attributes;
- The group is led by an independent chairman to steer the process, probe opinions, promote consensus and avoid prejudice; and
- The justification for the group's chosen weights and scores is fully explained.

The selected software are to be evaluated based on the weight and scale as given in Table 7 and the total score is to be calculated by using the following formula:

$$\sum_{i=0}^{n}\left(\text{CiWiSi}\right)\ \text{x100}$$

Table 6. Matrix analysis for various Open source GIS softwares

OGIS Software	Cross Platform Operation	Installation	Programming languages	Open source Operation System	Maturity	Level of Simplicity	Data format, database and interoperability	Functionalities	Geo-spatial simulation and visualization	Security	Statistical and Geostatistical Analysis
Diva GIS	+	+	+	*	--	--	+	*	*	*	-
gvSIG	+	++	*	*	+	*	++	++	+	++	+
MapWindow GIS	+	++	+	*	+	--	++	+	+	+	+
OpenJUMP	+	++	+	++	*	+	++	++	-	++	+
ILIWS	*	*	*	*	-	--	-	*	+	*	+
GRASS	+	*	++	++	++	*	-	*	++	++	++
SavGIS	+	*	*	+	*	-	-	*	*	*	+
GeOxygene	+	*	+	++	--	++	-	*	+	++	+
GeoServer	++	+	+	++	*	++	++	*	*	++	-

(Source: Modified from Chen et. al., 2010; Akbari and Rajabi, 2013).

Table 7. Selection criteria using weighted method based on the pair wise comparison

Selection Criteria	Weight (W)	Performance Scale (Very Poor-Poor-Average-Good-Excellent) (S)	Remarks
Internet connection and security (C1)	0.125	Scale (0-0.25-0.5-0.75-1)	Twice as important than Operating System and programming language as it is expected WKMP will be Internet based and therefore should be well secured
Operation System (C2)	0.063		Users are less bothered about the operating system.
Programming language (C3)	0.063		Given less priority as its final users are not expected do any programming or write any codes.
Level of simplicity (C4)	0.125		As important as Internet and security as more simplicity means more user friendliness.
Data format and database (C5)	0.125		As important as user friendliness as wider data access means more dissemination of information.
Maturity (C6)	0.125		As important as data format and database capacity as matured and well developed software are less likely to face any obstacles than those software still in its development phase
Installation and test running (C7)	0.125		Top priority given as easy installation and speed indicate wider acceptability
Functionalities (C8)	0.125		As important as data format and database as functionalities should be well defined and easy to operate

where, $C_{i\,=}$ criteria, $W_{i\,=}$ weight for that particular criteria, $S_{i\,=}$ scale factor or performance factor for that particular criteria.

However, a combination of both Matrix analysis and Multi-criteria analysis can be done where Matrix analysis can be used for primary criteria whereas Multi-criteria analysis using weighting and scoring can be applied for assessment of secondary criteria as shown in Figure 3.

CONCLUSION

The number of open source software is emerging very rapidly due to their popularity, interoperability and cross-platform performance. A number of assessment criteria such as cross platform operation, installation and test running, programming languages, level of simplicity, data format and database, maturity, functionalities are taken into consideration. Two methods such as matrix analysis and multi-criteria analysis can be used. Matrix Analysis provides a general overview of various criteria assessed over a performance scale (very poor-poor-average-good-excellent). While multi-criteria analysis compares various criteria using weight and assigned value ranging from 0-0.25-0.5-0.75-1 for the scale (very poor-poor-average-good-excellent). Therefore, multi-criteria analysis is more specific in comparing among potential OGIS software

Figure 3. Stepwise screening using primary and secondary criteria

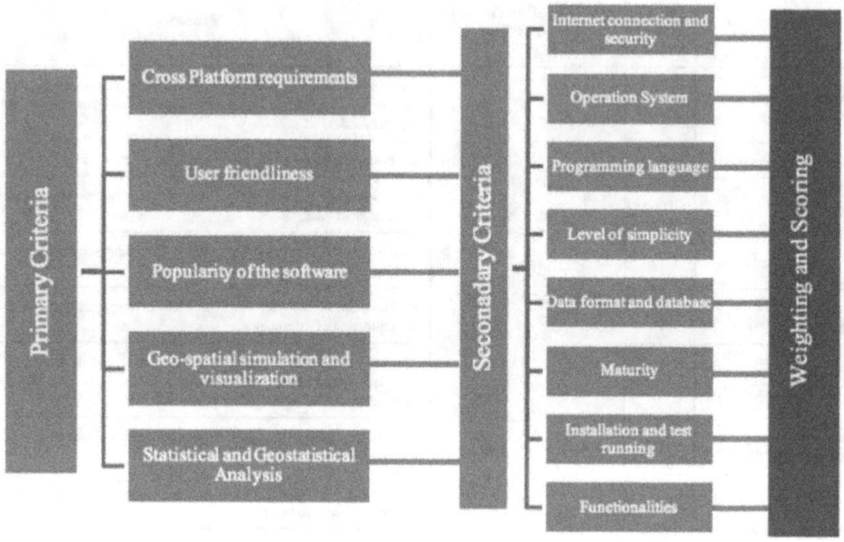

with scored assigned to each software during the assessment process and thus making it easy to select particular software. Another approach is to apply Matrix analysis to shorten the list of potential software from a large number of software and then use Multi-criteria analysis for the shortened list to come with the most suitable software assed based on the highest scoring.

REFERENCES

Akbari, M., & Rajabi, M. A. (2013). Evaluation of Desktop free/Open source GIS software based on functional and non-functional capabilities. *Technical Gazette, 20*(5), 755–764.

Câmara, G. (2003). *Why Open Source GIS Software?* Retrieved from http://www.directionsmag.com/entry/why-open-source-gis-software/123925

Chen, D., Shams, S., Carmona-Moreno, C., & Leone, A. (2010, October). Assessment of Open Source GIS Software for Water Resources Management in Developing Countries. *Journal of Hydro-environment Research, 4*(3), 253–264. doi:10.1016/j.jher.2010.04.017

Confino, J. P., & Laplante, P. A. (2010). An open Source Software Evaluation Model. *International Journal of Strategic Information Technology and Applications*, *1*(1), 60–77. doi:10.4018/jsita.2010101505

Donnelly, F. P. (2010). Evaluating open source GIS for libraries. *Library Hi Tech*, *28*(1), 131–151. doi:10.1108/07378831011026742

Free Software Foundation. (n.d.). Retrieved from http://www.fsf.org/licensing/essays/free-sw.html

Galli, P. (n.d.). *Open Source Is the Big Disruptor*. Retrieved from http://www.eweek.com/article2/0,1895,2186932,00.asp

Hartnett, P., & Bertolotto, M. (2003). Developing interoperable web-based geo-spatial Digital. GIRSUK 2003. *Session*, *5b*, 185–188.

Ca^mara, G., & Onsrud, H. (2004). Open-source geographic information systems software: myths and realities. In *Open Access and the Public Domain in Digital Data and Information for Science: Proceedings of an International Symposium*. National Research Council, U.S. National Committee for CODATA.

Morrison, K. W., & Purves, R. S. (2002). Customizable landscape visualizations. Implementation, application and testing of a web-based tool. *Computers, Environment and Urban Systems*, *26*(2-3), 163–183. doi:10.1016/S0198-9715(01)00033-3

Shekhar, S., Vatsavai, R. R., Sahay, N., Burk, T. E., & Lime, S. (2001). WMS and GML based interoperable web mapping system. *Proceedings of the ninth ACM international symposium on Advances in geographic information systems*, 106-111. 10.1145/512161.512185

Sherman, G. E. (2008). *Desktop GIS: Mapping the Planet with Open Source Tools*. The Pragmatic Programmers LLC.

Steiniger, S., & Bocher, E. (2009). An overview on current free and open source desktop GIS developments. *International Journal of Geographical Information Science*, *23*(10), 1345–1370. doi:10.1080/13658810802634956

Tiemann, M. (2009). *How Open Source Software can Save the ICT Industry One Trillion Dollars per Year.* White paper. Retrieved from http://www.opensource.org/files/OSS-2010.pdf

Chapter 3
Nurturing a Geospatially Empowered Next Generation

Derek Starkenburg
University of Michigan, USA

Christine F. Waigl
University of Alaska – Fairbanks, USA

Rudiger Gens
University of Alaska – Fairbanks, USA

ABSTRACT

For new generations of citizens in all countries, a level of proficiency in geospatial concepts and skills will be required to realize the potential of professional and developmental opportunities. The teaching of geospatial skills links into traditional science, technology, engineering, and mathematics (STEM) curriculum objectives, community-wide concerns and initiatives, and global citizenship. Therefore, by the pre-university and undergraduate level, it is desirable for each student to have acquired such competencies. Free and open-source tools that are accessible and affordable in most areas of the world, along with data availability, offer an opportunity to support teaching such a curriculum. Here, core geospatial concepts are introduced, along with available data and tools. Then, using three scenarios, it is shown how the core concepts can be applied to different settings for educational purposes.

INTRODUCTION AND BACKGROUND

In order to educate the next generation of geographic information system (GIS) users, this chapter organizes all learning objectives based upon the core concepts outlined by Kuhn (2012). These concepts are summarized in Table 1.

DOI: 10.4018/978-1-5225-5039-6.ch003

Table 1. Core concepts of spatial information from Kuhn (2012)

Concept		Question
Spatial	Location	Where are you? Where is "here"?
	Neighborhood	What is nearby?
	Field	What qualities are associated with "here"?
	Object	What properties define the things that are here?
	Network	On what links and pathways can physical goods or information travel?
	Event	What spatio-temporal change occurred?
Information	Granularity	On what scale and resolution is the spatial information provided?
	Accuracy	How correct is the spatial information?
	Meaning	How to interpret it? What does it imply?
	Value	What is its relevance? How does it relate to other societal values?

This chapter will explain these core concepts using three scenarios on different continents. These scenarios contain the basic structural underpinning for pedagogical units that can be developed by teaching staff for use with their classes and within the locally applicable curricular framework. The scenarios can be generally adapted to a large variety of settings. This approach demonstrates that the questions behind the concepts are the same regardless of the environment in which the learning takes place. However, depending on the context, these questions can be answered in very different ways. To begin, a comprehensive set of available data and tools are presented. Then, the application of some easily accessible tools are demonstrated via the scenarios.

DATA

Any educator who wishes to incorporate geospatial learning goals into his or her teaching practice will be confronted with the question of where to find, and how to select, the appropriate data and tools to bring the scenarios and activities to life. The data will be of a type for use with GIS or geospatially capable software. The availability of professional development opportunity for educators to acquire and develop the requisite skills is critical for a wide-spread rollout of a geospatial curriculum. Pioneers and innovators, though, will find a wealth of opportunities on the internet for independent study, thanks to the strong ethos open-source software and open-data among both developers and data providers (such as local, regional and national government entities, as well as public research institutions) in this area.

Geospatial data can be broadly divided into two categories: raster and vector data. Raster data are images composed of many picture elements known as pixels. Each pixel contains location information as well as a pixel value or pixel attribute (e.g. brightness, height, etc.). Vector data represent geographic features such as points, lines or polygons. The information content is linked with the geolocation as well as the attributes of the data set.

Geospatial data are stored in a number of formats. Most images, or so-called raster data, are stored in the GeoTIFF format that keeps both data and metadata – the information that describes the data, specifically its georeference information – in a single file. The equivalent for point information, also referred to as vector data, is the shapefile format. However, in the recent past the personal geodatabase format is increasingly used within the most widely used commercial GIS package, ArcGIS. Though a personal geodatabase provides flexibility and makes it easier to retrieve and analyze data, it is a proprietary database format and, therefore, not likely to become a generally accepted standard. A large variety of text file formats that can be converted into the common vector formats are used to store geospatial information. The comma-separated values (CSV) file format are taken for their ease of use, while the extensible markup language (XML) file format is known for its flexibility in storing complex information and for its power in being transformed into other formats. Common GIS data formats include ESRI™ Shapefile (proprietary), GeoJSON (open), Google's KML (proprietary) and the Global Positioning System (GPS) Exchange Format GPX, which is an XML schema. GIS desktop software is typically able to read and convert between all of these formats. For quick conversions and visualizations, an online tool such as GPSVisualizer offers great simplicity and convenience.

The Database of Global Administrative Areas (GADM, 2015) provides administrative boundary files on various levels from the global to the local scale in a number of geospatial formats. These files can be used as a reference for almost any geospatial project. Another source for reference maps is the collaborative project OpenStreetMap, providing a free editable map of the world, usable within a GIS but also on websites and GPS devices (Bennett, 2010). Global and national administrative and basic climatic and demographic datasets are also available via portals such as Diva-GIS and Natural Earth Data.

Wikipedia is a free encyclopedia that is open for people to contribute, available in 288 languages. Most pages about places contain geolocation information that is linked to GeoHack, a geolocation interface from the Wikimedia Tool Labs. The coordinates can be exported into a large number of formats and are also directly linked to a large variety of geospatial services. Wikipedia also contains climate summaries for many cities gathered from a variety of sources, which enables students to quickly investigate climatic data tied to their location.

TOOLS

One of the most widely used cross-platform free, open-source and multilingual (as of 2012 in 48 languages) GIS applications is QGIS (2015), originally developed as an incubator project of the Open Source Geospatial Foundation (OSGeo). The application is developed by an active community and updated by frequent releases. It has been adopted by public entities such the Austrian state of Vorarlberg and the Swiss Canton of Solothurn (QGIS, 2015). Another free and open-source Graphical User Interface (GUI)-driven GIS system is the User-Friendly Desktop Internet GIS (uDig). Like QGIS, it is a cross-platform application and reasonably well localized (14 languages as of the time of writing), but its set of features is considerably narrower. In the educational context, simpler software can, however, be an advantage.

The GeoServer is an example of an open-source server that allows the sharing of geospatial data (Youngblood & Iacovella, 2013). While it supports static file formats for raster and vector data, it is particularly useful in connection with geospatial databases such as PostGIS, an open-source geographic extension for the PostgreSQL object-relational database (Marquez, 2015). This allows the dynamic distribution of GIS layer information.

Most GIS and image processing tools are using the Geospatial Data Abstraction Library (GDAL) for reading and writing a large variety of geospatial data formats. GDAL also provides a number of powerful command-line tools to manipulate geospatial data. Coding languages such as Python allow users to overlay GeoTIFF files onto base maps to display geospatial information in figures and animations.

Google Earth™, especially the now freely available Google Earth Pro™, cannot be considered a comprehensive GIS. While Google Earth Pro™ is able to handle a number of GIS data formats, its main purpose is visualization, and not advanced geospatial analysis. This and other so-called virtual globes have very limited support for map projections and analysis tools, but can still be used by students for gathering qualitative information. Google Earth™ is also not an open-source application.

In addition to desktop software, there is a new class of geospatial tools to which the 21st century student has access. These are hand-held tools such as smartphones, e-readers, gaming, fitness and timekeeping devices (and indeed many more) which to-date are often routinely equipped with GPS receivers and capable of running small location-based applications ("apps") that are available for no or low cost. These are called geo-media devices or applications (SPACIT, 2015), and they can serve as affordable data acquisition devices to support learning activities.

Table 2 provides a summary of useful OSGS tools and data. The next section describes three scenarios that allow opportunities for students to investigate some of the geospatial data available. The scenarios are intended as springboard to stimulate the creative development of a variety of geospatial based learning activities tied to STEM curricula

Table 2. Summary of some useful tools and data

Online Resource	URL
QGIS	http://www.qgis.org/en/site/
QGIS tutorials	http://www.qgistutorials.com/en/
uDig	http://udig.refractions.net/
GPSVisualizer (online tool)	http://www.gpsvisualizer.com/
Geoserver	http://geoserver.org/
Openstreetmap	https://www.openstreetmap.org/
Python programming language	http://python.org
R programming language	https://www.r-project.org/
Wikimedia Tools Lab GeoHack	http://tools.wmflabs.org/
Diva-GIS free data	http://www.diva-gis.org/Data
GDAM	http://www.gadm.org
Natural Earth (raster &vector data)	http://www.naturalearthdata.com/

SCENARIOS

Urban Scenario

For an urban setting exemplified via the city of London, the focus is urban growth and land use (e.g., the design of a new building), within the context of the urban fabric and existing transportation network (Figure 1). In this scenario, the location denotes where the new building will be constructed. In its conception, a building project is the object and event of interest whose value is perceived as economic profit. Granularity describes the range of spatial scales over which one considers the surrounding context (e.g., everything within a 2 km radius, or only the surrounding block). Networks form the transportation that allows access to and from the site. The neighborhood contains other buildings and attractions which represent important field values that define the context. As the building design progresses, different drawings convey specific meaning about the new building.

The volume of data available for the London scenario is substantial. The Greater London Authority is the provider of a large number of geospatial data sets and is following an open data policy. Transport for London (TfL) exemplifies all the common ways open data can be distributed. Geospatial data that is static and does not change frequently over time is provided as shapefiles or text file formats. Furthermore, TfL offers a web user interface for searching the status and schedule of all modes of transport. An application programming interface (API) defines a

Figure 1. The city of London as an example for an urban scenario. The keywords on the right are potential axes of inquiry that can be developed into a variety of activities (e.g., field work, open source geospatial data analysis, etc.).

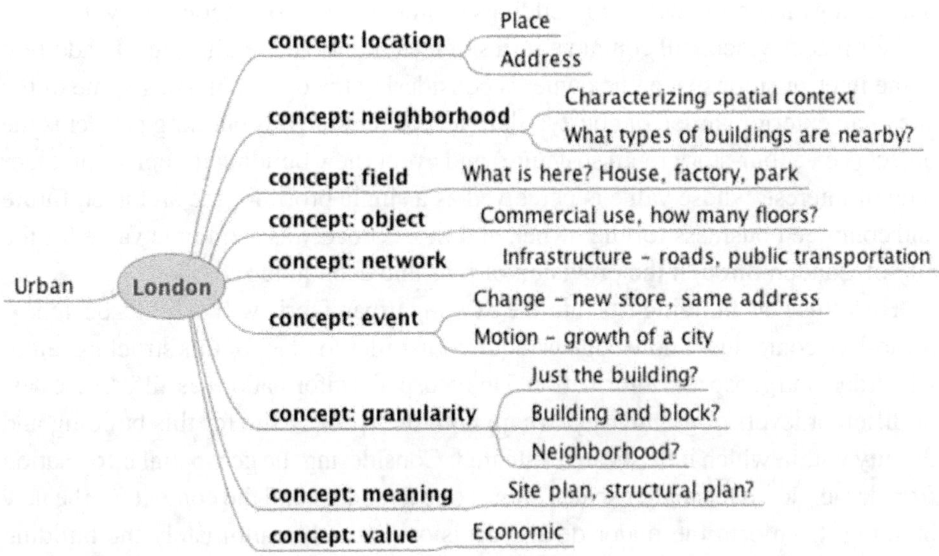

set of routines, protocols and tools for building software applications that distribute the geospatial information stored in databases in real time and that provide flexible access to the geospatial data. Alternatively, the databases also generate data feeds for the most common data searches that users can subscribe to. Table 3 highlights some useful websites to guide this urban scenario.

In this scenario, students will be asked to focus on a particular site location in downtown London. The location is the northwest side of the intersection of Chandos Place and William IV Street. At this location, there are a series of four-story buildings oriented in the southwest-northeast direction along Chandos Place. Download Google Earth™ (Table 3) and then locate this site by entering "Chandos Place and

Table 3. Helpful websites to guide the urban scenario

Online Resource	URL
Google Earth™	https://www.google.com/earth/
Transport for London (TfL)	https://tfl.gov.uk/maps
Google Maps	https://www.google.com/maps
CTBUH (Council on Tall Buildings and Urban Habitat)	http://www.ctbuh.org/TallBuildings/HeightStatistics/HeightCalculator/tabid/1007/language/en-US/Default.aspx

William IV Street, London." Think about the location specifically: what are some of the street addresses for the existing buildings at the NW corner of this intersection?

For this exercise, imagine that a developer (hereafter, "the owner") is considering purchasing this entire section of buildings in order to erect a new, four-story structure in their place, which will contain a series of restaurants and retail shops. In addition to the interior floor space, the owner is considering the option of using some of the site as an exterior garden or courtyard. In its conception, this building project is the object (i.e., a four-story retail structure) and event (new building to replace an older one) of interest, whose value is perceived as a single profit for the architect, future and continued business for the owner, and more broadly as economic value for the city of London through the provision of jobs and economic activity.

How this new building fits into the existing urban fabric will drive its design. A breadth of contextual knowledge is required in order to plan for this structure, all of which demand geospatial information. This geospatial information has different scales, or different levels of granularity, where both the specific site for this building and the city within which it exists are evaluated. Considering the geospatial information from larger down to smaller scales allows one to understand the context for the new building, to inform the major design decisions, and then ultimately the building itself in more detail.

The larger city scale involves analysis of the network of transportation that enables people to access the building. A map of the London public transportation system will enable one to see where different transportation routes exist, and where primary nodes (stations) are located. Navigate to the London public transportation page (Transport for London, Table 3), and enter "Chandos Place, London WC2N 4HS" in the empty field. Locations of various stations will appear (Figure 2). Click on one of the stations and its routes will appear on the map. Zoom out to get a sense of how many locations throughout London could use these routes to access the building site. Click on several stations and take note of where the routes go. How many different residential areas outside of downtown could use one of these routes to access the new building site? Connectivity plays a role in business success. Based on this study, can one confirm this site is well situated in terms of accessibility? Navigate to Google Maps (Table 3), and using this location (Chandos Place, London WC2N 4HS), obtain directions from one's own residence and place of work to the new site. How long will it take to get there by bus? Get a list of different residences and work places from other people, and try the same exercise.

Now consider that this site is located within an existing neighborhood that contains other buildings and outdoor spaces. Return to Google Maps and enter the location (Chandos Place, London WC2N 4HS). Once the map has zoomed to the site, clear the field and enter "attractions." The map will now display a myriad of other important places in your neighborhood, including theaters, restaurants,

Figure 2. Example of transportation routes to and from the study site

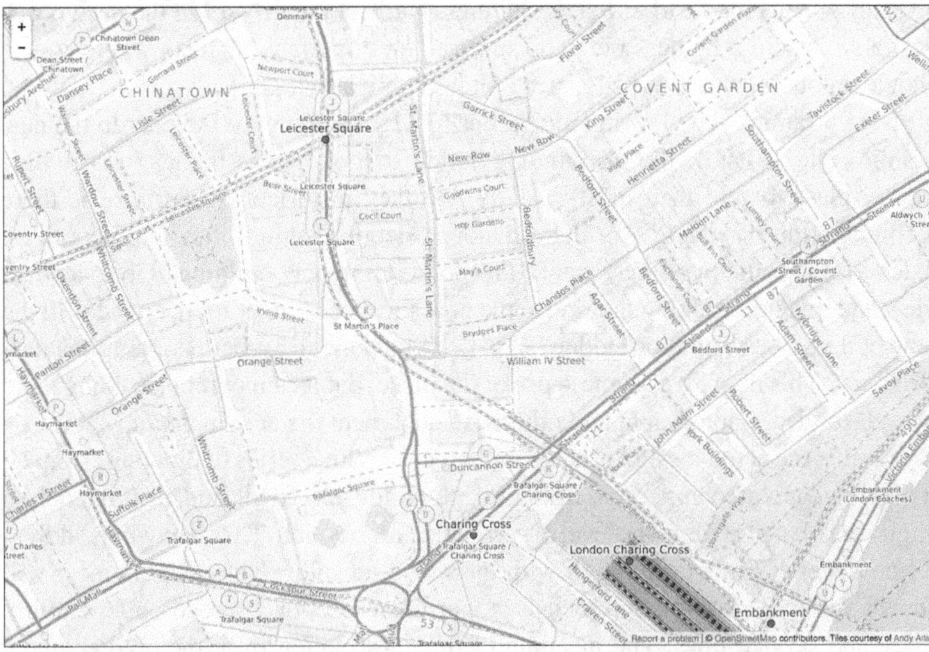

churches, along with parks and historical landmarks. Zoom out to a radius of about 2 km, and make a table showing how many of each type of attraction is within this range of the building site. In doing so, you are evaluating the attractions as a field of information, each point in the field being a particular place that someone may desire to go either on the way to, or from, the new building. Are these attractions complimentary to the intended function of the new building? Now use this field of information to further refine the design decisions for this structure. For instance, what type of retail spaces are lacking in this part of the city that should be included in the new building? Are there enough parks and open spaces in the neighborhood, or should this new building incorporate a public outdoor courtyard? Is there a major historical structure nearby to which the new building should respond with a view?

Now, return to this location on Google Earth™. Look at the location for this new building in a flat, plan view (from directly overhead). Make a sketch of the buildings and streets that surround the site. Now pan around in Google Earth™ to get a sense of how tall these buildings are. Students can do a crude estimation by adding up the number of stories and multiplying by 3.5 m (CTBUH, Table 3). Mark the heights of the buildings on the sketch. Based on this information, will there be opportunities to get southern light at the new building site? If there are certain places on the site that are more likely to have southern light, perhaps these locations should be preserved for important public spaces like lobbies and dining areas. If one decided earlier that including a garden or courtyard in the design was desirable, would it be possible to do so based on this analysis? If so, where on site should an outdoor garden be located? Should it be a roof garden?

Next, examine the view in Google Earth™ by zooming even closer to the new building site. What do the surrounding buildings look like aesthetically? Are they mostly glass, or are they mostly solid? Are they historical, modern, or is there a combination of styles? Are there major horizontal datum lines (e.g., cornices, balconies) to which you may want to respond in the new building? Go to a street view and imagine walking the site. Is the new site closed in, or more open? Will the new building be visible or hidden? What can be done to attract people to the new structure? This map gives some sense of the place, but it cannot replace going there oneself to visit, photograph, and otherwise document the area in detail.

Finally, the actual design process will proceed with a series of drawings from all trades (e.g., architectural, civil, structural, plumbing, electrical, and heating/cooling engineers) that provide some level of geospatial information. These drawings address a finer level of granularity, each with their own meaning. The civil engineer will show a variety of information such as property lines, set-backs, roads, alleys, utility lines, and service lines. The meaning of this information from the civil engineer helps to inform the building size and shape, and puts constraints on its design. As the design progresses, precise information about the location of individual structural elements, such as columns, will be documented by the structural engineer. The meaning of the structural drawing lies in its grid of coordinates showing where structural support will be constructed.

Ultimately, the finished building may become an event once again in the future, if for example expansion or additions are added. Externally, other changes to the neighborhood will also provide events that inform feedback responses to the building, such as changes in the retail services or number of residential units provided by nearby structures, which could render the building more or less relevant in its future context. This demonstrates that the entire geospatial context of building design within an urban fabric should be perceived as a living process.

Recreational/Cultural Scenario

This scenario demonstrates how geospatial data can be used for analysis of the Iditarod Trail Sled Dog Race that takes place annually in Alaska, USA. In this case, the object is the team of dogs and their musher (those who race the sled dogs). The value of the event is recreational and cultural. The neighborhood is the continually changing environment surrounding the traveling dog team, and the trail itself is the network. Events mark special occasions along the race route, e.g. stops at locations such as checkpoints, advances in standing, or a decision to abandon the (extreme) venture. Therefore, the events provide meaning to the race by generating interest and excitement. The events are also characterized by their granularity, which can have a broad spatial and temporal scale when the team is racing quickly over vast distances, or a smaller scale during stops (Figure 3).

The data for any Iditarod trail related analysis need to be collected from various sources (Table 4). The Alaska Department of Natural Resources keeps the information about the Iditarod National Historic Trail, on which the Iditarod Trail Sled Dog Race is mostly run, as a line shapefile (Alaska Department of Natural Resources, Table 4). The coordinates of the checkpoints are available on the Iditarod website (ITC Geolocation Information, Table 4). In order to generate a map of the race (Figure 4), the information needs to be edited and put into the right format. The parts of the Iditarod trail that are not used for the sled dog race need to be removed. The individual checkpoint coordinates are collected and used to generate a point shapefile. Historic weather information can be downloaded from various weather services (e.g. Weather Underground). The weather information is usually available from nearby airports where these measurements are acquired on a regular basis. We

Figure 3. Same as Figure 1, except for the Iditarod dogsled race as a recreational/ cultural example

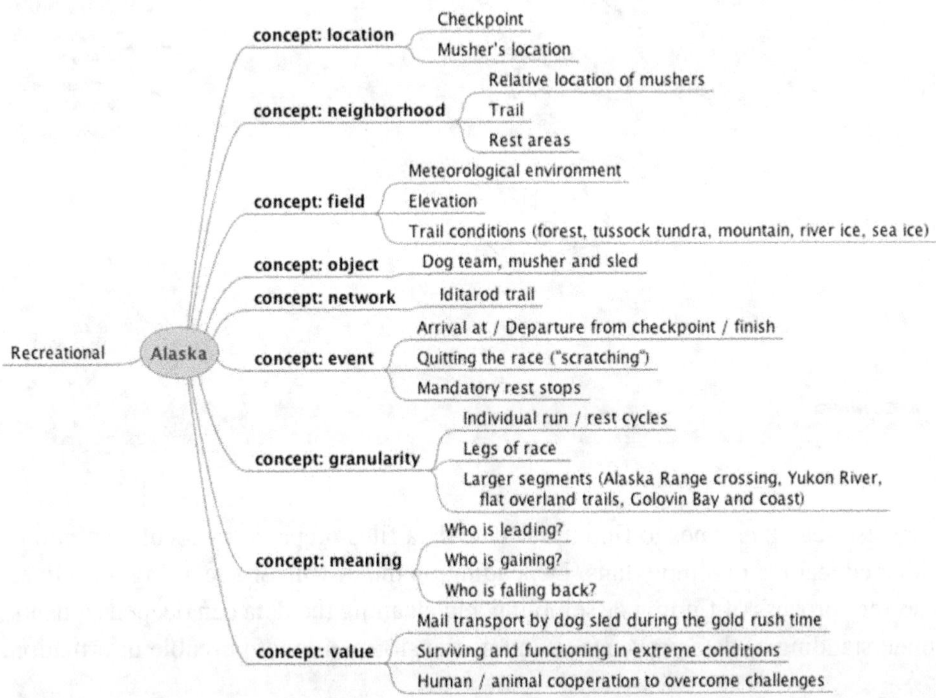

Table 4. Helpful websites to guide the cultural/recreational scenario

Online Resource	URL
Iditarod Trail Committee (ITC) History	http://iditarod.com/about/history/
ITC Education Portal	http://iditarod.com/teachers/
ITC Race and Checkpoint descriptions	http://iditarod.com/race-map/
ITC Geolocation Information	http://iditarod.com/race/checkpoints
ITC Today	http://iditarod.com/about/iditarod-today/
Alaska Dept. of Natural Resources: Iditarod Trail Shapefile	http://dnr.alaska.gov/mlw/mapguide/metadata/mdfiles/iditarod_trail.html
KML files from UAF GINA (2008 & 2009)	http://glink.gina.alaska.edu/public/iditarod/iditarod2008.kml
	http://kml.gina.alaska.edu/iditarod.kml
Mushing Tech Blog SHP and GPX	http://mushingtech.blogspot.com/2014/02/the-iditarod-track-file.html
1925 serum run information	https://en.wikipedia.org/wiki/1925_serum_run_to_Nome

Figure 4. Iditarod race map

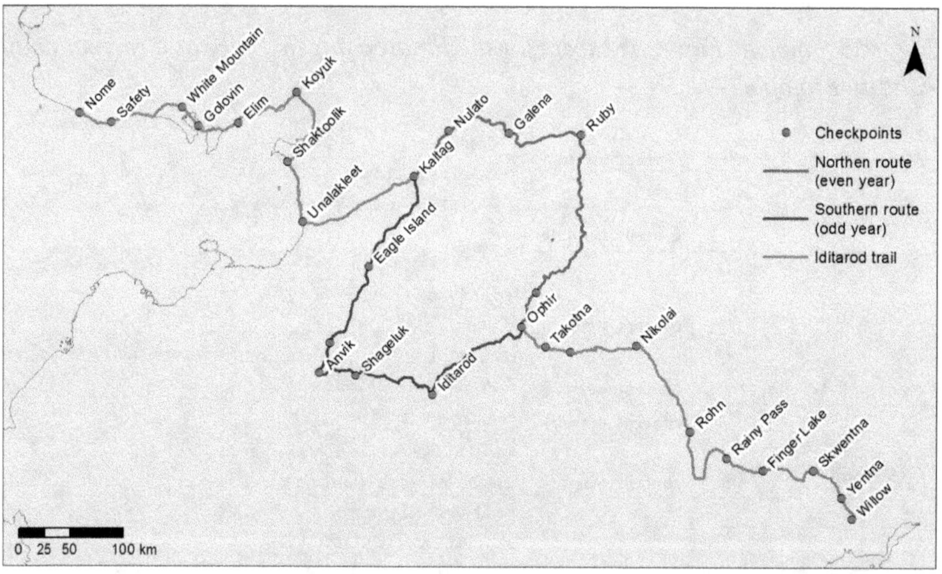

may use search engines to find up-to-date data files prepared by local institutions, other educators, or simply fans. Depending on the instructor's learning objectives, the very process of finding, assembling and cleaning the data can deepen students' understanding of data provenance and its transformation into useable information.

The Iditarod Trail Sled Dog Race (hereafter Iditarod) in Alaska is an example of a culturally situated recreational scenario. The Iditarod, organized by the Iditarod Trail Committee (ITC), is widely used to support pedagogical projects; the ITC education portal (Table 4) even provides collaboratively developed teaching material. Even though almost none of the existing lesson plans address geospatial learning, the existing wealth of documents can be used to generate introductory material and support a variety of activities.

Dog mushing remained a mode of travel for people in remote portions of Alaska well into the 20th century, until the snowmobile took precedent (ITC History, Table 4). To preserve dog mushing in the face of modernization and to commemorate the gold rush route from Seward to Nome known as the Iditarod Trail, the Iditarod race was conceived, and the first event was run in 1973 (ITC History, Table 4). The race also evokes an episode of the history of the circumpolar north, the 1925 serum run (Table 4), when diphtheria anti-toxin was transported by dog team to the remote coastal town of Nome, which was facing a rising, lethal epidemic. Racers of the Iditarod travel from Anchorage to Nome in March of every year (ITC Today, 2015). Typically, 70 to 80 participating dog teams cover the trail, stopping at 26 or 27 official, volunteer-staffed checkpoints, over a duration of 10 to 15 days approximately (ITC, Table 4).

For this scenario, one might imagine a musher while she is on the trail racing during an Iditarod. In this way, one considers the geospatial relevance as linear in time and place. At any moment in time, and even prior to the start of the race when the musher is preparing, she and her specific team of dogs form the object of interest. Both the musher and the dogs have specific levels of skill, ability, experience, and motivation that already precondition success. The value of the event is threefold: competitive for the musher and her team, culturally/historically relevant to Alaska, and profitable from the perspective of tourism (some businesses provide Iditarod package tours). Ultimately, the ability to experience the race requires knowledge of various geospatial concepts.

Since the Iditarod is a journey, the location and neighborhood of the musher are dynamic in that they continually change along the network. This is because the network is a single trail, along which the musher will have a specific location at any one point in time, as can be discerned using, for instance, GPS satellites (ITC, Table 4). The neighborhood puts the location in context because it is comprised of surrounding features such as such as topography, vegetation, and weather conditions. The neighborhood can therefore be described by its fields, or values of interest that determine comfort and safety. For instance, is the neighborhood relatively flat or is there an incline? Is the density of trees or a nearby hill enough to protect one from wind? What is the temperature distribution along the trail through a particular valley (are there cold pockets or milder locations on hilltops)? Is their deeply drifted snow on the other side of the upcoming ridge? The educator who chooses an activity based on this scenario during the time the race is taking place can elect for students to research the locations (start, finish, checkpoints), neighborhood, fields, and objects (dog teams and mushers) beforehand, thereby assembling and mapping the geospatial backdrop on which the events will play out.

Events occur along the network each time the musher stops at a checkpoint. Mandatory and voluntary stops provide the musher and their dogs with rest and sustenance, but they also mark progress on a map of the trail in time and space. Events also occur when a given musher advances beyond other teams that had been previously in the lead. The meaning of these events forms the heart of the race itself. Understanding where a musher is at any time, whether or not that musher is in motion or at rest, and knowing why, are all key to assessing who is in the lead and who is more likely to win the race. Is a musher at a certain checkpoint because she is on schedule and preparing to depart soon, or is she there for an extended time nursing an injured dog? Graphing the teams' progress, velocity along the trail and run/rest schedules opens opportunities for evaluation and comparison: qualitative (Can we discern differing strategies? Classify them by front-runners, mid-field, back-of-the-pack, or by rookies vs. veteran mushers?) or quantitative (Which teams appear to be progressing better or worse than others? What can we learn from a statistical analysis of run/rest times and velocities? What is the best way to evaluate the standing at a given moment?).

Events also possess granularity since the focus of the musher shifts from long stretches of upcoming trail (larger spatial scale), to primarily the local conditions of the check point and the routine procedure required to occur at that point. When in motion, updates on musher progress also provide a granularity in time, defining a cadence of the race by the frequency with which new information about the mushers and their progress can be discerned. From a spatial perspective, a spectator sustains some level of tension owing to accuracy, which determines the degree to which the musher location, and precise local trail conditions, can be defined. That is, some ambiguity in time and space mean that only the musher can know exactly his or her progress at every instant, which builds excitement for the onlookers.

Environmental Scenario

This scenario uses geospatial data at different scales to provide qualitative and quantitative information about climate and weather for a location, defined here as the city of Amritsar, Punjab, India. The neighborhood is comprised of other cities at a similar latitude to Amritsar. The network is the atmospheric transport of heat, moisture and momentum that defines the climate for a location. The field in this context includes local weather observations such as temperature and precipitation.

The granularity determines the scale over which the field values are averaged in time and space, and hence the scale of the atmospheric processes under evaluation. For instance, the atmospheric general circulation and the Asian monsoon characterize large-scale patterns that effect whole parts of the neighborhood, whereas a single hourly observation is specific to one location at one time. An event is a single weather occurrence that may differ substantially from the expected climate (e.g., a record high temperature). The accuracy of a field observation depends on the instrumentation and the data processing. Ultimately, the **value** of weather and climate analyses lies in our ability to use them for economic planning in such areas as agriculture and utility services (Figure 5). Table 5 lists helpful websites that guide this scenario.

First, locate the city of Amritsar on a map, noting its latitude and longitude (31.64° N, 74.86° E). Access a global composite of satellite imagery, for example, the Blue Marble images on Visible Earth from NASA (Table 5). Even if students

Figure 5. Same as Figure 1, except for meteorological observations as an environmental example

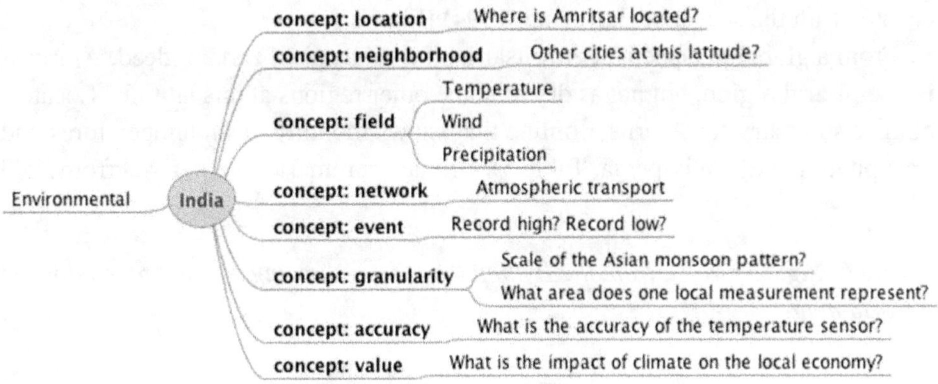

Table 5. Helpful websites to guide the environmental scenario

Online Resource	URL
NASA	http://visibleearth.nasa.gov/view_cat.php?categoryID=1484
Matplotlib	http://matplotlib.org/basemap/users/geography.html
Wikipedia (Main Page)	https://en.wikipedia.org/wiki/Main_Page
Wikipedia (Amritsar)	https://en.wikipedia.org/wiki/Amritsar#Geography_and_climate
Wikipedia (Cairo)	https://en.wikipedia.org/wiki/Cairo#Climate
Wikipedia (Ain Salah)	https://en.wikipedia.org/wiki/In_Salah
Wikipedia (Darbhanga)	https://en.wikipedia.org/wiki/Darbhanga#Climate
Weather Underground	http://www.wunderground.com/history/

have little computer programming experience, it is quite easy to use the Python language to plot a base map that locates Amritsar on the Blue Marble map. A code to do so is provided in the Appendix, and its results are shown in Figure 6. The code is a modification of that shown in Example 3 from Matplotlib (Table 5). Notice that over major land areas, a tan colored surface exists, roughly near about 30° N and 30° S (i.e., the subtropics). This is particularly evident over the continent of Africa. This tan color is indicative of a dry, reflective surface, common in arid regions, while the dark greens over central Africa indicate a wetter climate that is more heavily vegetated.

Distinct latitudinal bands of desert in the subtropics are related to the atmospheric general circulation (GC) which is the planetary scale movement of air driven by the uneven heating of the earth (Hartmann, 1994). Warm, humid air converges near the Equator and rises, cooling upon ascent until it reaches its dew point to produce heavy rains. As this air moves poleward, it descends around the subtropical zones. Descending air dries the atmosphere and quells storm formation in the subtropics. This large overturning of the atmosphere is a part of the GC called the Hadley cell. The region of warm, rising air and precipitation oscillates north and south of the equator with the seasons (Hartmann, 1994) (Figure 7).

From a global perspective, Amritsar is in a subtropical zone. Indeed, Amritsar is a semi-arid region, but not as dry as some other regions at this latitude. Locate a climate summary for Amritsar online that shows monthly mean temperatures and precipitation (e.g., Wikipedia, Table 5). Climate summaries are derived from field

Figure 6. Figure plotted from the Python language, showing Amritsar (red dot) on a world map

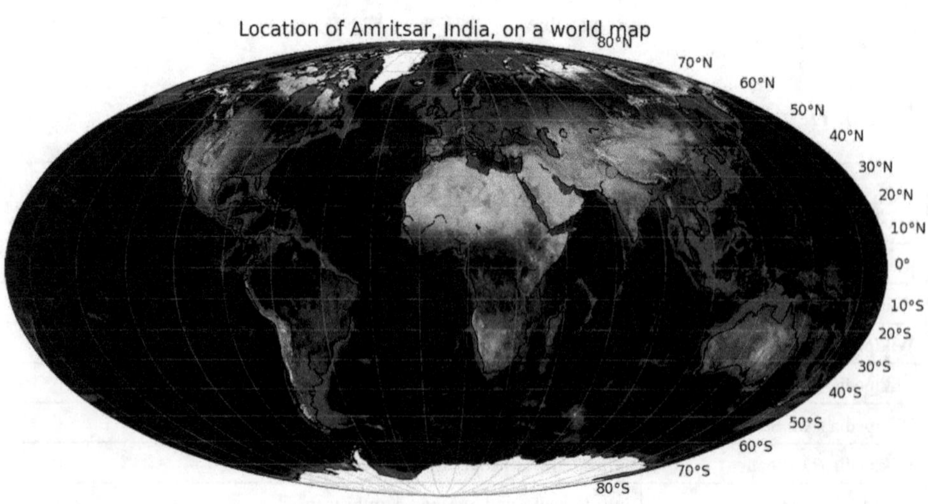

Figure 7. Idealized diagram of the Hadley cell, part of the atmospheric general circulation. Rising motion near the Equator results in low pressure (L) and precipitation, while sinking motion near 30°N results in high pressure (H) and dry conditions.

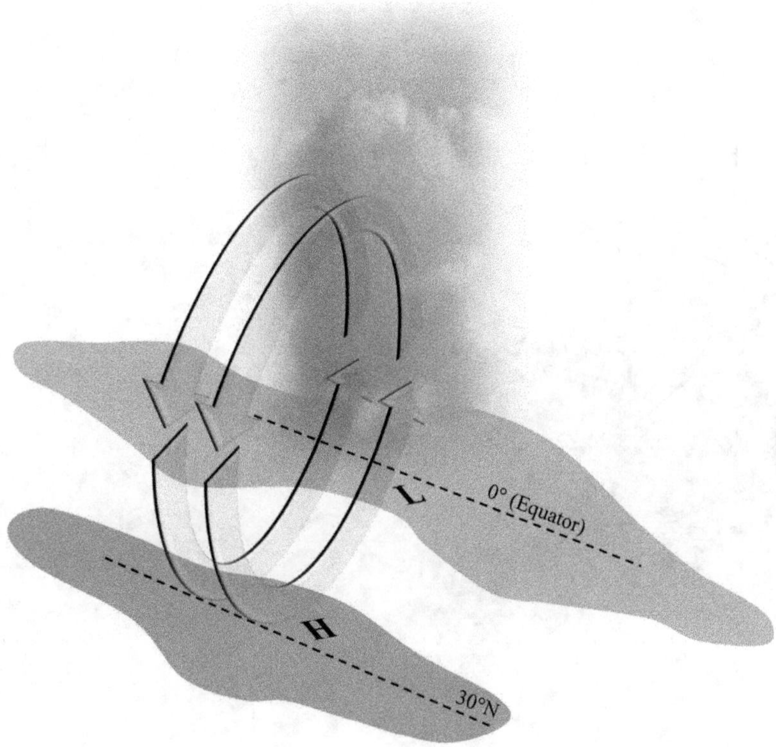

values that are averaged over long time periods. Take note of the average monthly high and low temperatures, as well as the average monthly and annual precipitation for Amritsar. Compare this to the same summary for other locations at roughly similar latitudes, for example Cairo, Egypt and Ain Salah, Algeria. What about other locations in India, such as Darbhanga? Clearly, Cairo and Ain Salah are much drier than Amritsar, and Darbhanga is the wettest of these cities.

To understand some of the reason for these differences requires consideration of the Asian monsoon, which is a seasonal shift in wind that brings precipitation into India during summer, and drier air in winter (Hartmann, 1994). The monsoon is a network influenced by the topography of India and its proximity to the ocean. Wintertime high pressure in the cold Himalayas pushes dry air southward into India (Figure 8), while the inverse of this condition occurs in summer when low pressure over the Tibetan Plateau draws warm, humid air northward from the ocean,

Figure 8. Conceptual diagram of the wintertime Asian monsoon, where high pressure (H) over the cold Himalayas promotes the southerly flow of dry air across India

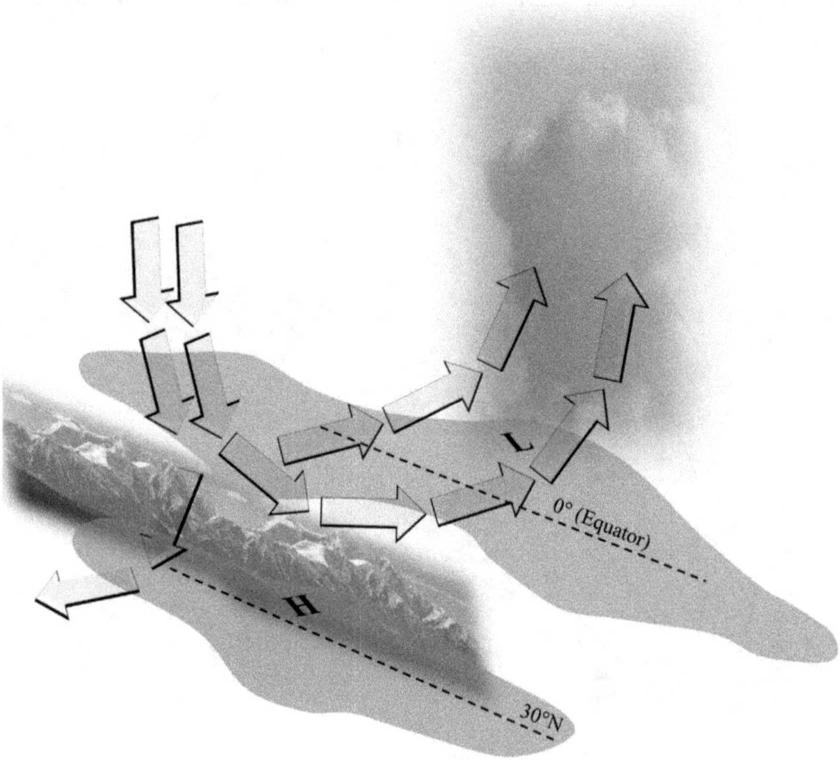

producing heavy rains throughout India (Figure 9) (Hartmann, 1994). Recall the climate summary from Wikipedia and note when the rainy season occurs in Amritsar and Darbhanga. There is an obvious spike in precipitation beginning in July due to the onset of monsoonal rains. Could this help explain why the warmest months of the year in these cities occur during May or June (note that Cairo and Ain Salah are actually warmest in July)?

With this broad picture, one can now examine a more local scale. Consider individual hourly field measurements of temperature, wind, and precipitation for Amritsar from a database such as the weather archives on Weather Underground (Table 5). Navigate to this website and enter "Amritsar, India." Select Amritsar when it appears below, and then select a date for which you want to see the historically archived weather data. Note that there is a table showing temperature, moisture, precipitation, air pressure, and wind. There is a list of hourly meteorological observations, available to download as a comma delimited file at the bottom of the page (be aware that this is a limited set of data having a number of missing

Figure 9. Conceptual diagram of the summertime Asian monsoon, where the warm Tibetan plateau generates low pressure (L) and rising air, bringing precipitation to India

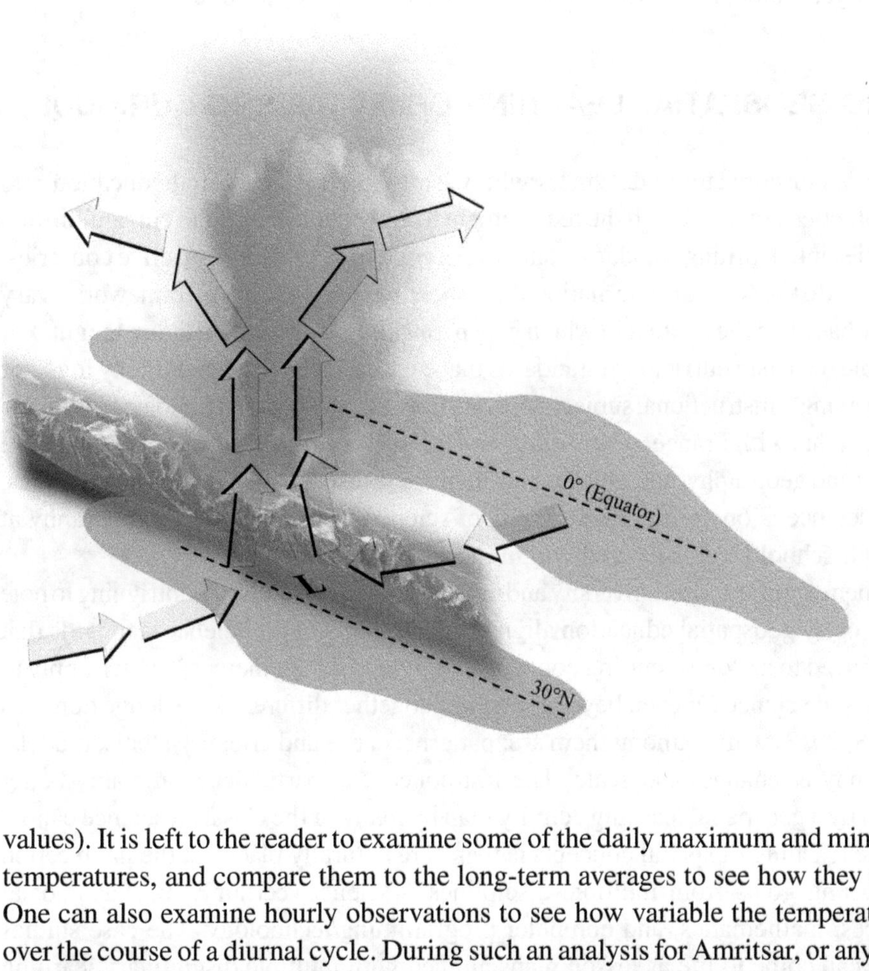

values). It is left to the reader to examine some of the daily maximum and minimum temperatures, and compare them to the long-term averages to see how they differ. One can also examine hourly observations to see how variable the temperature is over the course of a diurnal cycle. During such an analysis for Amritsar, or any other location, learners might discover an aberration from the expected temperature or precipitation regime. Such a weather event marks an unusual case which exemplifies the variability of a natural system that one attempts to quantify in the long term. Moreover, consider that the accuracy of hourly weather observations depends on the instrumentation being used to collect the observations, the person taking the observations from the instruments (and any data processing involved), and on the number of missing data values.

Finally, what value does an investigation of weather and climate data have for the society as a whole? Since climate defines the local conditions and hence the types of resources available (Hartmann, 1994), the analyses by atmospheric scientists help inform how humans can sustain themselves, and what hardships they must overcome to do so. For instance, utility companies can prepare for the seasons during which there will be a high demand. Growers can plan their season and determine what time of year to harvest, and what variety of crops can be sustained.

TYING GEOSPATIAL LEARNING OBJECTIVES TO CURRICULA

Educators or curriculum designers who wish to include geospatial education into their practice are faced with the requirement to embed the geospatial curriculum and activities into learning standards that govern education in their respective countries, states or districts. In an international perspective, such standard frameworks vary widely based on the context in which geospatial knowledge and skills are taught. For example, German curriculum standards place earth science and geography together within a single instructional subject, separate from the physical and biological sciences. Contrast this with France, where earth and life sciences are grouped together, as are history and geography, and physics and chemistry, respectively. In the United States, earth science is one of the three pillars of science instruction, while geography at the high-school level is located within the social sciences.

When examining pre-university and undergraduate curricula, it is intriguing to note how closely geospatial education aligns with the parts of the science standards that are referred to as "cross-cutting concepts" or "over-arching themes," which apply to all areas of science (or even beyond) and link together different knowledge domains (NGSS, 2013). Chief among them are: patterns; cause and effect; systems; models; constancy & change; and scale. The last concept in particular is arguably better served by a geospatial learning activity than by many of the classical science canon.

The teaching of spatial concepts is therefore naturally placed at the intersection of learning goals from multiple disciplines – science, certainly, but also social sciences, mathematics, and computer programming/technology. The case studies above can come to life as lesson plans aligned with multiple disciplinary learning goals. Students may:

- Collect, describe, analyze data (scientific practice).
- Carry out statistical analyses of the data and study their uncertainties (mathematics).
- Apply scientific methods to societal problems (applications of science and technology).
- Study the role of humans in their physical environment (geography, earth science).
- Learn about climate, landscape, ecological change (earth and life science).

- Make maps (geography).
- Write computer programs to visualize geospatial data (computer programming).

Geospatial learning is ideally suited to collaborative interdisciplinary project-based learning as exemplified in the National Research Council's study (NRC, 2006): "Spatial thinking is not an add-on to an already crowded school curriculum but a missing link across the curriculum. Integration and infusion of spatial thinking can help to achieve existing curricular objectives."

FUTURE RESEARCH DIRECTIONS

Learning opportunities such as those outlined above might lead to geo-spatial education taking roots in our secondary schools, reinforcing and extending existing curricula in the sciences and mathematics. The sample scenarios were designed to show how space can be apprehended on successively larger scales: from a snapshot of the local urban setting, via multi-week regional-level ventures, to achieving familiarity with a global nexus of causes, effects, and options to act. Today's young citizens also function on all these scales at once. It is our hope that the work outlined in this chapter serves as a springboard to further refine and extend the paradigms and develop their implementation.

CONCLUSION

In conclusion, this chapter has discussed a variety of data types and data sources which are available free and/or online, readily accessible to a wide audience around the globe. These data and tools are a source of information for instructors who want to create their own learning exercises geared towards engaging students in geospatial learning. To provide some direction to the instruction, a list of core concepts from the literature are presented and then exemplified via three scenarios. These scenarios are intended to show how the core concepts can be applied to a range of problem solving in different locations around the globe, and in different STEM subject arenas. The scenarios provide educational discovery opportunities that are intended as inspiration for generating classroom activities and projects centered around using some aspect of geospatial data, whether qualitative or quantitative, in order to solve real life problems. Many of these geospatial learning objectives can be merged with existing STEM curricula concepts, giving teachers an opportunity to address various international learning standards regardless of how the STEM subjects are organized within a given institution or governing region. In the future, it is going to be essential to have advanced geospatial skills in order for students to reach their full intellectual and professional potential.

REFERENCES

Bennett, J. (2010). *OpenStreetMap*. Birmingham, UK: Packt Publishing.

GADM database. (2015). *Global Administrative Areas.* Retrieved August 30, 2015, from http://www.gadm.org

Hartmann, D. L. (1994). *Global Physical Climatology*. San Diego, CA: Academic Press.

Kuhn, W. (2012). Core concepts of spatial information for transdisciplinary research. *International Journal of Geographical Information Science*, *26*(12), 2267–2276. doi:10.1080/13658816.2012.722637

Marquez, A. (2015). *PostGIS essentials*. Birmingham, UK: Packt Publishing.

National Research Council (NRC). (2006). *Learning to Think Spatially: GIS as a Support System in the K-12 Curriculum*. Washington, DC: The National Academies Press.

NGSS Lead States. (2013). *Next Generation Science Standards: For States, By States*. Washington, DC: The National Academies Press.

QGIS. (2015). *Background of QGIS.* Retrieved on August 30, 2015, from https://en.wikipedia.org/wiki/QGIS

SPACIT. (2015). Retrieved on August 30, 2015, from http://www.spatialcitizenship.org

Youngblood, B., & Iacovella, S. (2013). *GeoServer Beginner's Guide*. Birmingham, UK: Packt Publishing.

KEY TERMS AND DEFINITIONS

Accuracy: Accuracy is a key property of spatial information, capturing how information relates to the world. Information about accuracy answers questions about the correctness of spatial information.

Event: Events and processes are of central interest to science and society – capturing what happened in the environment, in a human body, in cells or molecules, in machines, or in financial systems. Event information answers questions about change.

Field: Fields describe phenomena that have a scalar or vector attribute everywhere in a space of interest (e.g., air temperatures on the earth's surface).

Granularity: Granularity characterizes the size of the spatial, temporal, and thematic units about which information is reported.

Location: A location expresses a relation between the figure to be located (e.g., your house) and a chosen ground (e.g., a named region, a street network, coordinate axes).

Meaning: Understanding what producers meant by some spatial information is crucial to its adequate use. Producing meaningful results and making sense of them involves determining whether the same things are called the same and different things differently.

Neighborhood: Neighborhoods are commonly thought of as regions, characterizing spatial context.

Network: Connectivity is central to space and spatial information. The concept of network captures binary relationships among arbitrary number of objects, called the nodes or vertices of the network.

Object: Objects provide a way of structuring spatial information. They describe individuals that that have an identity as well as spatial, temporal, and thematic properties.

Value: Information about values attached to or affected by spatial information answers questions about the roles played by spatial information in society.

APPENDIX

Python code for plotting a world map, with a red dot at the location of Amritsar. This code is a modification of the example source code from Matplotlib User's Guide which can be found at: http://matplotlib.org/basemap/users/geography.html as well as the Basemap Toolkit documentation available at http://matplotlib.org/basemap/:

```
from mpl_toolkits.basemap import Basemap
import matplotlib.pyplot as plt
import numpy as np

# latitude and longitude of Amritsar in degrees
lat_am = 31.64
lon_am = 74.86

# set up figure
fig1 = plt.figure(figsize=(20, 15))
# setup basemap.
m = Basemap(projection='moll',lon_0=0,resolution='c')
# draw coastlines.
m.drawcoastlines()
# draw parallels and meridians
# label parallels on right and top
parallels = np.arange(-80., 90, 10.)
m.drawparallels(parallels,labels=[False, True, True, False],
color="grey", fontsize=18)
meridians = np.arange(10.,351.,20.)
m.drawmeridians(meridians, color="grey")
# draw background
m.bluemarble()
# transform lat and lon into map coordinates
x, y = m(lon_am, lat_am)

m.plot(x, y, 'o', markersize = 15, color='red')
plt.title("Location of Amritsar, India, on a world map",
fontsize=24)
plt.show()
```

Chapter 4
Esri Association With Open Source:
Is It Free Too?

Faridedin Cheraghi
University of Tehran, Iran

ABSTRACT

Everyone involved in geospatial information systems has heard of Environmental Systems Research Institute (Esri)company. Most people have tried ArcGIS software at least for one time. Esri has been the pioneer in this industry for a long time; it also defines the future of geospatial trends. In this chapter, the author adopts a neutral position to discuss the relation of Esri and open-source community. They cover almost every aspect where Esri and open source co-occur. Even the latest achievement of Esri, which is R-bridge, is discussed here. Going into the details of everything is not the goal of the chapter; however, a minimum description is provided for each section. Proper references are given to the reader for further study.

INTRODUCTION

In the last few years Esri has contributed to the Open-source community and has announced to be excited being part of the evolving open-source community (ArcNews, 2011). Due to the successful and well-known proprietary software from this company (e.g. ArcGIS Desktop), the question that is brought to the mind is that whether Esri really cares about the open-source or it is only trying to stay in the business market of geospatial technology. Although we should distinguish between open-source and free; If a software is free, it isn't necessarily open-source and vice

DOI: 10.4018/978-1-5225-5039-6.ch004

Figure 1. Esri approach to open-source
(ArcNews, 2011)

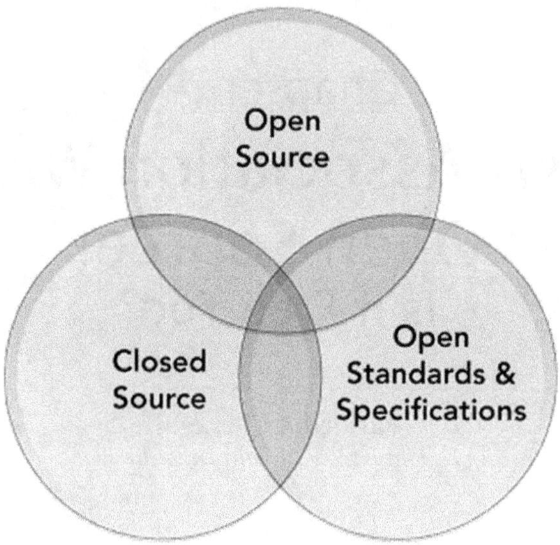

versa. In other words, an open-source software doesn't imply it is free. In fact, most of Esri's open-source projects belong to the second category and they are not free since they depend on a main commercial product. In the subsequent sections of this chapter, dependencies of individual project will be further discussed.

Plenty of other open-source projects from Esri exist that are not covered in this chapter. However, we have tried to compile the highlights here.

ESRI OPEN-SOURCE PROJECTS

There are so many opne-source projects from Esri that are available through GitHub website (Esri GitHub, 2015). Esri Geoportal Server, ArcGIS Editor for OpenStreetMap and the recent R-bridge are the most popular, which their capabilities will be highlighted in this section. Amongst too many open-source javascript projects from Esri, Terraformer and Esri Leaflet will be discussed either.

In addition, Esri has an initiative called ArcGIS for local government; "Maps and Apps" is another name that is in concert with it. ArcGIS for local government consists of several discipline based projects (e.g. Fire, Water and Sewer and etc). Each project contains a map, data, scripts, toolboxes and optionally an application (web or desktop) associated with it.

Esri Geoportal Server

Geoportal Server is a standards-based, open-source product that enables discovery and use of geospatial resources including data and services.

Geoportal Server allows cataloging the locations and descriptions of organization's geospatial resources in a central repository called a Geoportal, which one can publish to the Internet or intranet. Visitors to the Geoportal can search and access these resources to use with their projects. If you grant them permission, visitors can also register geospatial resources with the Geoportal. It gives an enterprise-level view of your geospatial resources regardless of their type or location. Resources are registered with a Geoportal using metadata, which describes the location, age, quality, and other characteristics of the resources. With access to this information about resources, an organization can make decisions based on the best resources available (Richard Kachelriess, Esri, 2015a).

Free Desktop Add-Ins

Esri Geoportal Server desktop add-ons add new levels of interoperability between Geoportals and ArcGIS for Desktop or ArcGIS Explorer. These clients are designed for data publishing, data search and discovery. The CSW[1] Clients let the user discover and use GIS resources available through a Geoportal directly from ArcGIS applications. They also provide the following functionalities:

- Search metadata catalogs directly from ArcMap or ArcGIS Explorer.
- View the title and abstract of search results and download the metadata XML.
- Add live map services to ArcMap or ArcGIS Explorer.

The Publish Client is a tool for ArcCatalog that helps the users to easily publish metadata from their local desktop to the Geoportal. The metadata can come from many sources including,

- Shapefiles
- Single-user Geodatabases
- Multiuser Geodatabases
- Any local data format that supports metadata in ArcCatalog.

Last but not the least is the WMC[2] client to open and view WMC file in ArcMap.

These three components come as free addin although the source code is not available to download. It is not crucial to install them to have an operational Geoportal, however they streamline the data management and some workflows to some extent.

Prerequisites

Esri Geoportal server source code is developed using many programming languages. The following list is compiled from GitHub web page of the project (Richard Kachelriess, Esri, 2015a):

Client Addins are written in C#. Geoportal web applications, web service and applets are mainly written in Java. JavaScript code is for embedding of ArcGIS JavaScript API. ActionScript is for embedding of Flex API. Web Ontology ("Web Ontology Language," 2015) Language is for definition of web services.

Java JRE must be installed prior to Esri Geoportal installation. Java is an interpreted language[3] and is platform independent; therefore, it can be installed on any operating system. The Geoportal web application is based on Java Servlet technology and therefore requires the use of a Servlet container. It supports Apache Tomcat, GlassFish and Oracle Weblogic as its web server. A mail server access is required to have functionalities like data download notification, harvesting notification and feedback function to contact administrator.

Esri geoprotal supports many database management systems as its backend. It supports Oracle, PostgreSQL, MS SQL Server and MySQL. Except MS SQL Server, all other DBMSs have a linux version.

For the full functionality of user-based roles, an LDAP-enabled Directory Server is also required for user authentication. Apache Directory Server-an open-source directory server-is an option for this purpose (ApacheDS, 2015). When no LDAP Directory Server is available, the Geoportal may be configured for a single administrative account. This is a quick option for an introductory setup, but is not recommended for production environments.

The Geoportal should reference a map service for its search page map. The Geoportal web application may consume ArcGIS Server map services (hosted locally or on another organization), a WMS service, or the OpenStreetMap map

Table 1. Esri Geoportal programming languages and their usages

Programming Language	Usage	Purpose
Web Ontology Language (OWL)	44.30%	Definition of web services
Java	19.40%	Geoportal web applications, web service and applets
C#	19.20%	Desktop client Addins
ActionScript	6.30%	ArcGIS Flex API
XSLT	5.60%	Translation of metadata
JavaScript	4.50%	Embedding of ArcGIS JavaScript map

for the Geoportal search page. For this purpose, ArcGIS JavaScript or Flex API must be installed locally or be used over the internet. However, the javascript API is not free for commercial use on publicly accessible websites. It is only free for the following use cases:

- Development and Evaluations.
- Non-commercial external use.
- Education (teaching purposes only).
- NGO or not-for-profit business.
- When used in conjunction with ArcGIS for Server license.
- When used in conjunction with ArcGIS Online Task Services.

This important consideration should be taken into account when using Esri Geoportal Server. Despite it is free and open-source, to have a map page in the Geoportal, at least one of ArcGIS web APIs (javascript, flex) must be used which is not free. Furthermore, if the user needs to consume local map services, an ArcGIS Server license must be bought. Therefore, Esri resembles an open-source supporter but this open-source project among others, is not free and it depends on one of its commercial products.

There are two optional features of Esri Geoportal that require ArcGIS Server locator service and/or geometry service. The locator service is necessary if you want to use the gazetteer drop-a-point-on-the-search-map functionality. The geometry service projects the bounding boxes of the metadata search results upon the search map. There are publicly available end points for these two services, although they are not for commercial use (Arc GIS Online, n.d.).

Web Service Protocols

The Esri Geoportal server catalog is exposed via a CSW service and as an OpenSearch endpoint. It is also accessible through the Geoportal REST API, which provides GeoRSS, KML, HTML, or JSON responses. One can access records that are hosted by the Geoportal through these protocols. The records themselves consist of metadata XML of a variety of flavors: ISO metadata XML standards, Dublin Core, FGDC, profiles of these, and service metadata. On top of that, user-defined schemas can be accommodated through customization (Richard Kachelriess, Esri, 2015b).

The Geoportal also supports connecting to a variety of interfaces to synchronize with resources at remote locations. Synchronization means that an endpoint is registered with the Geoportal to be revisited at a regular interval and retrieve updates of the resources at the endpoint (Richard Kachelriess, Esri, 2015b).

ArcGIS Editor for OpenStreetMap

This is another good example of Esri contribution to open-source community even though again, this one is dependent on the main product. ArcGIS Editor for OSM is an ArcMap plugin that allows to directly download OSM vector data and styles, store them locally, make the edits and then synchronize the edits with OSM online database (Christine White, Esri, 2015). This tool is a great tool for VGI enthusiast; it streamlines the updating process of OSM data. Furthermore people, who are looking for free dataset of road network, could utilize it to easily download OSM data accompanied by native OSM symbology.

ArcGIS Editor for OSM consists of three parts: an ArcGIS toolbox, with a set of geoprocessing[4] tools; a toolbar that hosts a button for conflict resolution; a feature class extension. The geoprocessing tools allows the user to download, symbolize and upload the data. The feature class extension[5] takes care of some auto-updates of geometry and attributes of features to better comply with OSM database schema. It also customizes attribute editor dialog of ArcMap[6] to further streamline feature attribute edition with tags. Once the data downloaded from OSM, the user should apply the feature class extension to the Layer via a geoprocessing tool that is available in the toolbox. If the user avoid this step, then the geometry and details of the edits will not be included when uploading to OSM server and leads to data corruption.

Conflict Resolution

Between the times a user download data from OpenStreetMap, make the edits locally and then submit the changes back to the OSM server, other OSM users could be working in the same geographic area and submit the changes just before the user. As a result, you might encounter a data state which is different from the time when you initially downloaded the data. In that case, the OSM server might reject your upload and indicate a conflict. The conflict resolution intuitive GUI helps the user with such cases.

Network Dataset

Using ArcGIS Editor for OSM, one can easily create a network dataset from the downloaded OSM data and solve routing problems. There is a geoprocessing tool that can be used to create the network dataset. Analyses such as the quickest route between two points, the area of service for a business, what barriers are in the path of a specific route, and more are possible when working with a network dataset. The important thing is that the network will only be as good as the underlying data. Some areas do not have adequate information to facilitate creating a useful network dataset; Therefore, It is not advised to use routing based on OSM data in such regions.

R-Bridge

R is a language and environment for statistical computing and graphics. R provides a wide variety of statistical (linear and nonlinear models, classical statistical tests, time-series analysis, classification, clustering and etc) and graphical techniques, and is highly extensible (R Project, 2015). What makes R so unique is the scientific community behind it. Many scientists around the world choose R as their primary development language and they implement their ideas and algorithms with R.

Esri unveiled R-Bridge in Esri User Conference 2015. R-bridge can be described by three main steps: set any ArcGIS supported data as the input; do the processing with R; output the data back to ArcGIS known formats. Esri has posted a working sample on GitHub that utilizes R-bridge in conjunction with R mclust (Chris Fraley et al, 2015) package to carry out point clustering. This sample is a great start for those interested in R and ArcGIS development (Shaun Walbridge, Esri, 2015b).

About 80% of R-bridge is written with ArcObjects C++ SDK and the other 20% is written in R. R-bridge can be installed and upgraded using a geoprocessing toolbox called r-bridge-install (Shaun Walbridge, Esri, 2015a). The minimum requirement for R-Bridge is ArcGIS Desktop 10.3.1 or ArcGIS Pro 1.1.

Creating a geoprocessing tool based on r-bridge resembles creation of a python script tool[7]. The only difference is the script source code. Python script tool uses a python script as its source but R-bridge geoprocessing tool uses one R file as its script source. R codes that are written on top of R-bridge can be run standalone; there is no need to use ArcGIS geoprocessing system and the user does not have to necessarily open Armap or ArcCatalog window.

Spatial Statistics Toolbox

The Spatial Statistics toolbox-being part of ArcGIS system-contains statistical tools for analyzing spatial distributions, patterns, processes, and relationships. While there may be similarities between spatial and non-spatial (traditional) statistics in terms of concepts and objectives, spatial statistics are unique in that they were developed specifically for use with geographic data. Unlike traditional non-spatial statistical methods, they incorporate space (proximity, area, connectivity, and/or other spatial relationships) directly into their mathematics (Spatial Statistics Doc, 2015).

The reason for mentioning this scientific toolbox here is due its transparency. Almost 90%[8] of the toolbox source code are written in pure python. One can easily take a look at the source and modify it accordingly.

JavaScript Projects

By catching a glimpse of Esri website on GitHub, and searching the "javascript " term, we face with tons of projects from Esri, which are all open-source. In the following section, the most notable projects are being explained.

Esri Leaflet

Leaflet is the leading open-source JavaScript library for mobile-friendly interactive maps. Weighing just about 33 KB of JS, it has all the mapping features most geo developers fancy about. Leaflet is designed with simplicity, performance and usability in mind. It works efficiently across all major desktop and mobile platforms, and can be extended with lots of plugins (Vladimir Agafonkin, 2015).

Esri Leaflet is a set of plugins for consuming ArcGIS Services; Esri Leaflet supports loading Esri base maps and feature services, as well as tiled map, dynamic map and image services. The primary purpose of Esri leaflet is to mashup data and services from ArcGIS Server and ArcGIS Online in Leaflet applications. It also Fosters an ecosystem of plugins for consuming Esri and ArcGIS services in Leaflet; Esri Leaflet Geocoder is an example. On top of that it provides wrappers for common tasks like visualizing and querying feature services.

The main question that users come across is whether Esri leaflet is a replacement of ArcGIS javascript API. The answer is, absolutely not. Esri Leaflet does not have nearly the level of API completeness as the JS API, in terms of covering all use cases and services that ArcGIS for Server provides. If the developer needs to have fully functional WebGIS which supports user authentication, feature editing and complex analysis, then ArcGIS JavaScript API is better suited.

Terraformer

Terraformer is a javascript modular toolkit for working with geographic data; it is divided into smaller modules:

- Terraformer Core Contains methods and objects for working with GeoJSON. With this module one can convert GeoJson objects (e.g. a Point) into Terraformer objects and use Terraformer functions such as point. within(polygon), convexHull or calculateBounds.
- WKT Parser: Parse Well Known Text into GeoJSON and vica-versa.

- ArcGIS Geometry Parser: Parse the ArcGIS Geometry Format into GeoJSON and vica-versa.
- GeoStore: A framework for persisting and querying GeoJSON features with pluggable indexes and persistant stores.

ArcGIS for Local Government

GIS is a multidisciplinary subject, which covers tones of disciplines. From electric, water and gas utilities to fire service, emergency management, law enforcement and election, they all take advantage of GIS and its services. Simply put, ArcGIS for Local Government[9] is ArcGIS for all of these disciplines. What can be done in each domain with GIS? How one would use GIS for its organization? AGLG[10] is the answer to these questions (AGLG Gallery, 2015).

Each AGLG project consists of miscellaneous components: a map template (mxd), referencing layers and symbology; data, mostly in a FileGeodatabase[11] format; optionally a desktop application or a desktop Addin, a WebGIS site based on ArcGIS for javascript API and a python toolbox.

The question is how AGLG is relevant to open-source? The point is AGLG offers a comprehensive informational model for organizations to build their geodatabase (Esri Informational Model, 2015). Local Government Informational Model[12] (LGIM) defines tables, feature classes, fields and their data type, relationship-classes and domains. Utilities data model, such as water, needs an extra geometric network that defines the network junctions, edges and connectivity rules. LGIM also describes all vital components of a geometric network. For example utilizing LGIM for water utilities incur several advantages (Howard Crothers, Esri, 2011):

Easier Deployment

It makes deploying the ArcGIS for Water Utilities maps and apps easier, faster and cheaper. For an organization to extend the model, there is no need to reinvent the wheel, it just make changes that are required to meet specific business needs of the organization.

Data Modeling Best Practices

The LGIM incorporates many best practices for water utility GIS. One of the most important best practices is how to represent a water, sewer or storm water system in GIS. In essence it modernizes the data models to reflect how water utilities have been deploying GIS over the past few years and also modifies the schema to fit the requirements of the ArcGIS for Water Utilities maps and apps.

Comprehensive Data Model

Older data models focused primarily on a data structure for the assets that comprise utility networks. Utilities want to model operational data (work orders, service requests, customer complaints, main breaks, capital improvement projects, etc) as well as base data (roads edge of pavement, road centerlines, elevation data, parcels, etc.). The LGIM solves this problem because it includes a complete schema for typical water utility base data and operational data

Easier Data Sharing

If organizations that commonly share data all adopt the Local Government Information Model, it will greatly reduce the time and resources spent establishing a common schema and migrating data to these schemas – thus allowing water utilities to focus on the maintenance and management of their authoritative data. In future it will become extremely easy to integrate all water utilities data in a unified Geoportal; as they share the same informational schema.

Best Cartographic Practices

Map templates from AGLG have best practice cartographic design for displaying water utility GIS data. In addition, Other map templates follow best practices for editing water utility data with ArcGIS Desktop.

Another example is APDM, which is the ArcGIS Pipeline Data Model designed for storing information pertaining to features and conditions found on or along gathering and transmission gas and/or liquid pipelines (Tom Coolidge, 2015).

In addition to data models, the source code of any application, desktop Addin or toolbox that is part of a project is included. This is very important if the user wants to make modification to those applications. One example is tax-parcel-viewer app; the source is available for windows which is written in C# and for web and smart phones written in javascript (Lindsay Thomas, Esri, 2014).

ESRI OPEN-SOURCE UTILIZATION

In this section, the way Esri utilizes many open-source software such as Python, GDAL and DOJO within its proprietary software stack, are being covered.

Python

Python is a widely used general-purpose, high-level programming language. Its design philosophy emphasizes code readability, and its syntax allows programmers to express concepts in fewer lines of code than would be possible in languages such as C# or Java. Python is cool and easy to learn; it is a good language for new programmers and a handy tool for experienced ones. Miscellaneous open-source libraries exist, both for scientific and commercial purposes. ArcGIS utilizes open-source python libraries such as numpy, matplot, xlrd and pyparsing in its software. These libraries are installed once ArcGIS Desktop gets installed.

ArcGIS has chosen python as its main scripting language due to its popularity. VBA is another scripting language, which Esri has deprecated since version 10; although VBA is accompanied with new ArcGIS Desktop media for backward compatibility.

Python is integrated with ArcGIS system. As soon as users start working with ArcGIS, they can feel the presence of python in every part of the system. Python makes ArcGIS very powerful; as if there is a lot of freedom to carry out almost any kind of computation and analysis in the system: field Calculator, label expression, display expression, model builder, python Addins and arcpy are examples of python in ArcGIS.

On top of python, Esri has developed ArcPy. ArcPy is a site package that builds on (and is a successor to) the successful ArcGIS scripting module. Its goal is to create the cornerstone for a useful and productive way to perform geographic data analysis, data conversion, data management, and map automation with Python. The additional power of using ArcPy within python is the fact that Python is a general-purpose programming language. It is interpreted and dynamically typed and is suited for interactive work and quick prototyping of one-off programs known as scripts while being powerful enough to write large applications. In ArcGIS, applications and tools written with ArcPy benefit from the development of additional modules in numerous niches of Python by GIS professionals and programmers from many different disciplines.

Arcpy is very popular across the world. At the time of writing this chapter, about 4000 questions are asked in GIS.StackExchange.com[13] website that are tagged with Arcpy (GIS StackExchange, 2015). Despite Python, Arcpy is not completely open-source. If someone digs into the main site package (arcpy), he/she will hit the dependencies on ArcGIS closed assemblies. Hence, to take advantage of ArcPy, one must install one of Esri main products like ArcGIS Desktop.

GDAL

GDAL/OGR is a translator library for raster and vector geospatial data formats that is released under an X/MIT style Open-source license by the Open-source Geospatial Foundation. Everybody in geospatial community is aware of the power of GDAL (GDAL Wiki, 2015).

Esri utilizes this powerful library to read/write many custom raster formats under the hood. In fact, many users aren't aware of this but to confirm this fact, one can search for GDAL in ArcGIS Desktop installation folder. Another proof is the wiki page for GDAL.

DOJO

Dojo is an open-source toolkit that helps to write robust and efficient JavaScript code. JavaScript is a language that runs within a web browser, and there are various flavors of web browsers that interpret JavaScript in slightly different ways. Toolkits such as Dojo, jQuery, YUI, Prototype, and many other javascript libraries are designed to abstract away the browser idiosyncrasies so that you don't have to learn them all and handle them in your code.

There are several reasons that Dojo was chosen as the foundation for the ArcGIS API for JavaScript (AGJavaScript Doc, 2015):

- The dojo comprehensive build system.
- An easy to use class based inheritance system which adapts to object-oriented programming.
- AMD[14] module loader for managing code in large applications.
- Vector graphics abstraction via dojox/gfx to simplify drawing graphics using SVG, Canvas or VML.
- Dojo and dijit are both fully accessible and open-source.
- Dojo fully supports internationalization.

Although other mature javascript libraries like jquery and ExtJS were available, Esri has chosen dojo, a robust and powerful open-source javascript library for extending its javascript API; Esri has made a delicate decision to opt dojo as the foundational library for its prominent javascript API, which is evident from comprehensive application that are developed with ArcGIS javascript API and dojo.

Operating System Compatibility

ArcGIS software packages are mostly platform dependent and they depend on Microsoft technologies such as Windows OS and Visual studio. However, from version 10 and above, some of the products like ArcGIS for Server[15] have become a product that is available to be installed on a Linux box too. Actually, this is one step forward to get closer to the open-source community.

The question is why a Microsoft oriented company like Esri, tends toward an open-source OS?

1. There are many empirical statements that say DBMSs like PostgreSQL (Magnus Hagander, 2015) and Oracle™ run faster and with a higher performance on a Unix-based OS. However, there is not any proven benchmark to evaluate this inference. The claim is just empirical and based on user's experience (Simo Kivistö, 2015).
2. ArcGIS Server core consists of Server Object Container (SOC) and Server Objects Manager (SOM) that are written in C++. As C++ codes compile to machine binaries, the execution are almost the same on Windows and Linux. However, ArcGIS Server manager (above 10) is written in Java with a built-in apache tomcat web server; there are claims that java runs faster on a linux box rather than a windows tailored system (Martin Kuba, 2015).
3. A software price is an important factor for choosing it. Despite a Linux OS that only incurs maintenance costs, deployment of ArcGIS Server on a windows server (2008 or above) needs an extra Windows server license which is not a beneficial choice for many customers.

There can be many other reasons and motivations for ArcGIS Server on Linux, but the reduction of overall cost of software deployment in an organization can be the main reason. In fact, Esri has made a good choice here to take this step to make a mature product like ArcGIS Server, platform independent.

OGC COMPLIANCE

The Open Geospatial Consortium (OGC) and the International Organization for Standardization (ISO) Technical Committee 211 (TC211) are the two leading geospatial standards organizations. Esri is an OGC-Compliant company and implements new standards and protocols proposed by OGC.

ArcGIS Desktop is the client to most of OGC standard services out of box. One can consume WMS, WCS and WMTS services with ArcMap. On the other hand, direct access to Web Feature Service (WFS) is supported through an extension called ArcGIS Data Interoperability extension, which incurs extra cost. Although a geoprocessing tool to convert a WFS service to an offline feature class resides in ArcToolbox (AGDesktop Doc, 2011).

Esri is a principal member of OGC and actively participates in implementation of open specifications and standards. OGC certifies ArcGIS Server for WCS, WMS, WCS and WFS compliance.

ArcSde for PostgreSQL spatial types and functions is certified for the implementation of OGC simple feature access (OGC, 2015) besides few other DBMSs[16] .

In addition, Esri Geoportal Server is certified for implementation of Catalog Service for the Web (CSW); Geoportal Server also implements Web Map Context (WMC), although it has not been certified for this implementation yet.

Open Specification

Shapefile is a widely used spatial exchange format first introduced by this company and then evolved and changed to an open specification binary format. There are also other open specifications that are suggested by Esri but none made its way to become an OGC standard format. One of highlights is Geodatabase XML format that models the whole geodatabase schema and data in one xml file (AGDesktop Doc, 2013).

File Geodatabase is the dominant format in ArcGIS software to store spatial data for single use. In addition to simple feature types such as point, line and polygon feature classes, more advanced types are supported as-well: annotations, dimensions, topology, networks, terrains, TIN and raster datasets. Unfortunately, File Geodatabase specification has not become open but due to its comprehensiveness may become in near future. However, File Geodatabase C++ API is a free library initiated by Esri to allow 3rd party software (both proprietary and open-source) to interact with this format. Using this API, one would be able to create, edit, and read File Geodatabases without using the ArcObjects API. Consequently, there would not be a need for an ArcGIS license.

MbTiles and Vector Tiles

Tile based web maps are made up of millions of tiles. Imagine loading all of those tiles; it would be inefficient and slow. That's where MBTiles comes in. The MBTiles specification is an efficient format for storing millions of tiles in a single SQLite database. Each SQLite database is self-contained and represented as a single .sqlite file. No external setup is required. You can copy a .sqlite file from a desktop computer to a mobile device and have all its rows, tables, and indexes ready to be used. It's a portable, single-file solution for storing and serving web maps (Mapbox, 2015a)

MBTiles takes advantage of utilities found in SQLite for solving problems like duplicate imageries. Maps that cover large areas of solid color like ocean or empty land can contain thousands of duplicate, redundant tiles.

At higher zoom levels this could lead to millions of solid (blue) tiles, all exactly the same. Instead of loading all of those look-alike tiles, MBTiles can reference tile coordinates to raw imagery. Thousands of tile coordinates can be paired to the same raw image, drastically reducing the file size required to serve a map of multiple zoom levels.

Esri has incorporated Mbtiles reader in its Run-times. For example with ArcGIS runtime SDK for Android, one can easily read local mbtiles by implementing some of the base classes of the SDK (Esri Android Doc, 2015).

Vector tiles are the vector data equivalent of image tiles for web mapping. They apply the strengths of tiling, developed for caching, scaling and serving map imagery rapidly, to vector data.

Consider an image tile at the z/x/y coordinate 14/4823/6160. Figure 2 is a PNG image that depicts the corner of lower Manhattan with roads, building footprints, and parks.

Figure 2. A sample vector tile

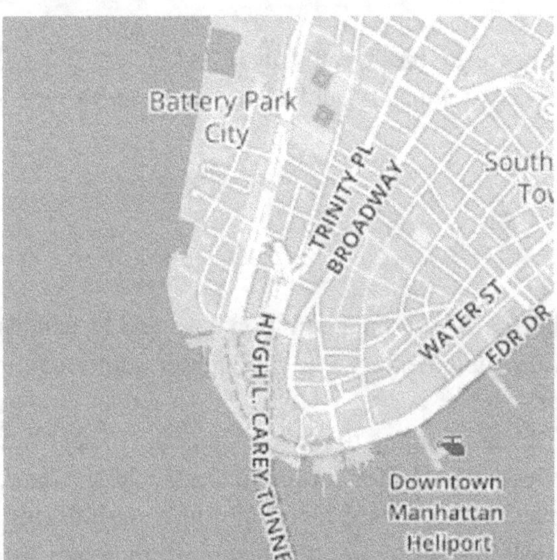

A vector tile at 14/4823/6160 would contain all the corresponding geometries and metadata like road names, area types, building heights in a compact, parsable format. Vector tiles are a highly performant format that provide greater flexibility in terms of styling, output format, and interactivity.

Vector tiles are designed to use the same tiling and coordinate scheme used by image tiles and take full advantage of HTTP protocol. Caching, high scalability and efficient distribution make vector tiles a much more attractive solution than running, say, an enormous cluster of database and application servers that respond to queries on demand (Mapbox, 2015b). Both Mbtiles and Vector tiles specifications are introduced by MapBox however they couldn't make their ways to become an OGC standard yet.

Esri has announced that ArcGIS Runtime SDKs for iOS, Android and Windows Phone as well as desktop applications are developing ArcGIS specific implementations of Vector tiles rendering technologies that follow this specification. They are building a hardware abstraction layer that seamlessly uses new OpenGL and DirectX graphics technology optimizing for the mobile platforms (Andrew Turner, 2015).

GeoPackage

GeoPackage is an open, standards-based, platform-independent, portable, self-describing, compact format for transferring geospatial information. Unlike Mbtiles and vector tile, it is an OGC open standard.

The GeoPackage specification describes a set of conventions for storing the following within a SQLite database:

- Vector features (resembles Vector tiles)
- Tile matrix sets of imagery and raster maps at various scales (resembles mbtiles)
- Schema
- Metadata
- Extensions

These capabilities are built on a common base and an extension mechanism is described to provide implementers a way to include additional functionalities with their GeoPackages.

It seems GeoPackage is going to replace shapefile in near future. It contains all suitability of Mbtiles and yet better because it is an OGC standard. We believe it can be extended to replace Vector tiles by incorporating some indexing algorithms into it.

Due to its extensibility, in future, it will support more advanced data types like those that are accessible in a File Geodatabase: topolgy, network, terrain and TIN.

Since a GeoPackage is a SQLite database, it supports direct use, meaning its data can be accessed and updated in a "native" storage format without intermediate format translations. GeoPackages are interoperable across all enterprise and personal computing environments, and are particularly useful on mobile devices like cell phones and tablets in communications environments with limited connectivity and bandwidth ("OGC GeoPackage," 2015).

There are many implementations of the standard by many software producers. To name a few, QuantomGIS 2.4.0 can read (but not write) GeoPackage Features as of version 2.4.0. GeoPackage Tiles support will be available soon.

GDAL supports GeoPackage Features as of version 1.11.0. Tiles support was recently developed and will be available in GDAL 2.0.

ArcGIS for Desktop and Server 10.2.2 and above support GeoPackage Features. ArcGIS Desktop 10.3.1 adds GeoPackage Tiles support, and ArcGIS Runtime SDK's for Android and Java 10.2.4 also support reading vectors and rasters, and writing vectors. And ArcGIS Pro as of version 1.1 will support reading vectors.

ArcGIS and SQLite

ArcGIS Desktop 10.3.x supports spatial data in SQLite databases that contain one of the following storage types ("SQLite and ArcGIS," 2015):

- **Esri ST_Geometry:** The database file must have the extension .sqlite
- **SpatiaLite:** The database file must have the extension .sqlite
- **GeoPackages:** The GeoPackage file must have the extension .gpkg

OPEN-SOURCE DATABASE MANAGEMENT SYSTEM

Among commercial DMBSs like Microsoft SQL Server and Oracle, Esri supports *PostgreSQL*, the free and open-source DBMS, as an ArcSDE technology backend. A special attention is paid to the PostgreSQL here, because the focus of this chapter is on open-source GIS.

OGC Data Types

Enterprise geodatabases utilize several different UDTs[17]. Esri has implemented st_geometry and st_raster types in PostgreSQL DMBS for vector and raster data storage. Although their implementation differentiates from the reference spec of OGC, storing spatial data with these data types, make the data more transparent to the users and 3rd party applications (e.g. open-source applications).

Direct SQL Access

Spatial types have associated functions or methods that are used to access and manipulate the data using Structured Query Language (SQL). Accessing the data via SQL allows external applications to work with the tabular data managed by the geodatabase or database. These external applications can be non-spatial database applications or custom spatial applications developed in an environment other than ArcObjects. For example QGIS can be a client to read directly from a st_geometry database.

With the types and functions in place, one can use SQL to create spatial tables; read and analyze the spatial data; and insert, update, and delete simple geometry data. SQL syntax varies slightly from one database to another, but all the spatial types that Esri supports are based on the International Organization for Standardization (ISO) SQL Multimedia (MM) spatial specification.

ST_Geometry

ST_Geometry is implemented as a user-defined type that can be used in databases and geodatabases such as IBM DB2, IBM Informix, Oracle, and PostgreSQL. ST_Geometry also can be used in SQLite databases (AGDesktop Doc, 2015). ST_Geometry is present by default in geodatabases created in Oracle and PostgreSQL. It seems Esri implies the preference of st_geomtry over other spatial types like Oracle sdo_geometry or PostGIS geometry.

Insertion

Users can use SQL to insert spatial data to a database or geodatabase table that has a ST_Geometry column. They should use ST_Geometry constructor functions to insert specific geometry types. They can also specify that the output of certain spatial operation functions be output to an existing table.

Edition

SQL edits to existing records often affect the non-spatial attributes stored in the table; however, users can edit the data in the ST_Geometry column using constructor functions inside SQL UPDATE statements.

If the data is stored in a geodatabase, there are additional limitations that the user should follow when editing with SQL:

- Do not update records using SQL if the data has been versioned unless SQL is used in conjunction with a versioned view.
- Do not modify any attributes that affect other objects in the database that participate in geodatabase behavior such as relationship classes, feature-linked annotation, topology, or networks.

If the user neglects the above precautions, it may leads to data corruption and ArcGIS may not be able to read the data properly.

ST_Raster

ST_Raster is a user-defined data type. It is stored in a single row, in a single column of object type ST_Raster in a user-defined table. ST_Raster provides full geodatabase support as well as SQL access to raster data. This allows writing SQL applications using database management system (DBMS) that can access and use raster operations and queries. In ArcGIS, ST_Raster is supported in Oracle, Microsoft SQL Server, and PostgreSQL databases.

To use ST_Raster type, it takes extra steps and it must be configured in the DBMS. The ST_Raster data type can be used within the geodatabase and ArcGIS; plus, it provides Structured Query Language (SQL) access to raster data for 3rd party applications. Two different categories of ST_Raster functions are used to access the ST_Raster type using SQL:

- **ST_Raster and ST_PixelData methods:** These are methods that instantiate and work with the ST_Raster object.
- **ST_RasterUtil stored procedures:** The ST_RasterUtil stored procedures support the creation and maintenance of the ST_Raster type.

Contrary to st_raster many advantages, it seems Esri doesn't tend to upgrade st_raster library to work with newer PostgreSQL versions; It is not updated to work with PostgreSQL newer than 9.0. Based on an Esri staff quote in Esri GeoNet website, st_raster is not the perfect data type for raster storage in modern databases:

While it's unfortunate that ST_RASTER wasn't released with PG 9.1 and 9.2 implementations, I think the root issue is that using databases to store rasters peaked in the ArcGIS 8.3-9.1 timeframe. Once ArcGIS Server 9.2 could fashion mapping caches, the need for a tiled basemap in the database waned. Modern raster support hinges around mosaic datasets, which are file-based, and are much faster and more flexible than database solutions. Mosaic datasets have the benefits of retaining the original images in their native format, provide overlays for tiled access, permit on-the-

fly calculation of alternate forms (e.g, hillshade from a DEM and reclassification), and they don't fill your database transaction logs with terabytes of static data (greatly simplifying the database backup situation). Rather than reverting to storing BLOBs in databases (which was a pre-8.0 solution), I'd recommend giving mosaic datasets a chance (Vangelo, Esri Staff, 2015).

So we shouldn't expect to see st_raster in the future releases of the ArcSde for Postgresql.

CONCLUSION

Esri is playing an important role in geospatial industry. In this chapter we have explained that despite many Esri open-source software offerings, none is free. By using any of its open-source software, the user is obliged to buy a commercial license, due to the dependency upon the main software. If an organization goes towards Esri product spectrum, it could achieve to almost any desire for Geo-data processing and visualization. Although this will results in a term called vendors lock-in[18].

If your organization task is very critical and is moving on the edge of geospatial technology, then Esri is perfect choice. On the other hand, if software price is the key criterion for your organization choice, then you should opt for open-source GIS.

Last not the least is the costs associated with the development of a GIS system. Esri software is popular and tons of developers may be accessible to hire to develop with Esri software SDKs[19]. Although this may not be the case with open-source GIS.

REFERENCES

Agafonkin, V. (2015, September 3). *Leaflet*. Retrieved September 3, 2015, from http://leafletjs.com/

AGDesktop Doc. (2011). *WFS To Feature Class Conversion Tool*. Retrieved September 6, 2015, from http://help.arcgis.com/EN/arcgisdesktop/10.0/help/index.html#//00120000000q000000

AGDesktop Doc. (2013). *ArcGIS Help 10.2 - Geodatabase XML*. Retrieved September 6, 2015, from http://resources.arcgis.com/EN/HELP/MAIN/10.2/index.html#//018r00000005000000

AGDesktop Doc. (2015, September 3). *SQL with ST_Geometry*. Retrieved September 3, 2015, from http://resources.arcgis.com/EN/HELP/MAIN/10.2/index.html#//006z0000001v000000

AGJavaScript Doc. (2015, September 3). *Why Dojo?* Retrieved September 3, 2015, from https://developers.arcgis.com/javascript /jshelp/why_dojo.html

AGLG Gallery. (2015, September 3). *Gallery | ArcGIS Solutions*. Retrieved September 3, 2015, from http://solutions.arcgis.com/gallery/#s=0

Apache, D. S. (2015). *Apache Directory Website*. Retrieved September 6, 2015, from https://directory.apache.org/

ArcNews. (2011). *Open Source Technology and Esri | ArcNews Online*. Retrieved August 28, 2015, from http://www.esri.com/news/arcnews/spring11articles/open-source-technology-and-esri.html

ArcObjects Doc. (2015, September 2). *Creating class extensions*. Retrieved September 2, 2015, from http://resources.arcgis.com/en/help/arcobjects-net/conceptualhelp/index.html#/Creating_class_extensions/000100000201000000/

Coolidge, T. (2015). *APDM.net | ArcGIS Pipeline Data Model (APDM) Clearinghouse*. Retrieved from http://apdm.net/

Crothers, H. Esri. (2011). *Benefits of the Local Government Information Model for Water Utilities | ArcGIS Blog*. Retrieved from http://blogs.esri.com/esri/arcgis/2011/10/11/benfits-of-the-local-government-information-model-for-water-utilities/

Doc, E. A. (2015, August 28). *Local MBTiles | ArcGIS for Developers*. Retrieved August 28, 2015, from https://developers.arcgis.com/android/sample-code/local-mbtiles/

Doc, S. S. (2015, September 3). *Spatial Statistics toolbox*. Retrieved September 3, 2015, from http://webhelp.esri.com/arcgisdesktop/9.3/index.cfm?TopicName=An_overview_of_the_Spatial_Statistics_toolbox

Does PostgreSQL perform on Windows as well as on Linux - Server Fault. (n.d.). Retrieved August 27, 2015, from http://serverfault.com/questions/674609/does-PostgreSQL-perform-on-windows-as-well-as-on-linux

Esri GitHub. (2015, August 27). *Esri on Github*. Retrieved August 27, 2015, from https://github.com/Esri

Fraley, C. (2015). *mclust: Normal Mixture Modelling for Model-Based Clustering, Classification, and Density Estimation* (Version 5.0.2). Retrieved from https://cran.r-project.org/web/packages/mclust/index.html

GIS StackExchange. (2015, September 3). *arcpy tag wiki*. Retrieved September 3, 2015, from http://gis.stackexchange.com/tags/arcpy/info

Hagander, M. (2015, August 27). *Is PostgreSQL suited to one OS? Is it better on Linux than Windows? - Server Fault*. Retrieved August 27, 2015, from http://serverfault.com/questions/222430/is-PostgreSQL-suited-to-one-os-is-it-better-on-linux-than-windows

Is PostgreSQL suited to one OS? Is it better on Linux than Windows? - Server Fault. (n.d.). Retrieved August 27, 2015, from http://serverfault.com/questions/222430/is-PostgreSQL-suited-to-one-os-is-it-better-on-linux-than-windows

Kachelriess, R. Esri. (2015a, August 28). *Esri/geoportal-server*. Retrieved August 28, 2015, from https://github.com/Esri/geoportal-server/

Kachelriess, R. Esri. (2015b, August 28). *Geoportal server as a broker · Esri/geoportal-server Wiki*. Retrieved August 28, 2015, from https://github.com/Esri/geoportal-server/wiki/Geoportal-server-as-a-broker

Kivistö, S. (2015, August 27). *PostgreSQL on Windows and Linux*. Retrieved August 27, 2015, from http://serverfault.com/questions/674609/does-PostgreSQL-perform-on-windows-as-well-as-on-linux

Kuba, M. (2015, August 27). *Java Performance On Various Platforms*. Retrieved August 27, 2015, from http://dior.ics.muni.cz/~makub/java/speed.html

Language, W. O. (2015, September 4). In *Wikipedia, the free encyclopedia*. Retrieved from https://en.wikipedia.org/w/index.php?title=Web_Ontology_Language&oldid=679385402

Mapbox. (2015a, August 28). *Mbtile | Mapbox*. Retrieved August 28, 2015, from https://www.mapbox.com/guides/an-open-platform/#mbtiles

Mapbox. (2015b, August 28). *Vector Tile | Mapbox*. Retrieved August 28, 2015, from https://www.mapbox.com/guides/an-open-platform/#vector-tiles

Model, E. I. (2015). *Local Government Information Model | ArcGIS for Local Government*. Retrieved September 6, 2015, from http://solutions.arcgis.com/local-government/help/local-government-information-model/

OGC. (2015, August 27). *OGC Compliant Products*. Retrieved August 27, 2015, from http://www.opengeospatial.org/resource/products/compliant#ESRI

OGC GeoPackage. (2015, August 28). Retrieved August 28, 2015, from http://www.geopackage.org/

Open Source Technology and Esri | ArcNews Online. (2011). Retrieved August 28, 2015, from http://www.esri.com/news/arcnews/spring11articles/open-source-technology-and-esri.html

Project, R. (2015, September 2). *R: What is R?* Retrieved September 2, 2015, from https://www.r-project.org/about.html

SQLite and ArcGIS. (2015). Retrieved August 28, 2015, from https://desktop.arcgis.com/en/desktop/latest/manage-data/databases/sqlite-and-arcgis.htm#ESRI_SECTION1_E4F82E510ED443088D87F6226E982EBA

Thomas, L. Esri. (2014). *Esri/tax-parcel-viewer*. Retrieved September 6, 2015, from https://github.com/Esri/tax-parcel-viewer

Turner, A. (2015, August 28). *Vector Tiles preview | ArcGIS Blog*. Retrieved August 28, 2015, from http://blogs.esri.com/esri/arcgis/2015/07/20/vector-tiles-preview/

Vangelo, E. S. (2015, September 3). *libst_raster_pg.dll for PostgreSQL 9.2*. Retrieved September 3, 2015, from https://geonet.esri.com/thread/114964

Walbridge, S. Esri. (2015a). *R-ArcGIS/r-bridge-install*. Retrieved September 6, 2015, from https://github.com/R-ArcGIS/r-bridge-install

Walbridge, S. Esri. (2015b). *R-ArcGIS/r-sample-tools*. Retrieved September 6, 2015, from https://github.com/R-ArcGIS/r-sample-tools

White, C. Esri. (2015). *Esri/arcgis-osm-editor*. Retrieved September 2, 2015, from https://github.com/Esri/arcgis-osm-editor

Wiki, G.D.A.L. (2015, August 14). GDAL. In *Wikipedia, the free encyclopedia*. Retrieved from https://en.wikipedia.org/w/index.php?title=GDAL&oldid=676030336

ENDNOTES

1 Catalog Service for the Web.

2 Web Map Context: This is a companion specification to the OGC Web Map Service. It states how a specific grouping of one or more maps from one or

more map servers can be described in a portable; a platform-independent format for storage in a repository or for transmission between clients.

3 An interpreted language is a programming language for which most of its implementations execute instructions directly, without previously compiling a program into machine-language instructions.

4 Geoprocessing is a known concept across ArcGIS platform. Geoprocessing is a GIS operation used to manipulate spatial data. A typical geoprocessing operation takes an input dataset, performs an operation on that dataset, and returns the result of the operation as an output dataset.

5 For a detail description on feature class extensions see (ArcObjects Doc, 2015).

6 Attribute Editor is a dialog in ArcMap for attribute manipulation.

7 A python script tool is a python script that imports Arcpy and is being used under a ArcGIS toolbox.

8 Except "Geographically Weighted Regression (GWR)", "ZScore Rendering" and "Count Rendering" which are not python script tool.

9 For the sake of clarity we will call it AGLG in the rest of this chapter.

10 Stands for ArcGIS For Local Government.

11 FileGeodatabase (FGDB) is a proprietary format from Esri. It is the dominant format in ArcGIS system.

12 Called LGIM from now on.

13 Stackexchange websites are user-driven Question/Answer portals to help answer question in many areas.

14 Stands for Asynchronous Module Definition.

15 Right now ArcGIS Engine Runtime and ArcReader are available to be installed on a Linux box, too. Note that virtualization is another option to install any software on a Linux OS.

16 ArcSde for Oracle, DB2, Infromix and Netazza are also certified.

17 User-Defined DataTypes.

18 vendor lock-in, also known as proprietary lock-in, makes a customer dependent on a vendor for products and services, and prevents him/her from using another vendor without substantial switching costs.

19 Software development kits.

Chapter 5
Spatial Data Mining, Spatial Data Warehousing, and Spatial OLAP

Amira M. Idrees
Fayoum University, Egypt

Mostafa Lamlom Ahmed Khaled
Institute of Statistical Studies and Research, Egypt

Amal Hassan Ali Talkhan
Institute of Statistical Studies and Research, Egypt

ABSTRACT

Data mining is one of the current vital fields for all types of data including spatial data. An example of useful extracted patterns from spatial data is to describe changes in metropolitan poverty rates based on city distances from major highways. Geospatial is a term that has recently been gaining in popularity; moreover, many applications on geospatial have different uses in different fields such as geo-tagging, geospatial technology, and geo-fencing. Analyzing spatial data is considered a complex task due to its details as it is related to locations with a special representation such as longitude and latitude. Other attributes are involved in the description of objects which can be analyzed using different data mining techniques. In this chapter, a demonstration of the basic information is performed considering spatial data and spatial data mining including all aspects such as the different type of data, different methods of analysis, different mining techniques, and other related topics.

DOI: 10.4018/978-1-5225-5039-6.ch005

AN INTRODUCTION TO GEO-SPATIAL DATA

GIS (Geographic Information System) simply considered a computer system that enables users to do some analytical processing by capturing, storing, retrieving, sorting, and comparing spatial data. GIS provides various types of patterns displayed on different kinds of generated map with different resolutions that allow users to see the map features more easily and clearly which enable them to more understand the relationships among these patterns (ESRI, 2015) .

GIS provides user with detailed gathered information about geographic places and their spatial relationships. It is considered as a framework concerned with collecting information, providing relationships and making some analytical image processing (ESRI, 2015) .

GIS provides a set of layers as in Figure 1 after interpreting and translating the captured features on a map. These layers represent the spatial data and the relationships among them that reflected from the map processing like for example roads, rivers and political boundaries as in Figure 1 (ESRI, 2015).

Spatial Data

Simply is the information about the locations and geographic features reflected on the processed maps stored as coordinates and topology (ESRI, 2015). There are two main types of spatial data Raster data and Vector data providing different ways in representing and storing data (ESRI, 2015).

Figure 1. Data layers

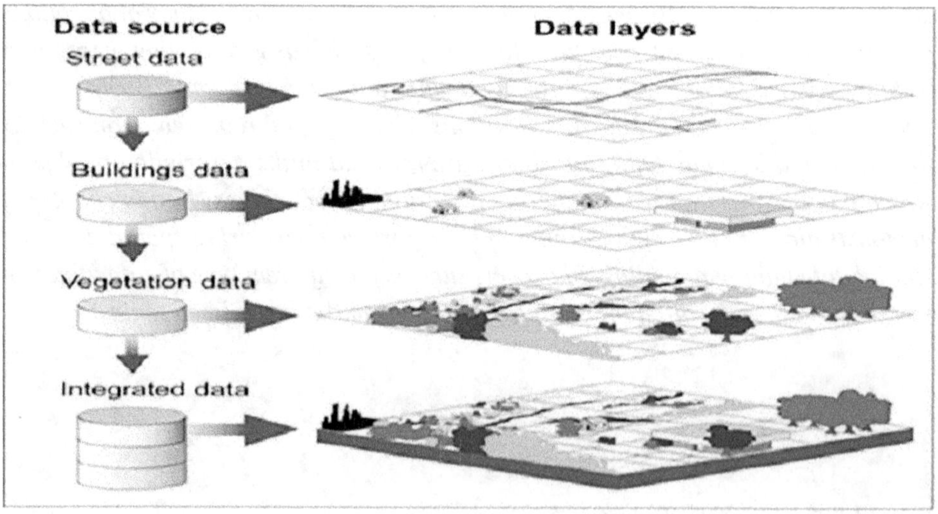

Vector Data

Represents data of the real world patterns as points, line and polygons. Streets, Land parcels and lighting poles are examples of vector data storing them as separated boundaries (ESRI, 2015).

The spatial objects are stored using data structures as point, Line and Polygon in the vector as the following (Kuba, 2001)

- **Point**: Simply defined by its coordinates.
- **Line**: The sequence of points connected together within a line.
- **Polygon**: The sequence of connected lines.

Raster Data

Unlike vector data model, the real world on a map is divided into grid of cells as in Figure 2.The advantage over vector data is storing.

Figure 2. Raster data

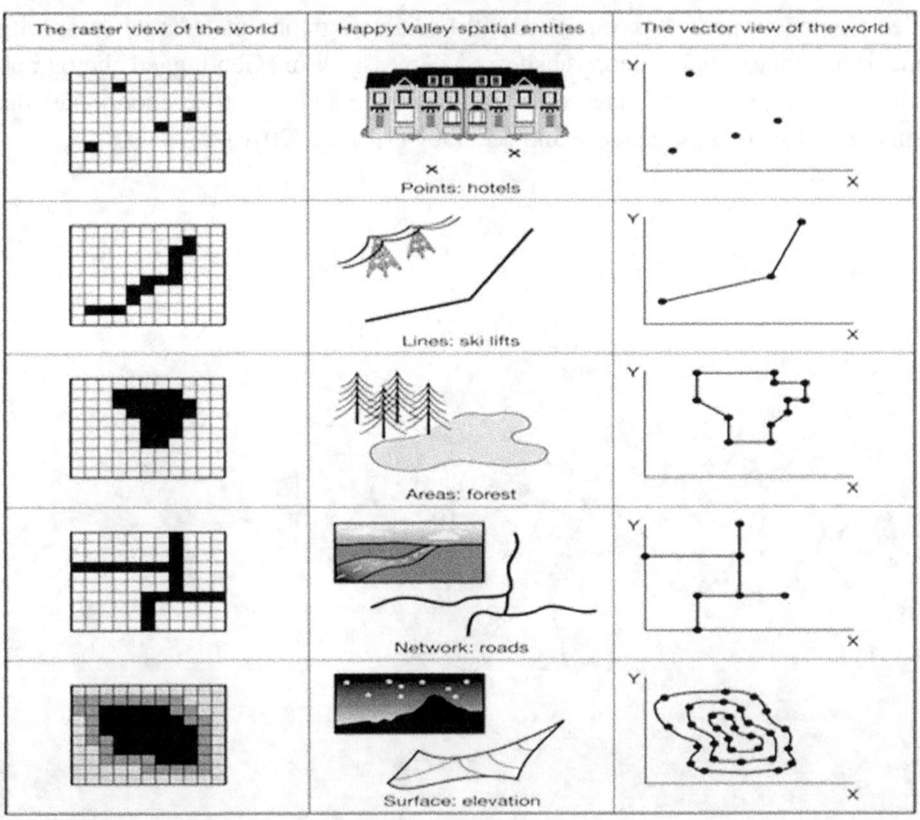

The main role of raster models is continuously storing different data like height surface, a satellite picture and chemical concentrations (ESRI, 2015). There are many users that can analyze and discover the relationships among spatial data sets like for example (IBEG, 2011)

- Advanced GIS users as in Figure 3 (researchers, experts).
- Statisticians, who do not use GIS tools but use statistical tools as in Figure 4.
- Ordinary users with no GIS experience.

What Is Geospatial?

Recently there was a need to integrate or merge the geographic information (spatial data with relationships) with new technologies in order to provide new geographic applications that what known as Geospatial (Lounge, 2014). Geospatial can be applied through many applications with different targets like for example Geo-tagging, Geo-matics, and Geo-fencing (Lounge, 2014).

Geo-Tagging

Geo-tagging as a process is concerned with adding geographic metadata about specific media like for examplevideo, websites, SMS messages or aGeo-tagged photograph. Metadata that has been added consists of latitude and longitude coordinates that supports place names, distance and accuracy (Worden, 2010).

Figure 3. Advanced GIS users

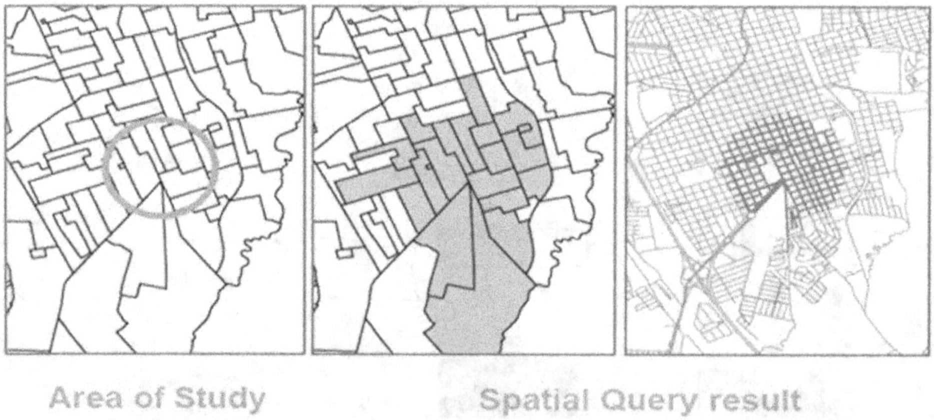

Figure 4. Statisticians users

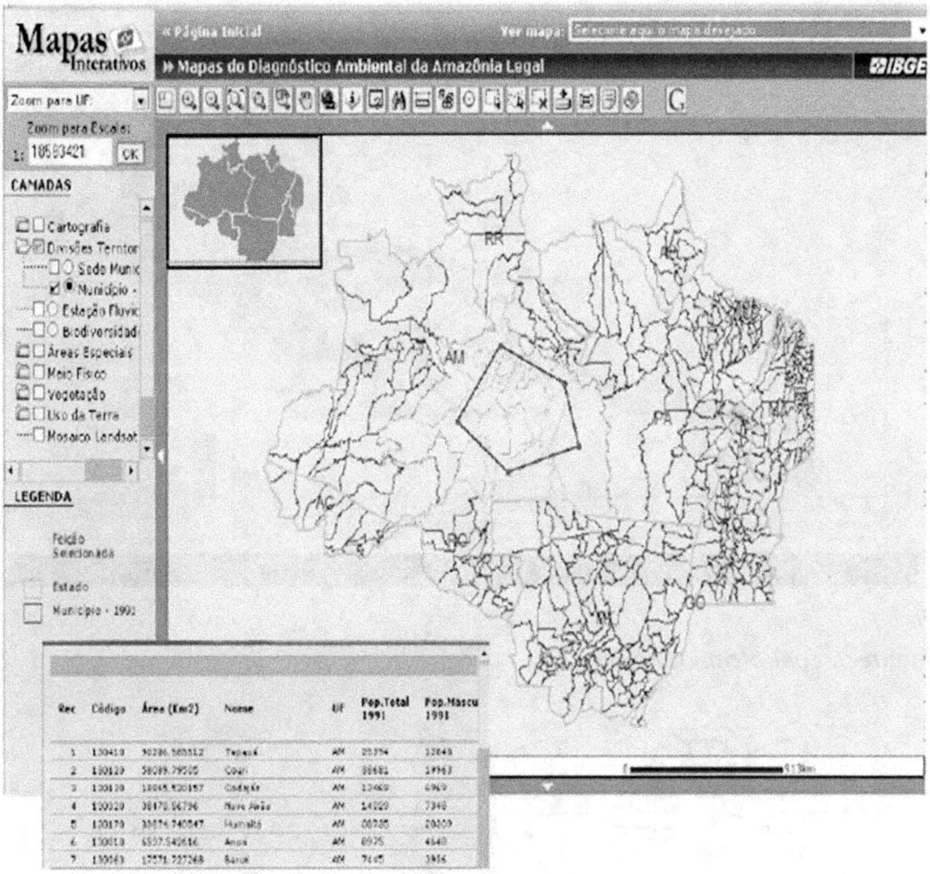

- Adam Savage in August of 2010, of MythBusters, a photo of his vehicle was taken by his smartphone as in Figure 6. After that he used his Twitter account to publish the photo with the phrase off to work.
- IPhone and Motorola are cell phones using GPS chip along with built-in cameras to allow users to geo-tag photos.
- Flickr and Panoramio are programs enable users to add metadata like Geographic coordinatesof a photograph to a map.
- Savage accurately determined the location of his home, his car and the time he goes to work.

Figure 5. Geo-tagging

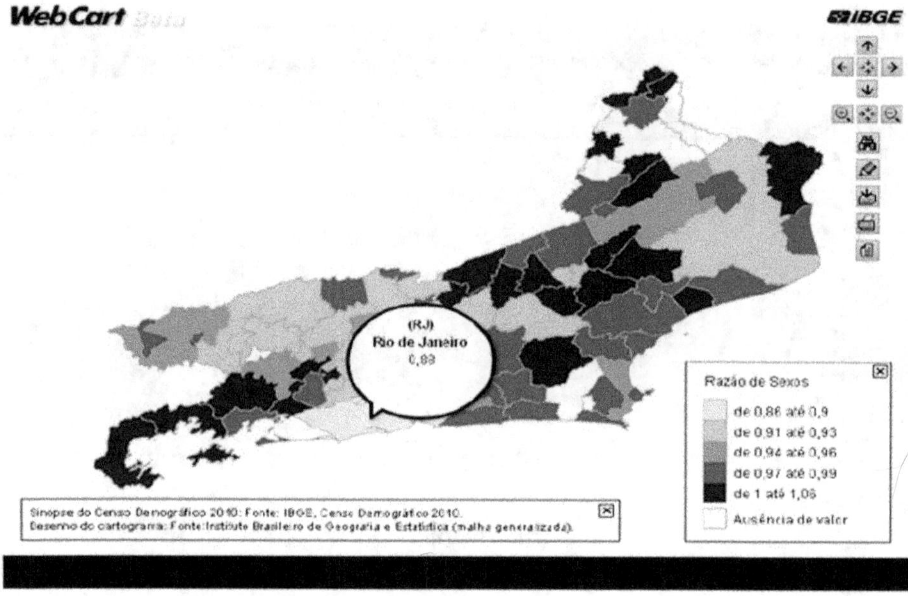

Figure 6. Spatiotemporal objects

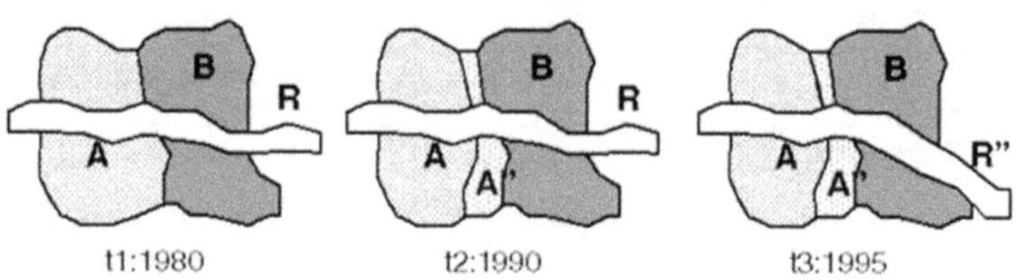

Geo-Matics

Geo-matics are considered Geo-spatial technologies concerning with get, manipulate, and store geographic information using technologies like remote sensing, and geo-fencing (Lounge, 2014).

Geo-Fencing

Geo-fencing is used to determine and calculate the locations on the earth using GPS depending on sent and received signals from satellites. There are many applications use this technology like sending SMS to customer's within some area as announcements to specific products (Kim2, 2013).

Geospatial Analysis

Spatial analysis concerned with examining the locations, attributes and relationships of features in spatial data usingsome analytical techniques to discover new information from spatial data. Geospatial analysis contains many techniques using different analytic approaches that applied in various fields likeastronomyto chip fabrication engineering using ‘place and route’ algorithms to construct complex wiring structures(ESRI, 2015).

What Is Spatiotemporal?

A spatiotemporal database is a database that manages not only the space information but also the time information related to thisinformation. Common examples include (Xiaoyu Wang, 2013)

- Tracking of moving objects with respect to given time.
- Wireless communication networks database within a geographic region existing only for a short time span.
- An index of different species within a geographic region.

Spatiotemporal Applications

There are three different types of spatiotemporal classified according to the type of data they manage (Xiaoyu Wang, 2013)

- Applications that mange objects with continuous motion. For example, applications that supports moving objects with changing in their positions as in a navigationalsystem. Also a car moves in a road network with unchangeable shape.
- Applications dealing with discrete changes among objects that located in space with some periodically changes in both shape and position. For example, “land parcels” or “rivers” are changingpositions in cadastral information system, or neighboring “land parcels” change theircommon border with discretely changes.
- Applications may manage objects integrating continuous motion as well as changes ofshape. For example, in environmental application, a ‘storm’ is measured as a movingphenomenon, which changes properties intensity and shape over time.
- In a cadastral system, spatiotemporal objects may be evolved as follows, see Figure 7.

Figure 7. The taxonomy of spatial and spatiotemporal data warehouse
(Mithal, Nipun Garg, Surabhi, 2010)

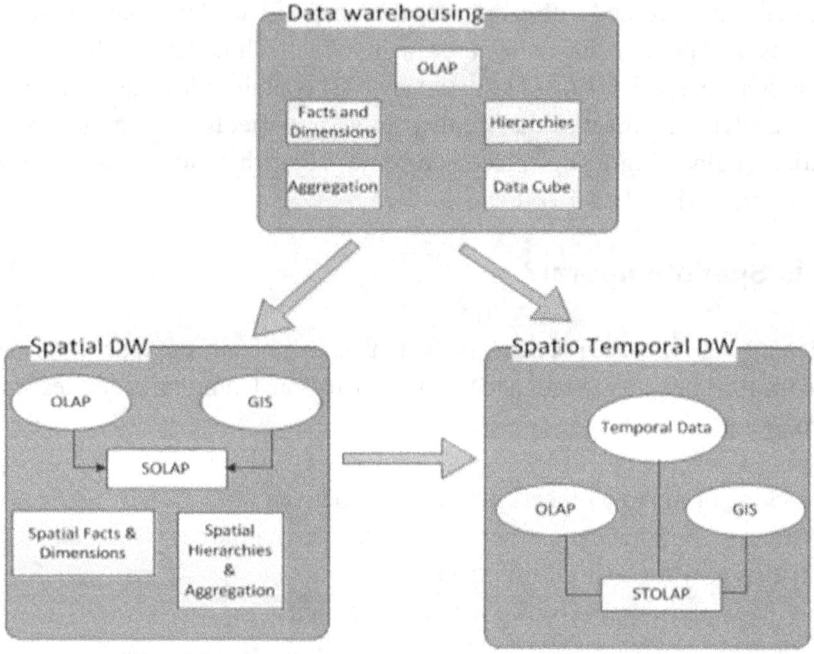

- During 1990, the system displays that land parcel A has common borders with Band River R runs through both Land parcel A with clay soil and Land parcel B with forest soil.
- In 1990, land parcel A was split into A and A”. Land parcel B has soil type “high-densityforest” and A” has soil type “sparse forest.”
- In 1995, finally river changed its position and become R”.

Geospatial Big Data

Geospatial big data refers to spatial data sets with hugecapacity that needs higher capacity than current computing systems. Large part of big data is actually geospatial data withgrowing sizereaching to 20% every year. Users can analyze huge amounts of geospatial data by big data analytics for geospatial data. (Lee, 2015). Geospatial data can be divided into three forms: Raster data, Vector data, and Graph data which considered as different forms of geospatial big data. (Lee, 2015)

Geo-Spatial Data Warehouse (GDW)

Spatial Data warehouse approaches recentlydiscoveredand used for analyzing large volumes of spatial data (Bédard, Y., T. Merrett & J. Han, 2001). Geo-Spatial data warehouse is concerned with hot to make effective and efficient querying of spatial data with historical information. While spatial databases are reasonable for answering transactional queries without enough historical data.

What Is the Relation Between Spatial Data, Spatiotemporal, and Spatial Data Warehouse?

A data warehouse can be divided into facts and dimensions represented in a star or snowflake schema generating what called data cube. The data cube contains cells thatare computed for accurate query processing. Slicing; dicing, Roll up and Roll down are main parts of common OLAP operations (Mithal, 2010). Figure 7 shows the relationship between spatial, spatiotemporal and a data warehouse.

Why Geo-Spatial Data Warehouse (GDW)?

Organizations with big spatial data are increasingly interested in GDW in order to provide more efficient representation of data for users. This merge provides an efficient service for investors using Business Intelligence System (ESRI, 2012).

GDW Approach Has Three Phases (Petek, 2011)

- Generating spatial data warehouse.
- Implementing ETL (extract, transform and load) procedure for data from different sources (spatial and non-spatial).
- Designing interface for performing data analysis.

The Main Features of a Spatial Data Warehouse

- Conceptual model: There are two basic schemas (Star and Snowflake) for spatial attributes.
- Spatially enable components: The main parts are spatial measures, spatial dimensions and spatial hierarchies.
- Spatial OLAP operations: The basic operations are roll up, drill down expanded to spatial attributes.

Spatial Data Warehouse (SDW)

The increasing availability of spatial data has made the necessary need to enhance OLAP with spatial analysis tools. OLAP systems were adapted to solve problems in different areas of application. New various kinds of decision support systems like Spatial Data Warehouses (SDW), and Spatial OLAP (SOLAP) tools have been made available. SDW is a collection of geographical information supports spatial analysis. SOLAP tools allow users to explore data, these tools combine OLAP analysis with capabilities of GIS systems. The relevance of SOLAP technology has been widely recognized in several application domains (e.g., agriculture, environmental risk management, etc.) (Bimonte, S., Pinet, F., Miralles, A., Papajorgji, P., 2014).

The queries needed to support the decision making process can`t be fully executed on spatial databases. This need helped emerging the field of spatial data warehouses. Spatial data warehouses use the concepts of Data warehouses and provide support to store, index, aggregate and analyze spatial data (Bimonte, S., Pinet, F., Miralles, A., Papajorgji, P., 2014).

From Spatial Data, Spatiotemporal to Spatial Data Warehouse

A data warehouse consists of facts and dimensions modeled in a star, snowflake, or fact constellation schema. A data cube represents hierarchies. Common OLAP operations include slicing, dicing, Roll up, and Roll down (Mithal, Nipun Garg, Surabhi, 2010). These concepts are extended to spatial data warehouse. Integration of spatial data and data cubes allow many interesting spatial aggregation queries to be executed.

Spatial Data Warehouse (GDW)

The field of Spatial Data warehouses has been emerging since the past two decades due to the need to analyze large amounts of spatial data. Spatial data warehouses aim for effective and efficient querying of spatial data (Bédard.Y,Han.J, 2009).

Spatial Data Cubes

Spatial data cubes are data cubes where some dimension members or some facts are spatially referenced and can be represented on a map. Two categories of spatial data cubes are presented: feature-based data cubes and raster-based data cubes (Figure 8).

Feature-based data cubes include facts that match discrete features having geometry (vectors or cells) or not having geometry (in which case dimension members must

Figure 8. Feature-based and raster-based spatial cubes
(Bédard.Y,Han.J, 2009)

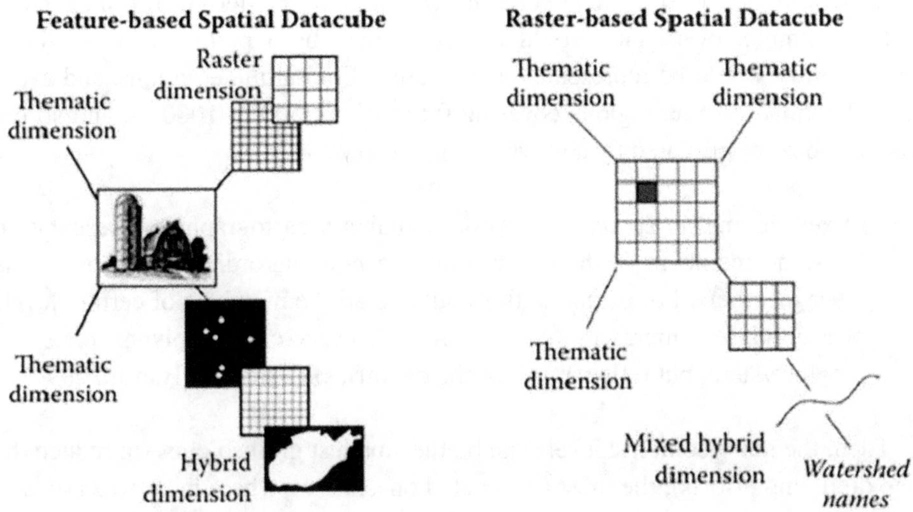

have a vector-based or raster-based geometry). Such fact geometry may be specific to this fact (in which case it may be derived from the dimensions) or it may correspond to the geometry of a spatial member.

Raster data cubes are made of facts that similar to regularly subdivided spaces of continuous phenomena, each fine grained fact being represented by a cell and every fine-grained cell being a fact. (Bédard.Y,Han.J, 2009)

Spatial Dimensions

There are spatial dimensions in addition to the usual dimensions of a data cube (in the multidimensional perspective, not the geometric perspective) that can be of three types according to the theory of measurement scales. These three types of dimension are as presented below:

- No geometric spatial dimension contains only nominal or ordinal locations of the dimension members, such as place names (e.g., St. Tahrir), street addresses (134 Main Street), or hierarchically structured boundaries (e.g., Cairo →Egypt → Africa). Neither shape nor geometry nor cartographic data are used. This is the only type of spatial dimension supported by non-spatial warehousing technology (e.g., OLAP).

Geometric spatial dimension contains a vector-based cartographic representation for every member of a dimension hierarchy to allow the cartographic visualization, spatial drilling, or other spatial operations (Bédard, Y., T. Merrett, J. Han, 2001).

For example, every city would be represented by a point, every province/state, country would be represented as polygons. Equi-altitude regions, and every generalization, such as regions covering 0 to 500 m, 500 to 1000 m, and so on, would also be represented by a polygonal geometry.

- Mixed geometric spatial dimension contains a cartographic representation for some members of the dimension, and nominal/ordinal locators for the other members. For example, this could be all the members of certain levels of a dimension hierarchy (e.g., a point for every city, a polygon for every province/state, but only names for the countries) (Bédard.Y,Han.J, 2009).

Then, the non-geometric levels can be the smallest grained ones (to reduce the map digitizing efforts), the most aggregated ones, or anywhere in between, in any number and/or in any sequence.

A mixed spatial dimension can also contain a cartographic representation for only some members of the same hierarchy level (e.g., all Canadian cities, but not all Mexican cities). The mixed spatial dimension presents some benefits of the geometric spatial dimension while suffering from some limitations of the non-geometric spatial dimension, all this at varying degrees depending on the mixes involved.

Furthermore, spatial dimensions relate to different ways to use geometry to represent a phenomenon: discrete feature-oriented vector data vs. continuous phenomena-oriented raster data. Depending on the type of geometry used, the users' potential to perform spatial analysis and geographic knowledge discovery changes significantly. As a result, users have the choice among seven categories of spatial dimensions as presented in Figure 9 below (Bédard.Y,Han.J, 2009).

The four additional categories of spatial dimensions are presented hereafter:

- **Raster Spatial Dimension:** The raster data structure is used for every level of the dimension hierarchy, the highest spatial resolution is used for the finest-grained level of the hierarchy. For example, one could use 100-km cells for a continent, 10-km cells for its countries, and 1-km cells for province/states. (Bédard.Y,Han.J, 2009)
- **Hybrid Spatial Dimension:** Some levels of the dimension hierarchy use the raster structure while other levels use the vector structure. For instance, this can be polygonal geometries for a continent and for its countries, while the raster structure is used for province/states. Inversely, this could be

Figure 9. Spatial data cubes with different examples of spatial dimensions.

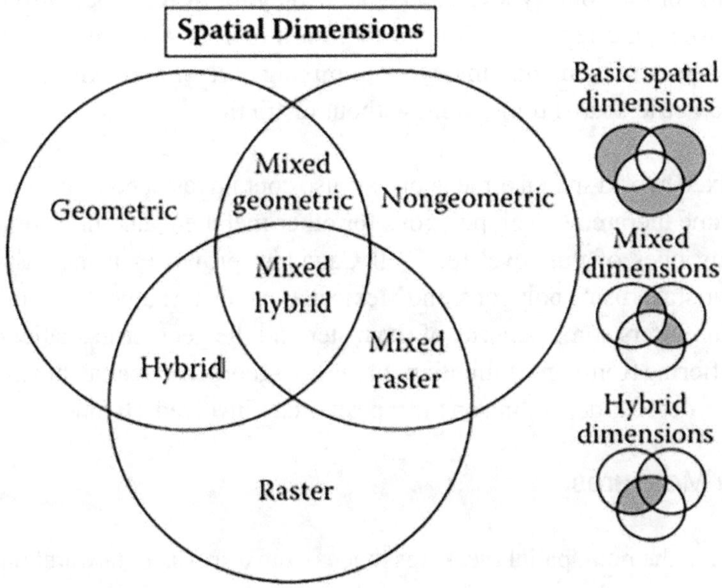

points for cities, polygonal geometries for provinces/states, 100-kmraster cells for countries, and 1000-km cells for continents. All levels must be cartographically represented. (Bédard.Y,Han.J, 2009)

- **Mixed Raster Spatial Dimension:** These dimensions contain raster data for some members of the dimension and nominal/ordinal locators for the other members (i.e., no geometry). For instance, this can be all the members of certain levels of a dimension hierarchy (e.g., cells for the province/state level, names for the country and continent levels, that is, no cartographic representation for these latter two levels of the hierarchy). A mixed raster spatial dimension can also contain raster cells for only some members of the same hierarchy level (e.g., all Canadian provinces, but not all American states). The mixed raster dimension offers some benefits of the raster spatial dimension while suffering from some limitations of the non-geometric spatial dimension, all this at different degrees depending on the type of mixes involved. (Bédard.Y,Han.J, 2009)
- **Mixed Hybrid Spatial Dimension:** These dimensions contain raster data for some members of the dimensions, vector data for other members, and nominal/ordinal locators for the remaining ones (i.e., no geometry). For instance, this can be all the members of certain levels of a dimension

hierarchy (e.g., names for cities, polygons for the province/state level, raster cells for the country level, and a name only for the continent level, that is, no cartographic representation for the finest and most aggregated levels of the hierarchy). Then, the same types of mixing possibilities exist as for the mixed geometric spatial dimension, without restriction.

A mixed hybrid spatial dimension can also contain raster cells for some members of the same hierarchy level, polygons for other members, and no geometry for the remaining ones of this level (e.g., all Canadian provinces using raster cells, all American states using polygons, and Mexican states using names). The mixed-hybrid dimension offers some benefits of the raster and the geometric spatial dimensions while suffering from some limitations of the non-geometric spatial dimension, all this at varying degrees depending on the type of mixes involved. (Bédard.Y,Han.J, 2009)

Spatial Measures

Addition to the non-spatial measures that exist in a spatial data warehouse, we may find three types of spatial measures in the multidimensional sense:

- **Numerical Spatial Measure:** Single value obtained from spatial data processing (e.g., number of neighbors, spatial density). Such measures contain only numerical data and are non-geometric spatial measures. (Bédard.Y,Han.J, 2009)
- **Geometric Spatial Measure:** Set of coordinates or pointers to geometric features that results from a geometric operation such as spatial union, spatial merge, spatial intersection, or convex hull computation. For example, during the summarization (or roll-up) in a spatial data cube, the regions with the same range of temperature and altitude can be grouped into the same cell, and the measure formed contains a collection of pointers to these regions.
- **Complete Spatial Measure:** Such measure is a combination of a numerical value and its associated geometry. For example, the number of clusters with their location.

Spatial ETL (Bédard.Y,Han.J, 2009)

Due to the special nature of spatial data, it rapidly becomes evident that integrating and aggregating/summarizing spatial data requires additional processing in comparison to non-spatial data. For example, one must make sure that the source of each dataset is topologically correct before integration and that it meets important spatial integrity constraints, that the overlay of these maps in the warehouse is also topologically correct (e.g., without slivers and gaps) and coherent with regard to updates, that the warehouse maps at the different scales of analysis are consistent, that spatial reference systems and referencing methods are properly transformed, that there is no mismatch problem between the abstraction levels of the dimension hierarchies and the cartographic generalization results. Spatial ETL requires an expertise about the very nature of spatial referencing (e.g., spatial reference systems and methods, georeferencing and imaging technologies, geoprocessing, and mapping) and this process can`t be executed 100% automatically.

Furthermore, there are issues related to the desire of users to clean and integrate spatial data from different eras. Trade-offs have to be made and different types of decision-support analyses have to be left out because the redefinition of basic spatial units over time, legacy systems were not documented according to good software engineering practices, spatial referencing of the fuzziness in spatial boundaries of certain natural phenomena that are re-observed at different eras, the spatial precision of measuring technologies has changed, and so on. One must realize that building and refreshing the multisource, multi-scale, and multi-epoch spatial data warehouse is feasible but costly that requires efforts, strategic trade-offs, and a high level of expertise. In some cases, properly dealing with metadata, quality information, and context-sensitive user warnings becomes a must.

Spatial ETL technologies are emerging as follows. One may use a warehouse-oriented non-spatial ETL tool or the built-in functions of OLAP servers with a spatial technology such as a GIS, an open-source spatial library, or a commercial transaction oriented spatial ETL. One may also use a fully integrated spatial ETL prototypes that are developed in research centers. This new category of spatial ETL tools may include spatial aggregation/summarization operators to facilitate the calculation of aggregated spatial measures. In spite of these difficulties, it is possible to develop simple spatial warehousing applications if one keeps the cartographic requirements at a reasonable level. Many applications are running today and successfully minimized the above issues.

Spatial Data Mining

During the last decade, due to the fast growing applications of GPS technology, web-based spatial data sharing and mapping, remote sensing, and services depending on locations, many research domains have created access to high-quality geographic data to incorporate spatial information in various studies, such as social analysis and business applications (Diansheng Guo, 2009).

Spatial Data Mining Architecture

Many architectures or models have been generated for data mining approaches. They include Han's architecture for general data mining prototype DB-Learn and DB-Miner architectures. This architecture is comparable to others as presented in Figure 10 (Hemalatha.M, Naga Saranya.N, 2011; Jassar, K. K., Dhindsa, K. S, 2015).

Data Mining Tasks and Techniques

Table 1 clarifies the kinds of data mining techniques and the classification methods for these techniques (Harvey, M.,Han, J., 2001).

Figure 10. Structure of spatial data mining
(Hemalatha.M, Naga Saranya.N, 2011; Jassar, K. K., Dhindsa, K. S, 2015).

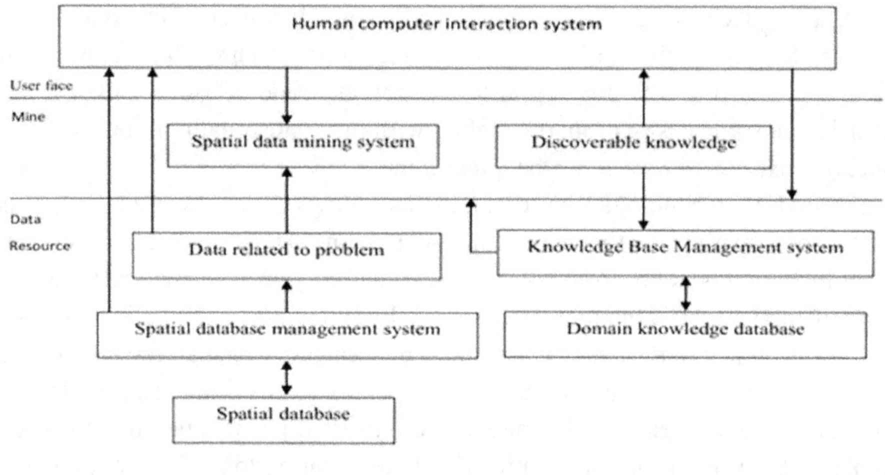

Table 1. Data mining tasks and techniques

Data Mining Task	Description	Techniques
Segmentation or clustering	Determine a finite set of implicit groups that describes the data.	Cluster analysis. Bayesian classification. Decisions or classification trees. Artificial neural network.
Dependency analysis	Predict the class label that a set of data belongs to, based on some training datasets.	Association rules. Bayesian networks.
Deviation and outlier analysis	Find relationships among data objects; predict the value of some attribute based on the value of other attributes.	Clustering mining methods. Outlier detection.
Trend detection	Lines and curves summarizing the database, often over time.	Regression. Sequential pattern extraction.

(Harvey, M.,Han, J., 2001)

Segmentation

Is responsible for classification input data into two main classes clustering and classification. The first class determines and describes relationships between data items. The second discovers rules in order to allocate data items to existing classes.

Dependency Analysis

It is the prediction of attributes values depending on other values. One of the most used graphical models algorithms is Bayesian networks that probabilistically examines and predicts relationships between items.

Deviation and Outlier Analysis

Concentrates on unexpected deviations by data items from the normal case.

Trend Detection

Aims to fast and simple estimation depending on linear and logistic regression with filtering techniques.

Data Clustering Techniques

Data Clustering is the classification of a data set into a group of members or patterns. Statistical classification technique in which objects are sub-divided into clusters with similar items that is different from the items in other clusters. We may classify clustering algorithms as in the proposed classification in Figure 11.

Hierarchical Methods

Tree of clusters in hierarchical form is built by these methods in which each cluster node (parent) has child clusters exploring data on different levels of divisions .There are two Hierarchical Methods categories:

- Agglomerative (bottom-up) (Jain & Dubes 1988); (JAIN, A. and DUBES, R., 1988).This method starts with one-point clusters and merges two or more most suitable clusters.
- Divisive (top-down) (Kaufman & Rousseeuw 1990); (KAUFMAN, L. and ROUSSEEUW, P., 1990).

This method starts with one cluster for given data points and splits the most suitable cluster in recursive manner. The process still continues until the requested number k of clusters is achieved.

- **Hierarchical Methods Advantages:**
    - Flexibility regarding the level of granularity.
    - Ease of handling of any forms of similarity or distance.
    - Consequently, applicability to any attributes types.
- **Hierarchical Methods Disadvantages:**
    - Vagueness of termination criteria.
    - Most hierarchical algorithms do not revisit constructed clusters for improvement.

Partitioning Methods (McLachlan & Basford 1988)

Data partitioning methods divide data into several subsets. Certain greedy heuristics are used because checking all possible subset systems is computationally infeasible in the form of iterative optimization with appropriate data these results in high quality clusters. (MCLACHLAN, G. and BASFORD, K., 1988).

K-Medoids Methods (Beckmann, 1990)

K-Medoids methods represent a cluster by one of its points. Clusters are outlined as subsets of points close to respective medoids and the objective function is outlined as the averaged distance between a point and its medoid.

Figure 11. Clustering algorithms classification

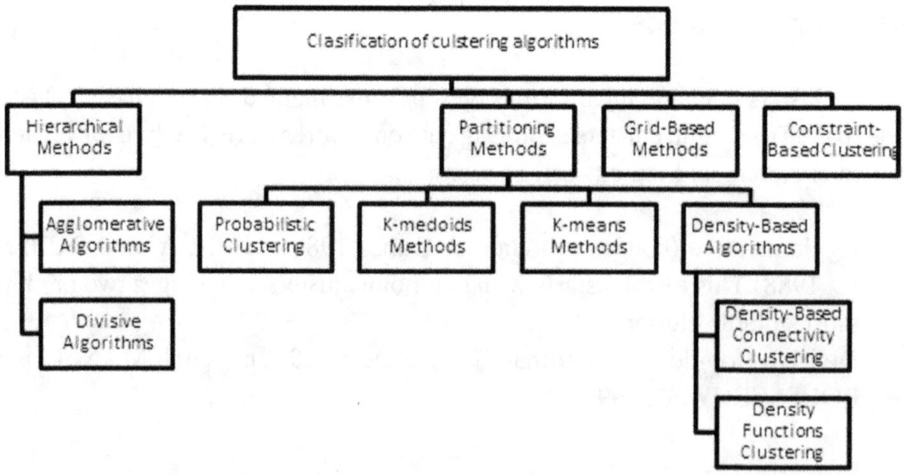

K-Means Methods (Hartigan 1975; Hartigan& Wong 1979)

K-Means is the most popular clustering tool used in scientific and industrial applications. It comes from exploring each of (k) clusters (C) by the mean(c) of its points that called centroid.

Density-Based Algorithms (Martin Ester, 2001)

A cluster is connected dense component that grows in any direction that density leads. So density-based algorithms are capable of discovering clusters of arbitrary shapes.

Density-Based Connectivity (Martin Ester, 2001)

Both Density and connectivity are measured in terms of local allocations of nearest neighbors.

Density Functions (Hinneburg & Keim, 1998)

"Hinneburg&Keim shifted the emphasis from computing densities pinned to data points to computing density functions defined over the underlying attribute space".

Grid-Based Methods (Schikuta & Erhart, 1997)

Grid-Based Methods provides another way of dealing with density, connectivity, and boundary inheriting the topology from the underlying attribute space.

Constraint-Based Clustering (Bradley et al., 2000)

The most frequent target is to bound number of cluster points from below. K-means algorithm sometimes provides a number of very small clusters. In real-world applications customers are rarely interested in unconstrained solutions as in Figure 12.

Dependency Analysis

- **Bayesian Networks (BNs) (Mihajlovic.V, 2012)**

BNs can be illustrated as directed acyclic graphs (DAGs) having nodes and edges. Where nodes represent random variables, and edges represent conditional dependencies between random variables that can be either continuous or discrete. (Daly.R, 2011)

Figure 12. Constraint-Based Clustering

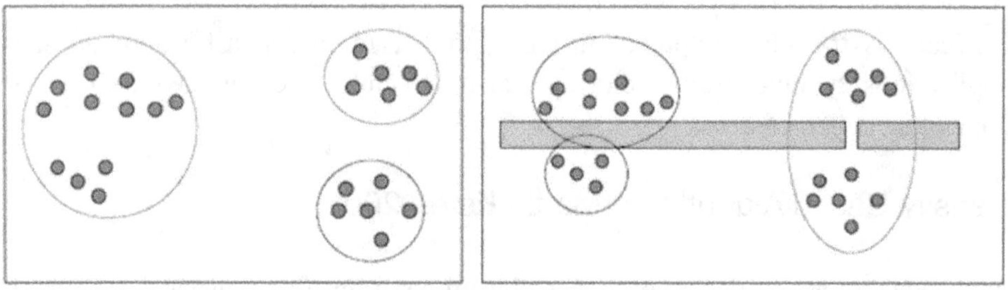

- **Inference in Bayesian Networks (Haipeng Guo, William Hsu, 2002)**

Bayesian networks (BNs) have become the key method for representation and reasoning under uncertainty in AI. BNs provide a normal way to encode exponentially sized joint probability distributions and a basis for effective probabilistic inference.

- **Spatial Bayesian Networks (Sanchez.LA, 2013)**

Having probabilistic relationships between spatial features related to each other and strongly related to each other.

- **Coupling Geospatial With Bayesian Networks (Jolma .A, Lehikoinen.A, Helle.I, 2011)**

These networks support both a methodological part and a software part. The first part is concerned with associating geospatial features with variables or their discrete values in BNs.

Most Used Bayesian Networks Algorithms

- **Genetic Algorithms (Larrafiaga, 1996)**

There are five approaches of heuristic search have discovered for solutions to combinatorial complex problems: evolutionary algorithms, neural networks, simulated annealing, tab search, and target analysis.

- **Estimation of Distribution Algorithms (Valdez.SI, 2000)**

EDAs suggest generating models that can explore the dependencies among the decision variables. Genetic Algorithm could use the available dependencies in case of the building blocks of the solution were correlated

- **Bayesian Optimization Algorithms (Xu.H, 2013)**

The Bayesian optimization algorithm generates a population of candidate solutions by constructing Bayesian networks. After the initialization of a population at random the population is then updated for a number of generations.

- **Simple BN Classifiers Naïve-Bayes (Greiner, Jie Cheng Russell, 2005)**

A Naïve-Bayes BN is a very simple structure as a tree having classification node (parent) and child nodes. A Naïve-Bayes does not allow connections among child nodes.

- **Tree Augmented Naïve-Bayes (TAN) (Cheng.J, Greiner.R, 2011)**

Assume that we have node set X={x1...xn, c} (c here represents classification node) of the given data. TAN classifiers use the conditional model $X \setminus \{c\}$ for tree learning as in Figure 13.

- **BN Augmented Naïve-Bayes (BAN) (Cheng.J, Greiner.R, 2011)**

BAN classifiers come as an improving on TAN classifiers in which an arbitrary graph can be constructed from the attributes. The BAN-learning algorithm is just like the TAN learner, but the Step 2 of the BAN-learner calls an unrestricted BN-learning algorithm instead of a tree learning algorithm see 4. (Cheng.J, Greiner.R, 2011).

Figure 13. A simple TAN structure
(Cheng.J, Greiner.R, 2011)

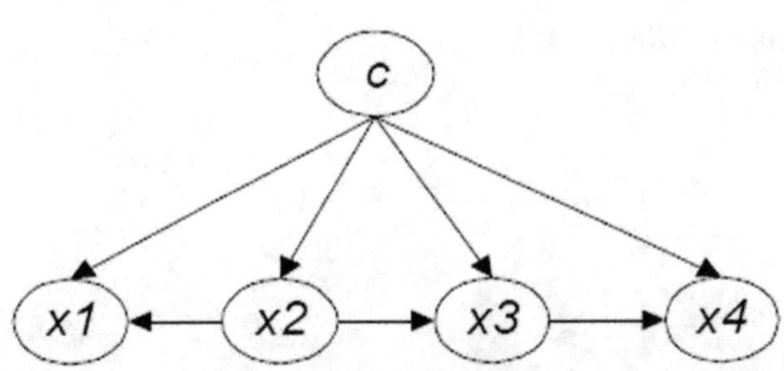

- **General Bayesian Network (GBN)**

Unlike the previous classifiers GBN learner treats the classification nodes as an ordinary node, see Figure 15.

Association (Girotra.M, Nagpal.K, 2013)

Due to its important and big role for marketers, association rule mining becomes most famous and important for them. Market basket is the most famous example for association rule. The role of association analysis is to discover frequently purchased items that have been purchased together. Sales Experts use this output in order to assign closed items together which has an implicit impact on the sales process.

Most Used Association Rule Mining Algorithms

- **AIS algorithm (Girotra.M, Nagpal.K, 2013)**

This algorithm is used to discover frequent item sets using candidate generation in order to detect them. The generated candidates are compared with the frequent item sets that already generated.

Figure 14. A simple BAN structure
(Cheng.J, Greiner.R, 2011)

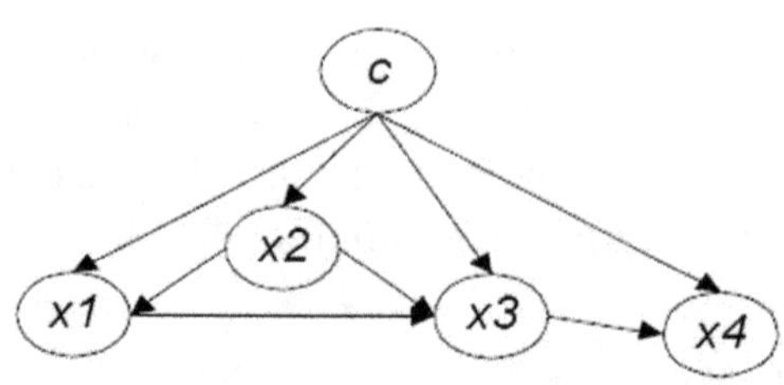

Figure 15. A simple GBN
(Cheng.J, Greiner.R, 2011)

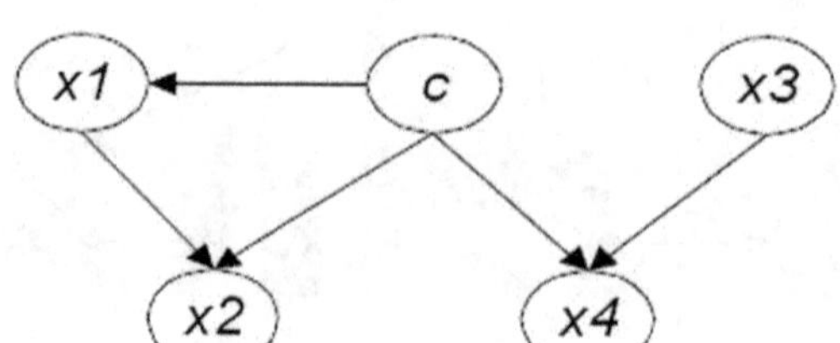

- **SETM Algorithm (Girotra.M, Nagpal.K, 2013)**

STEM algorithm like the AIS algorithm does on the fly counting based on the transaction read from the database. SETM uses relational operations that serve SQL operations. SETM generates candidates using standard SQL join operation after that separates candidate generation.

- **Apriori Algorithm (Zhao, Qiankun, 2003)**

One of the most used and known algorithms in mining association rules is A priori algorithm using a breadth-first search strategy for counting the support of item sets then exploring the downward closure property of support using a candidate generation.

- **AprioriTID Algorithm (Girotra.M, Nagpal.K, 2013)**

AprioriTID algorithm is similar to the Apriori algorithm using the generation function in order to determine the candidate item sets. AprioriTID algorithm differs from Apriori algorithm in the database that not referred for counting support after the first generation.

- **Apriori hybrid algorithm (Girotra.M, Nagpal.K, 2013)**

Apriori and AprioriTID use the same candidate generation procedure and therefore count the same item sets testing every transaction in the database. On the other hand, rather than scanning the database, AprioriTID scans candidate item sets used in the prior pass for obtaining support counts.

- **Eclat algorithm (Girotra.M, Nagpal.K, 2013)**

Eclat implementation represents the set of transactions as a bit matrix and intersects rows give the support of item sets. It follows a depth first traversal of a prefix tree

- **Recursive Elimination Algorithm (Girotra.M, Nagpal.K, 2013)**

Recursive elimination algorithm is dependent on elimination of items from the transaction database with recursively processing of transaction subsets.

- **FP-growth (frequent pattern growth) (Zhao, Qiankun, 2003).**

FP-growth stores the database in a compressed form using an extended prefix-tree (FP-tree) structure. FP-growth uses a divide-and-conquer approach to decompose both the mining tasks and the databases.

The Core Problems of Spatial Analysis (Xiaoshenga.L, Haifenga.R, 2010)

Spatial analysis is a function system which describes objects' spatial relations and spatial behavior, offers reference to objects' spatial query and spatial correlation analysis and services for spatial decision-support, is based on the definition and classification of geographical spatial objects' configuration. Having a given object o, spatial trend (Martin Ester, 2001) can be computed as an orderly variation of spatial attributes that moving away from *o*. Starting from *o* we use neighborhood channels to clear the movement of regression analysis on the particular attribute values for the objects of a neighborhood path in order to describe the harmony change

Algorithm Discovering Spatial Trends (Ester, M., Kriegel, H. P., Sander, J, 1997)

Linear regression is used because it is active and may the impact of some event to its neighborhood is either linear or transformed linear model. Returns all trends of the non-spatial attribute (attr) in the database of the length between (min_length && max_length) with the neighborhood defined by pred. If the correlation coefficient of the current trend is at least equal to (minconf), then incrementally tries to find significant trends of maximal length performing a linear regression on the (attr) values of the last objects on these paths are done.

- **Algorithm Global-Trend (Ester, M., Frommelt, 1998):** Beginning from an object *o*, the algorithm simultaneously creates all neighborhood channels of the same length beginning with (min-length) to (max-length).
- **Algorithm local-trends (Ester, M., Frommelt, 1998):** Providing a regression for each of the neighborhood channels with length (1 min-length) that can be expanded moreover having a significant trend only, see Figure 16.

Deviation and Outlier Analysis

Spatial outliers can be simply detected as they have distinct features from their surrounding neighbors. Spatial outliers discovery helps provide information from large spatial data sets.

Figure 16. Visualization of trends starting from the city of Regensburg
(Ester, M., Frommelt, 1998)

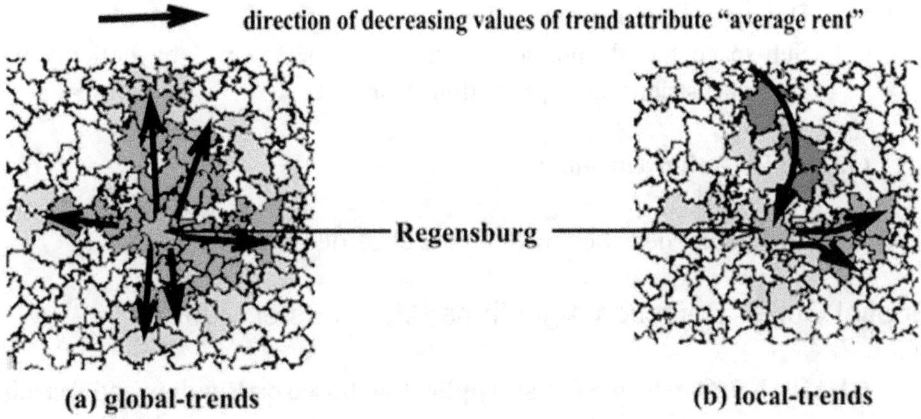

Spatial objects cannot be simply discovered as isolated points. These spatial properties affect the impact of a spatial object on its neighbors.

- **Outlier Detection Approaches, see Figure 17:**
    - Distribution-based approach: is defined as using standard statistical distributions.
    - Depth-based approach which binds data objects into a multidimensional information space.
    - Distance based approach which calculates the proportion of database.
    - Clustering-based approach: outliers can be obtained by products in the end of clustering.

Figure 17. Outlier detection approaches
(T. Surya, B. Azhagusundari, 2014)

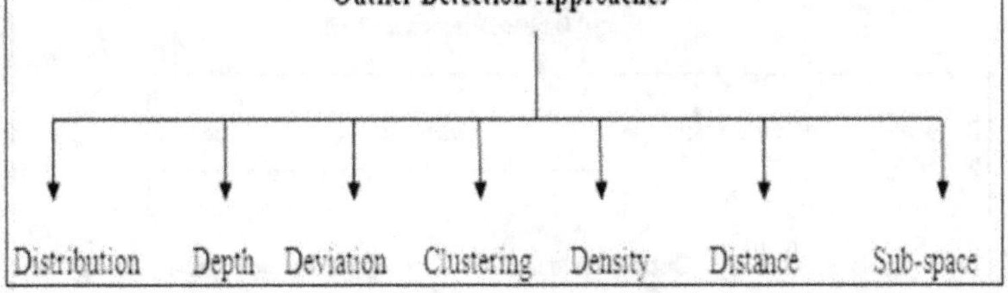

- Deviation-based approach: the characteristics of an object are cleared and outliers are defined as objects with deviations.
- Density based: depends on local outlier factor of each point.
- Sub-space based approach: outliers are determined by noticing the density distribution of predictions from data.

- **Outlier Detection Taxonomy**

Broadly the outlier detection methods are classified as in Figure 18.

Spatial Outlier Detection Algorithms (Martin Ester, 2001)

- **CLARANS (Clustering Large Application Based on Randomized Search)**

CLARANS is considered clustering based approach enhanced from partitioning method of clustering algorithms providing an identification of spatial structures.

- **DBSCAN (Density-Based Spatial Clustering of Applications with Noise)**

DBSCAN is considered density based approach discovering clusters of arbitrary shape.

- **BIRCH (Balanced Iterative Reducing and Clustering using Hierarchies)**

BIRCH is hierarchical clustering with agglomerative method that incrementally clustering multi-dimensional data in an attempt.

- **Index Based Algorithms**

Figure 18. Spatial outlier detection methods
(T. Surya, B. Azhagusundari, 2014)

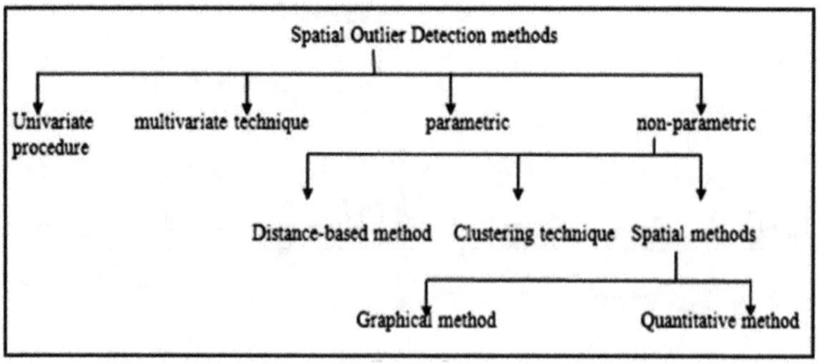

Having a dataset (T) and (N) is the number of objects then the function (F) used to measure distance that gives the distance between any pair of objects.

- **Moran Scatterplot Method**

Moran method is a graphical method with normalized attribute value that can be obtained

By equation (Z(f(i)=([f(i)-μf)σf]) against the neighborhood medial of normalized attribute values (w-z), that (w) is row normalized neighborhood matrix.

Spatial OLAP

Transactional systems are not designed to support the decision making processes. New types of systems have been developed to specifically meet decisional needs; they are called "Business Intelligence" (BI) solutions. BI systems usually have the data warehouse as a central component. These systems are optimized to facilitate and to improve complex analysis, and the performance of database queries involving thousands or more occurrences (Inmon, 2002). The most used BI solutions are OLAP (On-Line Analytical Processing) systems, which provide the capability to interactively explore the data warehouse.

GIS are very good candidates to support decisional needs. However GIS are not designed to support decisional applications or to support highly interactive navigation through spatial data at different levels of aggregation and through different eras. GIS are implemented as transactional systems, so alternative solutions must be integrated with them for analytical purposes (Bédard, Y., T. Merrett & J. Han, 2001). Among them, the coupling of GIS and OLAP into Spatial OLAP (SOLAP) is an interesting path to explore. SOLAP tools possess an interactive user interface allowing non-technical users to easily access, slice, dice, drill, swap, visualize and analyze their data.

Review of SOLAP Concepts

Experimenting the recent implementations of spatial data warehouses, they show that the common client tools used to exploit non-spatial data warehouses were not sufficient to fully analyze the geometric component of the spatial data, despite the fact that the geometric component is the heart of spatial data warehouses. A solution was then developed, which consists of merging the strengths of GIS, with the strengths of OLAP. This combination gave birth to Spatial OLAP or SOLAP, a term used for the first time by (Bédard, 1997).

A Taxonomy of SOLAP Operators

Proposed a taxonomy of SOLAP operators is composed of three levels. The first level of this taxonomic scheme consists of data cube elements. The idea is to group SOLAP operators according to what element of the cube they operate on. The second level of the taxonomy categorizes operators based on the type of operation they perform: multidimensional, spatial or spatiotemporal. Finally, the third level of the taxonomy is according to the operator's return type. Figure 19 shows the three levels of the proposed taxonomy along with an example. It is important to mention that the operators are subcategorized using the criteria of the second and/or third levels only if: i) the number of the operators in the category is large and ii) the group does not consist only of multidimensional operators.

As mentioned before, the first step for defining the operators' categories is to identify the elements of a cube. Therefore, the following categories are proposed using proposed taxonomy: i) Aggregate Functions on Facts, ii) Operations on Facts and Members, iii) Navigation Operations on Members, iv) Operations on Levels, and iv) Operations on Hierarchies and Dimensions. These categories will be explained in detail in the following subsections.

Figure 19. Levels of the taxonomy of SOLAP operators. (Ruiz, C. V., & Times, V. C., 2009)

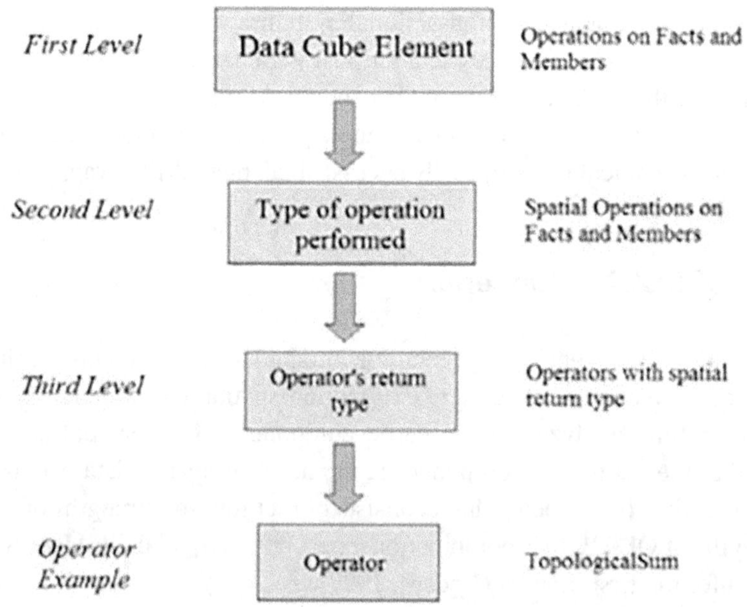

- Aggregate Functions on Facts

The aggregate functions are the special group of operations that can be applied to facts. When navigating through a cube, a new cube can be produced using an aggregate function. The second level of the taxonomy divides functions of this category into two subcategories: i) Spatial Aggregate Functions, which are applied to spatial measures; and ii) Multidimensional Aggregate Functions, which are applied to conventional measures (Ruiz, C. V., & Times, V. C., 2009).

- Operations on Facts and Members

The operations on facts and members are subcategorized using the second level of the taxonomy. The three subcategories emerged are:

1. Spatial Operations on Facts and Members: This group includes operations that can be applied to spatial members and measures. A large number of operators identified for this group. It is necessary to subcategorize them using the criteria of the third level of the taxonomy. The following three subcategories are proposed: i) Operations with Spatial Return Type, ii) Operations that Return Spatial Members and iii) Operations with Numeric Return Type.
2. Multidimensional Operations on Facts and Members: This group includes operations that are applied to conventional nonspatial members and measures.
3. Spatiotemporal Operations on Facts and Members: This group includes operators that facilitate performing queries about changes in geometries through time.

- Navigation Operations on Members

The navigation operations on members are categorized in: i) Spatial Navigation Operations, which allow navigating through spatial members; and ii) Multidimensional Navigation Operations, which allow navigating through conventional nonspatial members.

- Operations on Levels

This group includes operations that are applied to levels. As new operators are proposed, this group could be further subcategorized using the criteria of the second and third levels of the taxonomy. At the moment, only multidimensional operations are considered (Ruiz, C. V., & Times, V. C., 2009).

- Operations on Hierarchies and Dimensions

This group includes operations that are applied to hierarchies and dimensions. Researchers only considered the spatial objects that have predefined hierarchies for this division. There is a small number of operators identified so far, so it appears that the fusion of these two categories is a good alternative especially because hierarchies and dimensions are very related concepts. Figure 20 shows the complete proposed taxonomic scheme of SOLAP operators.

A Set of SOLAP Operators

Ruiz in (Ruiz, C. V., & Times, V. C., 2009) defined a set of SOLAP operators divided into three categories: multidimensional, spatial, and temporal. The combination between spatial and multidimensional domains is also considered.

- Spatial Domain

Topological operators are designed by (Ruiz, C. V., & Times, V. C., 2009) to return members instead of numeric values. These operators take a set of geographic members and return subsets of the members that meet the specified topological relation with respect to a specified geographic member or geometry. This allows queries like "show the cities (i.e., members) that are within a specified geometry" to be answered correctly. The topological operators considered are: Touches, Within, Crosses, Overlaps, Disjoint, Contains, Equals and Intersects. Metric operators take a set of geographic members and return, for each member, a scalar value that is consistent with the metric property being evaluated. The metric operators considered are: Area, Length, Perimeter, and Distance.

Figure 20. A proposed taxonomy of SOLAP operators (Ruiz, C. V., & Times, V. C., 2009)

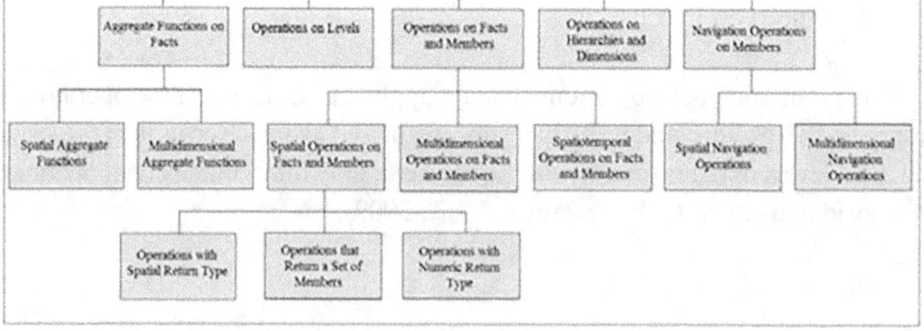

- Multidimensional Domain

The operators considered here are all the functions originally defined for the MDX language including the multidimensional aggregate functions. The aggregate functions supported by MDX are: Max, Min, Sum, Count and Distinct Count.

- Spatial Multidimensional Domain

One of the ideas presented in were used to define spatial multidimensional operators for the proposed set of SOLAP operators consists in the combination of aggregate OLAP functions and geographic operations. The OLAP functions considered can be classified into three categories: distributive, algebraic and holistic. The distributive functions taken into account are: Count, Min, Max and Sum. The algebraic functions considered are: Average, Standard Deviation, MinN and MaxN. Finally, Median, Mode and Rank are the functions considered for the holistic group. This combination leads to the six combinations of functions: Distributive Numeric, Distributive Spatial, Algebraic Numeric, Algebraic Spatial, Holistic Numeric and Holistic Spatial. these groups will be demonstrated in the next paragraphs.

The *Distributive Numeric group* includes the operators that result from the combination of the distributive OLAP function *Count* and topological operators.

In, Ruiz defined only one operator for this group: TopologicalCount. It returns the number of members from the set passed as a parameter that meet the topological relation indicated. Another set of operators belonging to this group is the one that results from the combination of distributive OLAP functions *Max, Min and Sum and metric* operators Area, Perimeter and Length. 's approach defined only three operators: *MetricMin*, *MetricMax* and *MetricSum*. The idea is similar to the one presented above, *MetricMin* operator returns the member from the given set that has the minimum value of the metric property (area, length or perimeter) indicated. *MetricMax* has a similar behavior to *MetricMin*. *MetricSum* operator returns the result of adding the values of the property. Finally, two operators result from the combination of *Max*, *Min* functions and the *Distance* operator: *MinDistance* and *MaxDistance*. The *MinDistance* evaluates the distance from the members of the given set to the member or geometry and returns the member with the minimum distance. *MaxDistance* has a similar behavior but it returns the maximum distance.

Theoretically, the operators of the *Distributive Spatial group* are obtained from the combination of distributive OLAP functions and geographic operations that have a spatial return type. In reality, topological operations are used in combination with the distributive OLAP function Sum. When applied to a set of geographic members, the Sum function performs the union of the given geometries returning the resulting geometry. For our approach, just one operator is defined: TopologicalSum. This operator returns the geometric union of the geometries belonging to the set passed as the first parameter that meet the topological relation indicated on the third parameter with respect to the geometry or geographic member given as the second parameter.

The operators of the Algebraic Numeric group include the operators generated from the combining *Average and Standard Deviation* functions *and metric* operators. In this case, two operators are proposed: *MetricAvg* and *MetricStdev*.

MetricAvg operator returns the average value of the metric property (area, length or distance) indicated. *MetricStdev* has a similar behavior but it returns the standard deviation instead of the average. Another operator that belongs to this group is *AvgDistance*. It returns the average distance from a group of geographic members to a specific geometry or geographic member. Other operators considered for this group are the ones generated after combining the *MinN* and *MaxN* functions with *metric* operators. These operators are: *MetricTopCount* and *MetricBottomCount*. They evaluate a set of geographic members and return the members with the highest (or lowest in the case of *metricBottomCount*) values for the geometric property given. Finally, the operators *TopCountDistance* and *BottomCountDistance* are proposed. They evaluate the distance from the members of a given set to a member or geometry and return the members with the highest (or lowest in the case of *bottomCountDistance*) distances.

The operators of the Algebraic Spatial group are obtained from the combination of algebraic OLAP functions and geographic operations that have a spatial result. No possible combinations were found so no operators are proposed. No operators are proposed for the Holistic Spatial group for the same reason.

Lastly, the operators of the Holistic Numeric group include: *MetricMedian*, *MetricMode* and *MetricRank*. *MetricMedian* evaluates a set of members and returns the median value for the metric property also indicated. *MetricMode* has a similar behavior but it returns the mode value instead. *MetricRank* returns the rank of a given member in a specified set by evaluating the specified metric property.

Besides the previously mentioned operators, another type of spatial multidimensional operators is considered. These are navigation operators that allow performing *drilldown* operations using topological and *metric* properties. *TopologicalDrilldownMember* allows the user to perform a *drilldown* on members of a specified set that meet a specified topological relation with respect to the members of a second specified set. *TopologicalDrillDownLevel* drills down the members of a set to one level below the lowest level represented in the set but only showing those members that meet a specified topological relation with respect to a member or geometry indicated. In the case of *MetricDrilldownMemberBottom*, it sorts, in ascending order, the children of each member in the first set, according to the value of the specified metric property (i.e., area, length or perimeter) as evaluated over the set of child members, then returns a set that contains the parent members and the number of child members, with the lowest value and that are contained by both sets. *MetricDrilldownMemberTop* has a similar behavior but it sorts the elements in ascending order. Finally, *MetricDrilldownLevelBottom* sorts the children of each member in the specified set in ascending order according to the specified metric property, as evaluated over the set of child members. MetricDrilldownLevelBottom returns a set that contains the parent members and the number of child members with the lowest value. *MetricDrillDownLevelTop* has a similar behavior but it sorts the elements in descending order and returns the topmost number of child members.

Finally, the last kind of spatial multidimensional operators considered are the ones originally specified for MDX that needed to be redefined in order to manipulate members of geographic dimensions. They are: Ancestor, Descendants, Ascendants,Members, Children and Siblings.

Categories of SOLAP

There are various basic approaches to GIS and OLAP tool integration to be considered: GIS-dominant, OLAP-dominant and total integration (SOLAP technology). The first approach offers full GIS capabilities and a GIS graphical user interface, but only offers limited simplified access to OLAP data sources and practically no OLAP functionality. The second approach is the opposite; it offers full OLAP capabilities and an OLAP graphical user interface, but limited GIS functionalities. Implementing a SOLAP system without using SOLAP technology requires major development to implement a custom front-end for every application, to integrate OLAP and GIS and use them as back-end services. The time required to implement a SOLAP functional system is reduced by using a fully integrated SOLAP technology such as researchers proposed in (Rivest S.,Bédard .Y,M. Proulx, T. Badard, 2005) (Rivest.S, Bédard .Y, Proulx .M.J., Nadeau. M., 2003) (Ruiz, C. V., & Times, V. C., 2009).

CONCLUSION

This chapter demonstrated basic information of spatial data, spatial data warehouse, spatial data mining, and spatial OLAP. The chapter presented an overview for different algorithms and approaches which are applied on spatial data with demonstrating the issues faced in the field. The chapter also demonstrated aome of the targets for applying different techniques on spatial data such as discovering associations and trends. However, the chapter can be further extended to demonstrate the previous research problem in the field.

REFERENCES

Frommelt. (1998). Algorithms for Characterization and Trend Detection. *KDD: Proceedings / International Conference on Knowledge Discovery & Data Mining. International Conference on Knowledge Discovery & Data Mining*, 44–50.

Bédard, Y., Gosselin, P., Rivest, S., Proulx, M. J., Nadeau, M., Lebel, G., & Gagnon, M. F. (2003). Integrating GIS components with knowledge discovery technology for environmental health decision support. *International Journal of Medical Informatics*, *70*(1), 79–94. doi:10.1016/S1386-5056(02)00126-0 PMID:12706184

Bédard, Y., & Han, J. (2009). *Fundamentals of Spatial Data Warehousing for Geographic Knowledge Discovery*. Taylor & Francis Group. doi:10.1201/9781420073980.ch3

Bédard, Y., Merrett, T., & Han, J. (2001). Fundamentals of Spatial Data Warehousing for Geographic Knowledge Discovery. Geographic Data Mining and Knowledge Discovery, 2, 53-73.

Bédard, Y. B. (1997). *Spatial OLAP*. Academic Press.

Bimonte, S., Pinet, F., Miralles, A., & Papajorgji, P. (2014). Spatial data warehouses and SOLAP. *GeoInformatica*, *18*(2), 269–272. doi:10.100710707-013-0196-9

Cheng, J., & Greiner, R. (2011). *Learning Bayesian Belief Network Classifiers*. University of Alberta.

Daly, R. S. Q. (2011). *Learning Bayesian networks: approaches and*. Glasgow, UK: Cambridge University Press.

ESRI. (2015). Retrieved 2015, from http://support.esri.com

Ester, M., Kriegel, H. P., & Sander, J. (1997). Spatial Data Mining: A Database Approach. *Symposium on Large Spatial Databases*, 47-66. 10.1007/3-540-63238-7_24

Fernando. (2000). *Geomatic Methods for the Analysis of Data in the Earth Sciences New*. Academic Press.

Girotra, M., Nagpal, K., Minocha, S., & Sharma, N. (2013). Comparative Survey on Association Rule Mining Algorithms. India. *International Journal of Computers and Applications*, *84*(10), 18–22. doi:10.5120/14612-2862

Greiner, J. C. R. (2005). *Comparing Bayesian Network Classifiers*. Academic Press.

Guo, D. (2012). *Spatial data mining and geographic knowledge discovery—An introduction*. ESRI.

Guo, H., & Hsu, W. (2002). *A Survey of Algorithms for Real-Time Bayesian Network Inference*. Kansas State, AAAI Technical Report WS-02-15.

Harvey, M., & Han, J. (2001). Geographic data mining and knowledge discovery: An overview. In Geographic Data Mining and Knowledge Discovery (pp. 3-32). Academic Press.

Hemalatha, M., & Naga Saranya, N. (2011). A Recent Survey on Knowledge Discovery in Spatial Data. *International Journal of Computer Science Issues, 8*(3), 473-479.

Inmon, W. (2002). *Building the Data Warehouse* (3rd ed.). Wiley.

Jain, A., & Dubes, R. (1988). Algorithms for Clustering Data. Prentice-Hall.

Jassar, K. K., & Dhindsa, K. S. (2015). Comparative Study of Spatial Data Mining Technique. *International Journal of Computers and Applications*.

Jolma, A., Lehikoinen, A., & Helle, I. (2011). Coupling Bayesian networks and geospatial software for. *19th International Congress on Modelling and Simulation*, Perth, Australia.

Kaufman, L., & Rousseeuw, P. (1990). Finding Groups in Data: An Introduction to Cluster Analysis. New York, NY: John Wiley and Sons.

Kim, B. J. (2013). A System for detecting the Stray of Objects within User-defined Region using Location-Based Services. *International Journal of Software Engineering and Its Applications*.

Kuba, P. (2001). *Data structures for spatial data mining*. Masaryk University Brno.

Lee, J.-G. A. (2015). *Geospatial Big Data: Challenges and Opportunities*. Elsevier Inc.

Lounge, G. (2014). *What is the Difference Between GIS and Geospatial?* Academic Press.

Martin Ester, H.-p. K. (2001). *Algorithms and applications for spatial data mining*. Geographic data mining and knowledge discovery .

McLachlan, G., & Basford, K. (1988). Mixture Models: Inference and Applications to Clustering. Marcel Dekker.

Mihajlovic, V. P. (2012). *Dynamic Bayesian Networks:A State of the Art*. University of Twente.

Mithal, N. G. (2010). *Spatial Data warehouses*. Academic Press.

Mithal, N., & Garg, S. (2010). *Spatial Data warehouses*. Academic Press.

Petek, T. (2011). *Geospatial Data Infrastructure for Economic Development.* Academic Press.

Rivest, S., Bédard, Y. M., & Proulx, T. B. (2005). SOLAP technology:Merging Hypermedia GIS with Spatial On-Line Analytical Processing: Towards Hypermedia SOLAP. *Geographic Hypermedia: Concepts and Systems.*

Rivest, S., Bédard, Y., Proulx, M. J., & Nadeau, M. (2003). *SOLAP: A new type of user interface to support spatio-temporal multidimensional data exploration and analysis.* Academic Press.

Ruiz, C. V., & Times, V. C. (2009). A Taxonomy of SOLAP Operators. *SBBD*, 151-165.

Sanchez.LA. (2013). *A Bayesian network-based method for spatial statistics.* Madrid: Computational Intelligence Group School of Computer Science, Technical University of Madrid.

Schneider, G. V. (2011). *On the Requirements for User-Centric Spatial.* Academic Press.

Surya & Azhagusundari. (2014). Spatial Outlier Detection Approaches and Methods: A Survey. *International Journal of Innovative Research and Development, 3*(4), 24-29.

Valdez, S. I. H. (2000). *Efficient Estimation of Distribution Algorithms by using the Empirical Selection Distribution.* 2Centre for Research in Mathematics A.C.

Wang, X. X. Z. (2013). Spatiotemporal Data Modeling and Management: A Survey. The University of Queensland.

Worden, G. (2010). *Computer Security Awareness at Home.* Academic Press.

Xiaoshenga, L., & Haifenga, R. (2010). *Extract spatial association rules by the method of spatial analysis.* Academic Press.

Xu, H. Y. J. (2013). *Effective Structure Learning for Estimation of Distribution Algorithms.* Creative Commons Attribution License.

Zhao, Q. (2003). Association Rule Mining: A Survey. Singapore: Technical Report, CAIS, Nanyang Technological University.

Chapter 6
Investigating the Impacts of DEM Type, Resolution, and Noise on Extracted Hydro-Geomorphologic Parameters of Watersheds via GIS

Vahid Nourani
University of Tabriz, Iran

Safa Mokhtarian Asl
University of Tabriz, Iran

Maryam Khosravi Sorkhkolaee
University of Tabriz, Iran

Aida Hosseini Baghanam
University of Tabriz, Iran

Masoud Mehrvand
University of Stuttgart, Germany

ABSTRACT

Water resources management is dependent on knowledge and understanding of water quantity and quality information with the latest developments in information technology such as geographic information system (GIS) to develop effective hydrological modeling within the water-based systems. The efficiency of such hydrological modeling relies on the accuracy of applied data. In this way, the application of low-quality data in developing models for integrated management

DOI: 10.4018/978-1-5225-5039-6.ch006

of water resources can impose irreparable financial and human resources and environmental costs in the catchment area. Thus, in regions that shortage of data is the issue, semi-distributed modeling is a useful tool. In this chapter, three aims are followed: (1) effect of digital elevation model (DEM) type and resolution on extracted hydro-geomorphologic parameters, (2) effect of wavelet-based de-noising method on extracted hydro-geomorphologic parameters, (3) determination of the optimal cell size to extract topographic attributes with good agreement to the real features.

INTRODUCTION

Within the past few years, the analysis of factors affecting the hydrological processes within basins has been one of the most significant areas of research in the field of hydrologic modeling. One of the most important prerequisites of developing the map of watersheds and knowing their hydrologic characteristics is the knowledge of rainfall-runoff simulating at the outlets of basins or sub-basins which is one of the most significant challenges of hydrologists, especially in the least developed countries that suffer from the lack of adequate data required for the hydrologic modeling.

Hydrologic models have been generally classified into two categories: lumped and physically-based models. Lumped models act as a black-box model and estimate runoff only at the catchment outlet. These models cannot provide any information about the distribution of saturated areas within the basin; therefore, they are unable to describe how saturated areas distributed within the basin and what their role in evapotranspiration and runoff production is, such as Stanford IV (Crawford & Linsley, 1966) and ARNO (Todini, 1996) models.

Since physically-based models have been developed based on spatially distributed digital and remotely sensed data sets of features such as precipitation, elevation, vegetation, etc., the quality of applied data plays an important role in obtaining efficient models. The application of low quality data in developing models for integrated management of water resources can impose irreparable financial and human resources and environmental costs in the catchment area. To overcome this problem, the physically-based models are applied. Physically-based models that are divided into two categories (i.e., distributed and semi-distributed models) can be utilized in regions that shortage of data or inapplicability of high quality data is the issue (Chen et al., 2004). Due to the various parameters involved in the distributed models, they are considered completely variable on the watershed area, these models require huge data quantities of watershed characteristics to have proper hydrological models such as WetSpa (Water and Energy Transfer between Soil, Plants a) model Atmosphere (Wang et al., 1996). In order to obviate the complexity of distributed models, semi-distributed models are posed. Since the parameters of semi-distributed models have relative spatial changes in the watershed area, so these models need less data than distributed models in hydrological modeling. The semi-distributed models such as TOPMODEL (Topographic Model) that have less complexity in hydrological modeling are a good trade-off between the lumped models and the distributed models (Beven & Kirkby, 1979; Nourani & Mano, 2007). Semi-distributed hydrologic models are created according to hydro-geomorphologic characteristics of the watershed. Terrain analysis and extraction of hydro-geomorphologic characteristics are usually performed by employing the Digital Elevation Model (DEM), as a means of topography information representation of the watershed via software of Geographic Information System (GIS). Indeed GIS is a system designed to capture, store, manipulate, analyze, manage, and present all types of spatial data. GIS is a relatively broad term that can be applied in various branches of science and technology. It is undeniable that the emergence of GIS in hydrology has improved hydrological models to a great extent since the accessibility to spatial elements such as slope, aspect and watershed or catchment area, etc., became easier.

Many hydrological models utilize GIS to edit the input data and to display output data. According to Maidment (1993) GIS applications in hydrology are classified into four categories:

1. Hydrological assessment and classification; GIS is utilized to extract the overall data of watershed, while the manually extraction of these data is so difficult.
2. Determination the hydrological parameters; commonly, GIS is applied to extract the parameters of hydrological models, especially in physically-based models.
3. Linking the hydrological models with GIS.
4. Integrated hydrological models in GIS.

According to Maidment (1993) since several hydro-geomorphologic parameters are extracted via GIS, in order to extract these parameters digital elevation data (DED) which represents the terrain, using numeric or digital elevation information of real land surface (Moore et al., 1991; O'Callaghan & Mark 1984) is utilized. There are lots of variety in the generation sources of DEDs such as photogrammetry, LiDAR (Light Detection and Ranging), IfSAR (Interferometric Synthetic Aperture Radar), land surveying and etc. (Li et al., 2005). Moreover, three methods to generate DEDs are classified as (Moore 1988; O'Callaghan & Mark, 1984): (1) DED based on regular square grid structure (DEM); (2) DED based on Triangulated Irregular Network (TIN) structure; (3) DED based on contour-based structure. DEM as the first type of DEDs issues some advantages and disadvantages (Moore et al., 1991). DEMs that have widely used in researches, are composed of square grids with the elevation values of each grid squares. Advantages of DEMs are including: simplicity in their producing, ease of computer implementation, and computational efficiency (Moore et al., 1991). One of the disadvantages of DEMs is that, the square grids cannot easily handle abrupt changes in topography of land surface (Moore et al., 1991). A DED based on TIN structures that is generated from a continuous surface of triangles, is composed of nodes and edges. The information about the location (x,y) and elevation (z) are located in nodes that are considered as vertices of triangles. Although TINs structure easily handles abrupt changes in topography of land surface, the production and relevant computations are more complex than square grid structure (Moore et al., 1991). Finally, contour-based structures are composed of lines that these lines represent the equal elevation on the land surface. Contours gives readily visualized of land surface topography. Lines are closer together in areas of steep topography, and further apart in areas with less relief (Moore et al., 1991).

The simplicity in produce, processing ease, and computational efficiency are the most important advantages of DEMs, which decrease the importance of its disadvantageous. Therefore it can be resulted that the stated advantages and disadvantages of DEM structures are relative, in the sense that what is considered an advantage for data storage may be a disadvantage for data processing. In this case, due to the dominant advantageous, DEMs based on regular square grid have been used widely in hydro-geomorphologic studies. In the following context, DEM word is referred to DED based on regular square grid structure in order to be concise.

Owing to the usage of hydro-geomorphologic parameters in hydrological models, it can be indicated that DEMs are used in hydroinformatics. Development of the flood prediction systems based on DEM is one of the most important challenges in hydrology (Wood et al., 2011). The extraction of geomorphologic parameters such as slope, specific catchment area (SCA), topographic index (TI) and etc. is possible by DEM. The geomorphologic parameters, which are also called topographic properties, have been generally classified into two categories: initial and secondary topographic characteristics. Initial topographic characteristics are extracted directly from DEM such as, slope, SCA, main channel length, area and perimeter of watershed and etc. (Moore et al., 1991). Secondary topographic characteristics are calculated via combination of some initial topographic characteristics (Jensen & Domingue, 1988). Area and perimeter of watershed, slope, SCA and main channel length as initial topographic characteristics with TI and stream power index as the secondary topographic characteristics are considered in this study.

TI, which shows the most wet points with lowest elevation near to main channel, is one of the most important and basic hydrologic parameters in physically-based models such as TOPMODEL (Beven & Kirkby, 1979; Nourani, 2011). O'Loughlin (1986) and Moore (1988) used TI for different purposes; estimation the saturated points of watershed and explanation of the spatial distribution of soil-water in the watershed. TI, also, is utilized in order to indicate the spatial changes of soil characteristics in the watershed (Moore et al., 1993).

The stream power index assesses the value of the water erosion power (Wilson and Gallant 2000). This parameter is used to determine the first order stream in models to reduce the soil erosion (Wilson & Gallant, 2000).

Since topographic characteristics (e.g. TI) are fundamental parameters in semi-distributed hydrological modeling, the investigations on the relevant factors (e.g., DEMs) can lead to appropriate modeling outcomes. Studies proved that DEM resolution has significant effect on extracted initial and secondary topographic characteristics such as slope, SCA, TI and etc. (Kienzle, 2004). Panuska et al. (1991) compared DEMs in range of 15 to 90 meters and discovered that, the watershed area changes via change of DEM resolution. Zhang and Montgomery (1994) studied mean value of slope in the Oregon and California states via 2 to 90 meters DEMs. The results showed that, the mean values of slope for Oregon and California states with 2 meters resolution DEMs are 65% and 34% and 41% and 25%with 90 meters resolution DEMs. Brasington and Richards (1998) discovered that, the mean value of slope and SCA decrease via increase of the DEM cell size. Kienzle (2004) showed that the mean value of slope and soil erosion raise via decrease of the DEM cell size. Saulnier et al. (1997) observed that the value of TI raise via increase in the DEM cell size. Nourani (2011) compared DEMs in range of 1 meter to 32 meters in order to study the DEM resolution effect on TI distribution via the application of continues wavelet transform. The outcomes showed that the value of TI, increase while DEM resolution decrease. Similar effects of DEM resolution on TI have been reported by Quinn et al. (1995) and Wolock and Price (1994).

In addition to the DEM resolution effects DEM type can also affect the extracted hydro-geomorphologic parameters. In order to show the effect of DEM type on hydro-geomorphologic parameters extraction, Jiang and Wong (2010) compared SRTM (Shuttle Radar Topography Mission) and LiDAR DEMs, and observed that, when LiDAR DEM with 2 meters resolution aggregates to 10 and 30 meters resolution, optimum results in comparison with SRTM DEMs were achieved. Falorni et al. (2005) compared SRTM and USGS (United States Geological Survey) DEMs. Through the study it is observed that drainage networks, watershed boundaries, values of slope change with changes on DEM type.

Over recent years, the application of high resolution DEMs such as LiDAR DEMs has increased by developments in computer field. In spite of progression in producing high resolution DEMs and application of them in various researches, DEMs with small cell size often have not shown satisfactory results (Bian & Valsh, 1993; Jiang & Wong, 2010). Meanwhile, some researches proved that application of high resolution data (i.e., LiDAR DEM) due to the application of mega data set would be involved noises (Nourani, 2011) and noises would affect the appropriate modeling results. Pan and King (2012) stated that when finer resolution in DEM data are used for computing the TI, some artificial features such as buildings, roads and etc. could add noise to the terrain information and produce errors in calculation of TI.

Thus it can be stated that the application of a high resolution DEM in hydrology from one hand enhances the precision of the models in case of huge data, on the other hand the possibility for more noise raises and quality of DEM reduces which leads to the appearance of errors in DEM-based models. It seems that application of a technique which can reduce the noises from high resolution DEMs might be effective to have more precise models. In this way, in order to have sufficient quality of DEMs it is essential to have pre-processing and noise removal operations prior to any usage of DEMs. Accordingly, suitable selection of proper de-noising method plays a crucial role in DEM noise removal. Wavelet transform (WT), as a powerful mathematical tool for data analyses based on scale or resolution, can be used as a de-noising method. Among divers application of WT in various branches of science it has been widely used in geophysics (Foufoula–Georgiou & Kumar, 1995), topographic feature extraction (Lashermes et al., 2007; Passalacqua et al., 2010; Wang & Cai, 2006), hydrological simulation (Labat, 2005; Nourani et al., 2009), image processing such as edge or singularity detection, image compression and image de-noising (JPEG2000; Jansen 2001; Mallat & Hwang, 1992).

Since DEM noise removal which is subjected in this paper can be considered as an image de-noising method, it is essential to review the literature about image de-noising which conduces to wavelet technique as a proper DEM de-noising method. In this light, Nasser Nahi in early 1970s was the pioneer who first went through image de-noising, later in 1970s, Tom Huang investigated the similar subject in his edited book. Then Lee in 1981 proposed digital image enhancement and noise filtering by use of local statistics for the first time. Following years of research on image de-noising was coincided with the invention of WT in late 1980s and dramatic progress on it in the future years. One of the seminal works which combined image de-noising with WT approach was Shapiro (1993), who used Zero-Trees of wavelet coefficients to eliminate the effect of noise from the image. Later, Donoho and Johnstone (1994) presented the theory of error by wavelet soft thresholding schemes for de-noising. A year after, Simoncelli and Adelson (1996) investigated the noise removal via Bayesian wavelet coring. Grace Chang et al. (2000) proposed an adaptive, data-driven threshold for image de-noising via wavelet soft-thresholding. Bjorke and Nilsen (2003) used the WT as a DEM pre-processing tool incorporated into a GIS framework. Lashermes et al. (2007) utilized wavelet-based methodology for DEM de-noising to introduce locally filter LiDAR elevation data and detect thresholds in topographic curvature and slope direction. Nourani (2011) utilized high resolution LiDAR data and Continues WT to show the resolution effect of DEM on TI distribution. As well, Nourani and Zanardo (2013) are used soft threshold wavelet transform for noise reduction of high resolution DEM. They used entropy of topography index as a criterion for determination of threshold in DEM. In a recent study, Biswas et al. (2012) removed the underlying noise of DEM via Curvelet method and as a consequence found that geomorphologic parameters such as slope and wetness index improved after DEM de-noising.

One of the important tasks in de-noising processes is the suitable selection of threshold value. A small threshold may yield a result close to the input, but may still be noisy. A large threshold on the other hand, produces a signal with a large number of zero coefficients. This leads to a smooth signal and may destroy topography details and cause blur and artifacts. Entropy as a measure of the uncertainty in a random variable can be used to determine the suitable threshold values. Entropy is variably described; examples include "a measure of the amount of chaos" or "of the lack of information about the system" (Koutsoyiannis, 2005). In 1948, Shannon formulated the concept of entropy, which can be regarded as a measure of information, disorder, chaos, uncertainty, or surprise, since these are different types of information. The Shannon information content has already been used as a measure of spatial variability and uncertainty in a few hydrologic studies (e.g., Farajalla & Vieux, 1995; Ibbitt & Woods, 2004; Kuo et al., 1999; Mendicino & Sole, 1997; Niedda, 2004; Vieux, 1993).

To quantify deviations from a frequency distribution, use is usually made of variance, but entropy is more powerful and general (Maruyama et al., 2005). Kagan et al. (1973) stated that, if the entropy of a random variable is maximized under the constraint that the variance and the mean are known for a set of values of the variable, the probability distribution function for the set is reduced to a normal distribution.

The reliability and efficiency of the information content (i.e., entropy) application to the de-noising of one dimensional time series have been discussed by Sang et al. (2009) and referred the details. Moreover, Singh (2011) has provided a review of entropy applications in hydrology and water resources.

Determination of optimal cell size of DEM as one of the important goals of this study is investigated through fractal concept. Therefore, beside the broad literature, which is reviewed by far, the fractal concepts also needed to be studied since it is used to find the optimal cell size in this research. It is noted that the optimal cell size is applicable to extract the hydro-geomorphologic parameters. Fractals, simply explain the irregular and fragmented patterns that are existing in the nature, since DEMs due to their structures also include the irregular and fragmented patterns so they can be easily managed by Fractal concept. According to Mandelbrot (1982) fractals can be utilized to variety of many geological, ecological and geophysical problems in the watershed and landscape. Following is the review of various applications of fractals. Mandelbrot (1967), for the first time determined the coastline length of Britannia by fractal concept. Later, fractals have been used to explain the spatial patterns governing on phenomena of earth surface (Burrough, 1986; O'Neill et al.,

1988). De Cola (1989) classified the land-cover via fractal dimension. Lam (1990) discussed on the complexity of three land types by application of fractal concept. Later then, Kuchment and Gelfan (2007) simulated the snow melting in watersheds of Sosna River and Kolyma River via based on self-similarity properties of fractal dimension. Recently, fractals have been used to describe spatial patterns in many landscape-level applications such as measurement of the geometric complexity of landscape features (Burrough, 1986; Mandelbrot, 1982). Fractal dimension that is the characteristic to determine the attributes of fractal patterns or spatial patterns governing on a set is applied to explain the similarity between phenomena (Mandelbrot, 1982). Burrough (1986) via natural log of one-fourth the perimeter against the natural log of the area calculated the fractal dimension of phenomena. Spatial scaling to one branch of the river determined with self-affinity properties of fractal dimension (Tarboton et al., 1988), whereas, drainage networks (River networks) are universal features of landscapes on the earth that lead surface water on the earth land flow into the oceans. Rodriguez-Iturbe et al. (1992) observed that the fractal geometry of the river networks in the watershed impressed the discharge masses and energy dissipation. Afterwards, Tokunaga (2003) utilized theory of fractal geometry to produce the new structure of stream ranking. Based on Literature review in field of fractal geometry that their expression is beyond the scope of this discussion, DEMs because of square grid structure can be easily managed by Fractal concept, so that it is used in this study to determine the optimal DEM cell size.

Overall, this study investigates the effect of DEM type and DEM resolution in hydro-geomorphologic parameter extraction. Firstly, in order to probe the DEM type effect on the issue, two types of DEMs are applied in current study; a) USGS (US Geological Survey) DEM with resolution of 10 meters and b) LiDAR DEM with resolution of 1 meter for the Elder Creek sub-basin of South Fork Eel River basin located in California. Moreover the 10 m USGS and 1m LiDAR DEMs are aggregated into (30, 50,70 and 90 meters) and (10, 30, 50,70 and 90 meters) resolutions, respectively. Secondly, the effect of wavelet-based soft thresholding de-noising approach on extracted hydro-geomorphologic parameters is carried out. De-noising approach is subjected for the high resolution (1m) LiDAR DEM for the Skunk Creek sub-basin of South Fork Eel River basin. Subsequently, 1m LiDAR DEM is aggregated into 1, 2, 4, 8, 16 and 32 meters resolutions to investigate the effect of resolution on entropy of TI. Finally, in order to extract the topographic attributes with good agreement to the real features, the optimal cell size for both USGS and LiDAR DEMs is investigated through two evaluation criteria as; (a) Adaptation of extracted channels with actual channels, and (b) comparison the fractal dimension of extracted channel with fractal dimension of actual channel.

In the following sections of the book chapter, material and methods including the concepts of wavelet transform, entropy and fractal are reviewed. Study area and data sources come after in the subsequent section. The results obtained using the proposed methodology are then presented and discussed, followed by some concluding remarks.

MARTIAL AND METHODS

Wavelet De-Noising Procedure

Wavelet analysis is multi-resolution analysis in time and frequency domain. The wavelet transform decomposes signals into different resolutions by controlling scaling and shifting. It provides a good localization properties obtained in both time and frequency domains.

In the case of a 2-D discrete signal, the decomposition results in a low-pass filtered field (called "approximation") and three high-pass filtered fields (called "details" or "wavelet coefficients"), which capture the high frequency components in the horizontal, vertical and diagonal directions in the frequency domain. This decomposition is repeated at increasing levels resulting in smoother (lower-resolution) approximations and details. Note that to each decomposition level, l, corresponds a certain signal resolution (i.e., the resolution goes as 2^l), therefore the analysis allows for considering signal features at different spatial scales. This procedure is useful to smooth many natural signals, such as topography and rainfall, which contain a large amount of redundant information. Such redundancy, which is due to spatial coherency or correlation, can be easily depicted as the majority of the wavelet coefficients with a value close to zero.

The process is initialized through the decomposition of the high resolution DEM, (a 2-D signal), $f(x,y)$, using a suitable orthonormal discrete 2-D wavelet as (Mallat, 1998):

$$f\left(x,y\right) = p_{L,n,k}\varphi_{L,n,k}\left(x,y\right) + \sum_{l=1}^{L} \sum_{j=H,V,D} \sum_{n=0}^{N-1} \sum_{k=0}^{K-1} d_{l,n,k}^{j}\psi_{l,n,k}^{j}\left(x,y\right) \tag{1}$$

where, $N\times K$ is the size of the signal, $p_{l,n,k}$ and $d_{l,n,k}^{j}$ are respectively approximation and wavelet (or detail) coefficients at level l; (n,k) are translation indices in x,y directions and j stands for the sub-band (H=Horizontal, V=Vertical and D=Diagonal directions). The functions $\psi_{l,n,k}^{j}\left(x,y\right)$ and $\varphi_{l,n,k}\left(x,y\right)$ are the mother wavelet (or wavelet function) and the scaling function, respectively. The decomposed signal can be inversely reconstructed using the inverse DWT. It is worth noticing that the wavelet transform should be applied to an extended area started from outside of the studied region in order to cope with the edge effect (Addison, 2002).

Most of the noise of the DEM tends to be represented by the smaller wavelet coefficients; therefore, to remove the noise, a natural filtering procedure is that of discarding a certain fraction of such wavelet coefficients. To this end, we use a so-called soft-thresholding scheme (Donoho, 1995), which, not only discards all the coefficients smaller than a certain threshold, but also reduces the remaining coefficients of a quantity equal to the threshold value. The new coefficients *d'* can then be obtained as (Donoho, 1995):

$$d' = sgn\left(d\right).\max\left(\left|d\right| - \tau, 0\right) \tag{2}$$

where *d* is the original coefficient (without reporting the indices for simplicity) and τ is the selected threshold. After applying the soft-thresholding procedure the signal can be reconstructed through the inverse DWT. The schematic diagram of wavelet based image de-noising is shown in Figure 1.

It is clear that the selection of a suitable threshold value is an important question in the de-noising process. A small threshold may yield a result close to the input, but may still be noisy. A large threshold on the other hand, produces a signal with a large number of zero coefficients. This leads to a smooth signal and may destroy topography details and cause blur and artifacts. Through this study the selection of the threshold by entropy is focused on.

Present your perspective on the issues, controversies, problems, etc., as they relate to theme and arguments supporting your position. Compare and contrast with what has been, or is currently being done as it relates to the chapter's specific topic and the main theme of the book.

Figure 1. The schematic diagram of wavelet based image de-noising

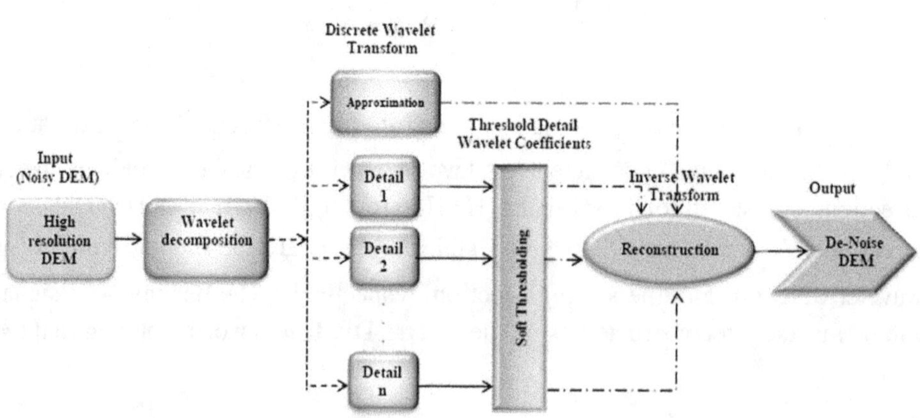

Entropy

Entropy is variably described "a measure of the amount of chaos" or "of the lack of information about the system" (Koutsoyiannis, 2005).The Shannon information content, *I*, (Shannon, 1948) has already been used as a measure of spatial variability and uncertainty in a few hydrologic studies (e.g., Farajalla & Vieux, 1995; Ibbitt & Woods, 2004; Kuo et al., 1999; Vieux, 1993), and is defined as:

$$I = \sum_{i=1}^{N_b} P_i \ln\left(P_i\right) \tag{3}$$

Where N_b is the number of discrete intervals (bins) in which the variable (in this case TI) range is divided; and P_i is the probability of the variable occurring within the bin *i*. The natural logarithm used in Eq. 3 yields information content in unit of *Nats*. The information content is an indication of how much information loss may occur during the filtering process. For topography with high variance a fine grid is needed to capture the information content, whereas a much coarser grid will be sufficient for a surface with almost constant elevation. The reliability and efficiency of the information content (i.e., entropy) application to 1-D time series de-noising have been discussed by Sang et al. (2009) and may be referred for details.

Fractals

A fractal is a mathematical set that has a fractal dimension and usually exceeds its topological dimension and may fall between the integers. A fractal is created by repetitive patterns and generally is self-similar and independent of scale. The self-similarity properties occur when the object growth scale is equal in all dimensions. Generally, fractals are divided into two categories: mathematical fractals and natural fractals. Mathematical fractals can be exactly self-similar. Some of the most famous mathematical fractals are the Sierpinski gasket, Cantor set and Koch curve (Figure 2). As, it is observed in Figure 2 while zooming the images no new detail appears; nothing changes and in fact, it is exactly copy of the whole image at smaller and smaller scales.

Natural fractals can be statistically self-similar, where the statistical properties of the object are consistent across scale, the example of such natural fractals are coastlines, cracking, tree branching and lightning. Figure 3 exhibits Brownian motion as a natural fractal object with self-similarity properties in its creation pattern.

Fractal dimension is the characteristic to determine the attributes of fractal patterns or spatial patterns governing on a set. There are various methods to calculate the fractal dimension. In this study, two methods are applied to determine the fractal

Figure 2. Exactly self-similar fractals (a) The Sierpiniski gasket: a fractal object which is self-similar over all scales, (b) The Cantor set: a fractal object which is self-similar over all scales, (c) The Koch curve: a fractal object which is self-similar over all scales (each of the circles contains self-similar parts of the whole object at different scales.)

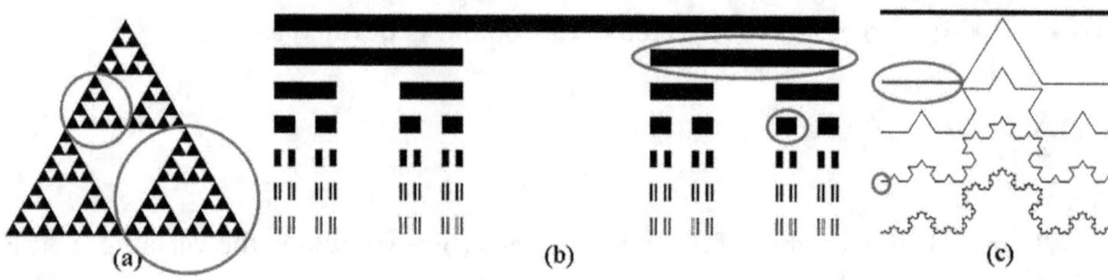

Figure 3. The two-dimensional trajectory of ordinary Brownian motion (The right-hand plot contains the first 1/16th of the trajectory of the left-hand plot enlarged to maintain the same degree of resolution between each plot.) After Addison 1997 Fractals and Chaos: An Illustrated Course, P S Addison 1997, Institute of Physics Publishing, Bristol and Philadelphia.

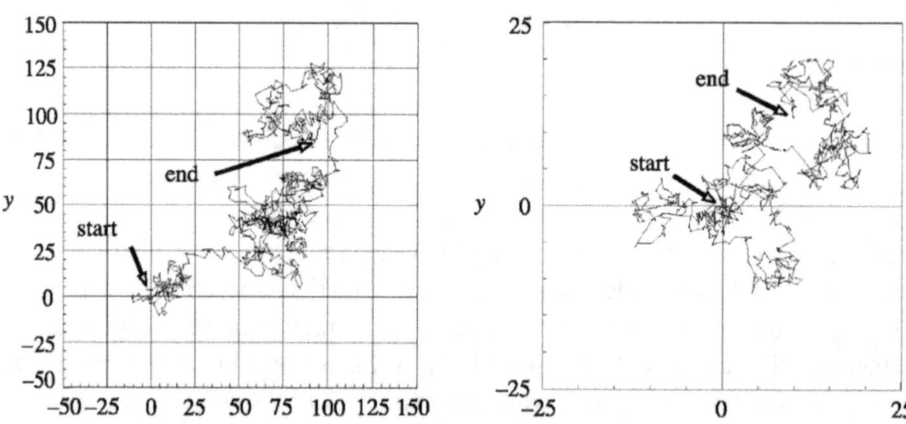

dimension; Box dimension method, and Horton's law method (Burrough, 1986; Ichoku et al., 1996).

Based on box dimension method, fractal dimension is calculated by computing the slope of a regression line between the logarithm of perimeter or area of phenomena to logarithm of cell size (Burrough, 1986).

According to Horton's law method, at first, bifurcation ratio of drainage network and length ratio of drainage network is calculated. Then the fractal dimension is calculated as:

$$D = \frac{Log\,R_B}{Log\,R_L} \tag{4}$$

where D is fractal dimension, R_B is bifurcation ratio of drainage network and R_L is the length ratio of drainage network (Ichoku et al., 1996).

Study Area

The South Fork Eel River basin in northern California, USA drains an area of approximately 1,800 km2. The climate is generally Mediterranean type and characterized by long, warm summers and cool, wet winters. Mean annual rainfall ranges from 1,500 mm to 1,800 mm most of which falls between October and April (James, 1983). Approximately 20 percent of the basin is owned by State Parks and the Bureau of Land Management and a small portion is owned by large timber industries, while the remainder is owned by small landholders, ranchers and residential communities. This research focuses on two sub-basins of the South Fork Eel River, they are called, Elder Creek and Skunk Creek that are largely undisturbed basins. The length of the main channel in the Elder Creek sub-basins is about 8 km and with an average stream gradient of 80 m/km. Elder Creek and Skunk Creek sub-basins have 16.84 km^2 and 0.54 km^2areas, respectively (Figure 4).

Applied Data

The applied data in this study are DEMs, which are utilized to extract the hydro-geomorphologic parameters. Two types of DEMs with various generation methods are utilized in this research i.e., USGS and LiDAR DEMs.

USGS DEM

The USGS has offered three DEM products for the conterminous United States for several years through the EROS Data Center (http://eros.usgs.gov):

1. Large-scale 7.5-min DEMs (approximately 1:24,000 scale) in both 30-m and 10-m resolutions,

Figure 4. Study area

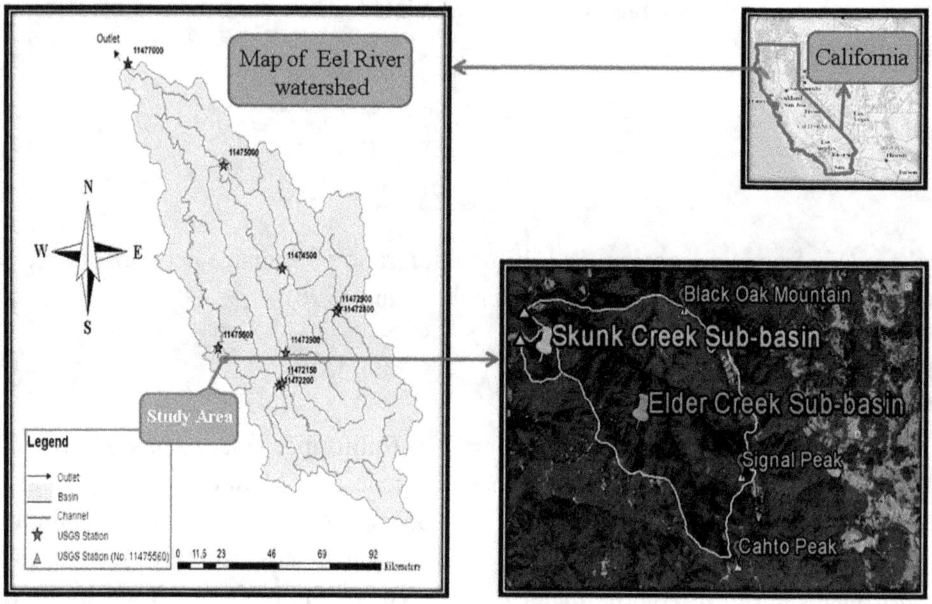

2. Intermediate-scale 15-min DEMs (approximately 1:100,000 scale) in a 60-m resolution,
3. Small-scale 1° DEMs (approximately 1:250,000 scale) in a 90-m resolution.

Since these data are offered in blocks, problems often arise in edge matching of adjacent blocks of DEM data, giving rise to slivers and seams when the blocks are merged together. To overcome these problems, the USGS has developed the National Elevation Database (NED) as a seamless mosaic of the best-available elevation data (http://ned.usgs.gov). Although 7.5-min elevation data for the conterminous United States are the primary data source, efficient processing and filtering methods have been applied to provide a consistent datum and projection, edge-matching, filling slivers of missing data at quadrangle seams, and conversion of all elevation values to decimal meters. The resolution data at the level of 1 arc-second (approximately 30-m resolution) are available nationwide, with approximately 70% coverage for one-third arc-second (10-m resolution). Through the current study the USGS DEM with 10-meter resolution is applied.

LiDAR DEM

LiDAR is an acronym for Light Detection and Ranging. This technology can be used to determine distance, determine chemical composition of the atmosphere, or for determining the velocity of a moving object (Cracknell & Hayes, 1991). LiDAR operates in much the same way as traditional RADAR, but instead of emitting radio energy and awaiting a reflective response, the LiDAR sensor emits short bursts of laser light towards an object or a surface. As the emitted light interacts with objects and surfaces during its travel, it is reflected back towards the sensor. The time it takes for the light to be reflected back to the sensor reveals the distance of the object or surface from the sensor. The intensity of the return is also measured. The distance from the light emitting sensor to the object being detected is determined from mathematical equations using the speed of light (NC Flood maps, 2003; http://www.ncfloodmaps.com/lidar.htm). Finally, the height of each point on the ground surface are calculated based on the obtained information from GPS (Global Position System) and IMU (Inertial Measuring Unit).Subsequently, elevation values in cellular networks are assembled and the outcome is LiDAR DEM. The applied LiDAR DEM in this is in resolution of 1 meter.

METHODOLOGY

GIS concept and technology are being used extensively in water resources engineering, planning and design. In this research GIS is utilized to extract the hydro-geomorphologic parameters for investigation the watershed via the digital map. Burrough (1986) introduced the GIS as, set of tools to collect, store, convert and show the spatial data. One of the most important ability in GIS is to reference the real-world spatial data elements to a coordinate system.

Data modeling in GIS is classified into two categories: raster modeling and vector modeling. In this study in order to generate various coarse resolutions by aggregating initial fin DEMs (i.e., 10-meter USGS and 1-meter LiDAR DEMs) raster modeling is used. Afterwards, in order to generate drainage network of sub-basins which is used in developing fractal dimensions vector model is used. It is noted that vector models in GIS, which represent the terrain attributes, are divided into three layers: (1) a layer includes points to display outlet location at basins and sub-basins, (2) a layer includes lines to display the flow path and drainage networks, (3) a layer includes polygons to display the basins and sub-basins (Figure 5).

One of the basic raster data that is produced via D8 method (i.e., unit flow direction algorithms) is flow direction raster, which is performed in Arc Hydro tool. Indeed, Arc Hydro is an ArcGIS-based system that provides a template for the

Figure 5. (a) Separating the features layer in GIS, (b) display merges main three vector layers in GIS

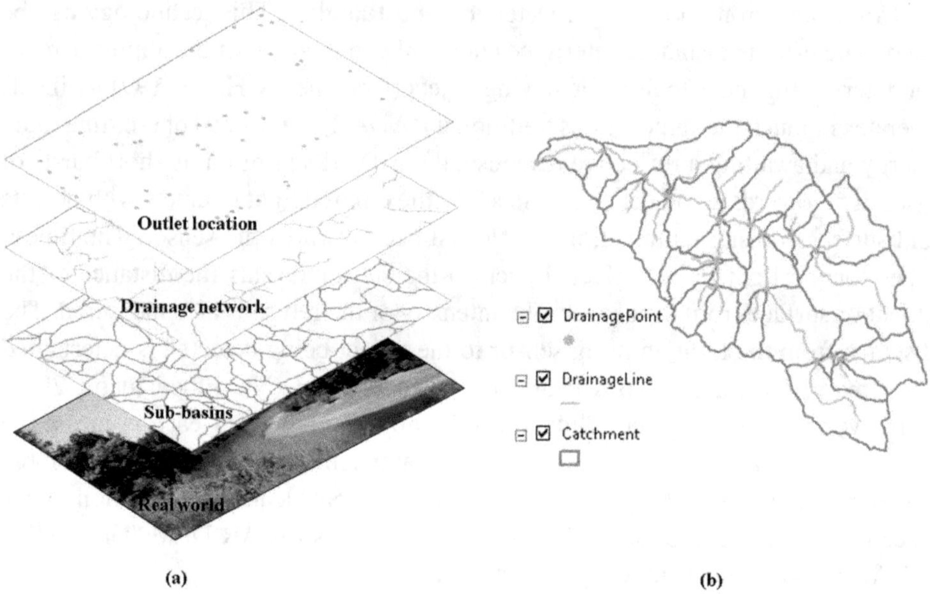

**For a more accurate representation see the electronic version.*

creation and manipulation of a wide variety of hydrologic and water resources related objects and their associated attributes, including the appropriate rules governing their topological interactions.

In order to produce the flow direction raster via D8 method, the errors in DEM which are usually classified as sinks should be eliminated firstly. A sink is an area surrounded by higher elevation values and is also referred to as a depression or pit, which is an area of internal drainage. Although some of sinks in DEMs may be natural many others are imperfections and are created due to the interpolation errors. Since sinks are areas of internal drainage, prevent down slope flow routing of water, thus, in order to generate continuous flow networks in the watershed, it is necessary to remove the sinks from a DEM. The scheme of an assumed sink that has been generated via interpolation is shown in Figure 6-a.

After elimination of the sink, via D8 method (i.e., unit flow direction algorithms), which is an algorithm to determine the flow direction (O'Callaghan & Mark, 1984), flow direction raster is generated. Flow direction for any cell of DEM, is a key function to extracted hydro-geomorphologic characteristics of the watershed. In fact, flow direction raster is the base of subsequent terrain analyzing. Performance of the D8 algorithm is based on 3×3 matrix. In fact, flow direction function is a central function. In order to initiate D8 algorithm, a processor locates in the central

Figure 6. Scheme of assumed sink, (b) Codes of flow direction from central cell to adjacent cells, (c) Elevation values of earth surface, (d) Flow directions based on D8 method, (e) Codes of flow direction based on D8 method

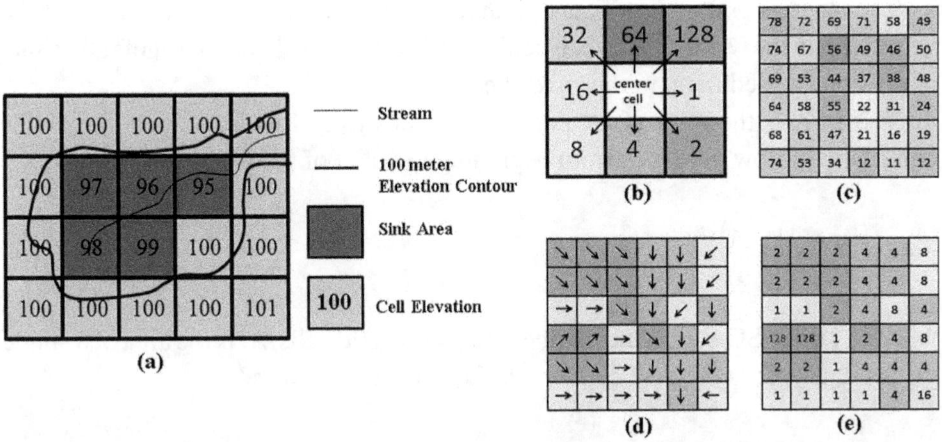

cell (Figure 6-b) and determines flow direction through an adjacent cells which has the minimum value of elevation. Then, according to direction of the flow regard to the selected adjacent cells, one of the eight flow directions is assigned; 1(east), 2 (southeast), 4 (south), 8 (southwest), 16 (west), 32 (northwest), 64 (north), 128 (northeast). In Figs. 6-b, 6-c, 6-d and 6-e the codes of flow direction from central cell to adjacent cells, elevation values of earth surface, flow directions based on D8 method and codes of flow direction based on D8 method are shown, respectively.

After determination of the flow direction in a basin the topographic characteristics are extracted. As stated before, the topographic properties have been generally classified into two categories: initial and secondary topographic characteristics. In this research, two initial topographic characteristics i.e., slope and SCA and one secondary topographic characteristic i.e., TI are used to investigate the effect of DEM resolution on extracting hydro-geomorphologic characteristics.

Slope measures the rate of change in elevation on the ground surface and is typically measured as a percent rise or in units of degrees. To calculate the slope for each cells of DEM, firstly, the elevation values of a 3x3 cell neighborhood around the center cell are considered. Then, based on the maximum downward slope method the slope values for each cells of DEM are calculated based on the slope between the center processing cell, and its lowest neighbor as (Jones, 1998):

$$S = \Delta Z / \Delta L \tag{5}$$

where, S; is the slope value, ΔZ; is value of the height difference between central cell and its lowest neighbor cell and ΔL; is value of the distance between center of central cell to center of its lowest neighbor cell.

The slope is one of significant parameters to determine the landscape and the vegetation in the watershed also, it could be effective on velocity of runoff and value of erosion and sediment in the watershed.

The SCA is the ratio of area at upslope cells to cell width, which these cells contribute in flow guidance from beginning to outlet of the watershed, as:

$$A_s = \left(\left(n+1\right)\times A\right) / C \tag{6}$$

where A_s; is the SCA, n; is the number of upslope grid cells, A; is the area of a single grid cell, and C; is the grid-cell length. Eq. (6) simplified as:

$$A_s = \left(n+1\right)\times C \tag{7}$$

since $A = C \times C$.

TI is a fundamental component of some semi-distributed hydrologic models such as TOPMODEL that measures the potential for soil saturation (Beven & Kirkby, 1979). The points in the watershed that have equal TI values have the same moisture conditions. It is calculated as the natural logarithm of the SCA divided by the tangent of the local slope in degrees (Wilson & Gallant, 2000) as:

Figure 7. Show the slope by degree and percent

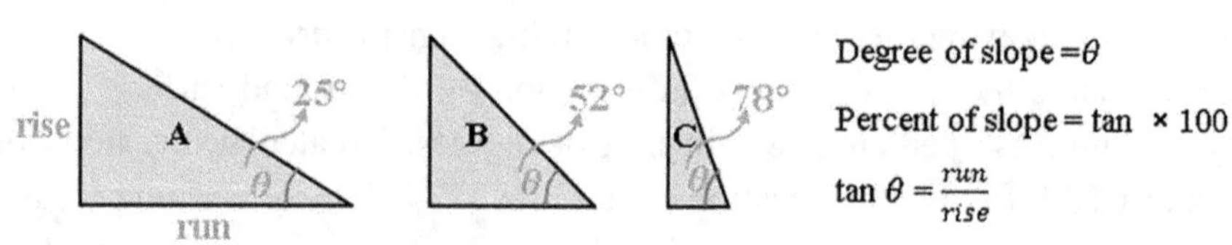

	A	B	C
Degree of slope:	25	52	78
Percent of slope:	47	128	470

$$TI = Ln\left(\frac{A_s}{tan\beta}\right) \quad (8)$$

where A_s; is the SCA, β; is the local slope in degree.

Beside the geomorphologic parameters (i.e., aforementioned topographic characteristics) there are hydrological parameters (i.e., time of concentration (TC) and drainage density) assessed in this study as well. TC is the time required for runoff to travel from the hydraulically most distant point in the watershed to the outlet (Haan et al., 1994). The hydraulically most distant point is the point with the longest travel time to the watershed outlet, and not necessarily the point with the longest flow distance to the outlet. TC varies depending upon slope and character of the watershed and the flow path. TC is generally applied to surface runoff and can be computed by different methods, such as SCS, Kirpich, California method and etc. Due to the California method TC is calculated as (California Culvert Practice, 1955):

$$Tc = \left(\frac{0.885\,L^3}{H}\right)^{0.385} \quad (9)$$

where *Tc*; is the time of concentration in *hr* unit, *L*; is the length of longest flow patch in *km* unit and *H*; is the elevation difference between divide and outlet in *m* unit. Among existing methods in order to calculate the time of concentration the California method is selected because the study area is located in California State.

Drainage density as a fundamental concept in hydrologic analysis is defined as the length of drainage per unit area. The term was first introduced by Horton (1945) and is determined by dividing the total length of streams within a drainage basin in *km* unit by the drainage area in km^2 unit, as (reference):

$$D_d = \frac{L_T}{A_T} \quad (10)$$

where D_d; is the drainage density in km^{-1} unit, L_T; is the total length of streams within a drainage basin in *km* unit, A_T; is the drainage area in km^2 unit.

Drainage density is dependent on the weather and the geomorphologic attributes of the watershed. A high drainage density reflects a highly dissected drainage basin with a relatively rapid hydrologic response to rainfall events, while a low drainage density means a poorly drained basin with a slow hydrologic response (Melton, 1957). If in the watershed, the Soil permeability is low, and the mean slope is high, drainage density increases. For example, in mountainous areas due to high mean slope and low Soil permeability, the drainage density is higher than foothills region.

Both aforementioned geomorphologic and hydrologic parameters are applied together in this study as hydro-geomorphologic parameters. These hydro-geomorphologic parameters are used for the comparison purpose between USGS and LiDAR DEMs as well as various cell sizes.

In the final approach of this study the information content of high resolution topography captured in the TI, proposes a methodology of localized smoothing (regularization) of the LiDAR data which leads to a more physically relevant TI distribution. The regularization methodology uses a soft-thresholding reconstruction in the wavelet domain, which removes the perceived noise and systematically reduces the very high gradients by an amount dictated by the selected threshold. This procedure leads to regularized field on which the TI is then computed. Through the study, the TI distribution from the regularized field exhibits information content, as expressed by an entropy measure, with a characteristic step-shape trend, and allows identifying a critical threshold value for the wavelet coefficients.

RESULTS AND DISCUSSION

In this study the data of two study areas (i.e., Elder Creek and Skunk Creek) are utilized to investigate the behavior of hydro-geomorphologic parameters. To this end, various resolutions of USGS DEM and high resolution LiDAR DEM were applied.

In the first approach which includes sub sections 5-1 and 5-2, the effects of DEM type and resolution on extraction of hydro-geomorphologic parameters were studied for Elder Creek watershed. The sub-section 5-1 goes through the extraction of sub-basins, drainage networks, and spatial distribution of TI as well as hydro-geomorphologic parameters (i.e., TC, drainage density, SCA, slope, TI). Moreover, the behavior of all above mentioned parameters in sub-section 5-1 were compared based on DEM type and resolution. The results of sub-section 5-1 revealed that both DEM type and resolution effects the values of extracted hydro-geomorphologic parameters. So that, the optimal DEM type and resolution is needed to have the most appropriate hydro-geomorphologic parameters in order to develop a high performance hydrologic model. The subsection 5-2 demonstrates the method to figure out the optimal DEM type and resolution via fractal concept. Through this sub-section the adaption of drainage network and fractal dimension of various DEMs were compared with the actual values derived from real image map of Elder Creek watershed, and ultimately the optimal DEM type and resolution for Elder Creek watershed was LiDAR DEM with resolution of 10 meter.

Since LiDAR DEM is a high resolution DEM and involves redundant information, in extraction of hydro-geomorphologic parameters can expose noise. Thus, a pre-processing approach seems to be useful in noise removal of LiDAR DEM. In this regard, in the second approach at sub-section 5-3, the wavelet-based de-noising technique was applied for Skunk Creek watershed.

The methodology of localized smoothing (regularization) of the LiDAR data via information content was conducted to identify a critical threshold value for the wavelet coefficients and then a soft-thresholding reconstruction in the wavelet domain by de-noising purpose was carried out.

Investigation the Effect of DEM Type and Resolution on Extracted Hydro-Geomorphologic Parameters

In order to investigate the effect of DEM type and resolution on extracted hydro-geomorphologic parameters the extraction of sub-basin and drainage network as well as TI spatial distribution is needed. After wards the hydro-geomorphologic parameters for each DEM type and resolution were extracted and compared.

Sub-Basin and Drainage Network Extraction

Firstly, sub-basins and drainage networks of Elder Creek watershed is extracted. The extraction of sub-basins and drainage networks demonstrated that watershed properties (e.g., mean elevation, area and perimeter), shape and number of extracted sub-basins vary with the change of DEM resolution. Application of each DEM type (i.e., USGS and LiDAR DEMs) with different resolutions led to different division of sub-basins and subsequently generated the different drainage networks with various lengths in the watershed (Figure 8). Owing to selected regions in the sub-sections of Fig 8, it is evident that the shape and numbers of extracted sub-basins differ while DEM resolution (compare Figure 8-a and b) and DEM type change (compare Figure 8-b and c). Tables 1and 2 indicate the extracted information about properties of Elder Creek watershed and number of the sub-basins for USGS and LiDAR DEMs in different resolutions, respectively.

According to Tables 1 and 2 decrease of DEM resolution could not lead to constant trend of decrease or increase over the mean elevation, area and perimeter values of the watershed. The reason might be detected on smoothing phenomenon, which occurs while DEM resolution decreases .The smoothing phenomenon is the merge of several adjacent cells and conversion to one cell, so that the mean elevation value over the merged cells is allocated to the new cell as the elevation value. It is obvious that the change in DEM resolution leads to alteration of flow direction in the sub-basins, in such case some cells which were allocated to a specific sub-basin

Figure 8. (a) Extracted drainage network of LiDAR DEM in 1 meter resolution, (b) extracted drainage network of LiDAR DEM in 50 meters resolution, (c) extracted drainage network of USGS DEM in 50 meters resolution

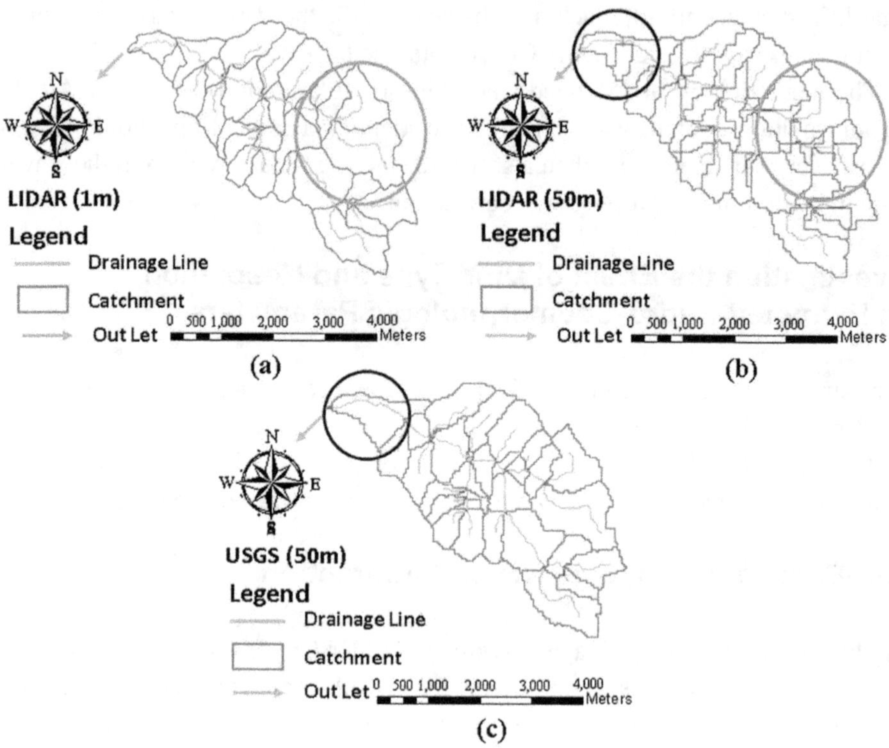

**For a more accurate representation see the electronic version.*

in the initial resolution might be allocated to the neighbor sub-basin by decrease on DEM resolution. This might affect the area and perimeter of new established sub-basins, and ultimately, area and perimeter values of watershed which are extracted based on flow direction raster, might vary along with DEM resolution decrease. Moreover, it should be noted that, the increase in watershed area, will not necessarily lead to increase in the watershed perimeter (e.g., Table 1, in USGS DEM from cell size 50 to 70 meters). Because the perimeter is calculated based on interpolation of peripheral cells and the variation in configuration of peripheral cells from diagonal form to direct form decrease the length of perimeter.

TI Spatial Distribution Extraction

TI value was calculated for each cell of DEMs by application of the Eq.8, and then the spatial distribution of TI which is highlighted by drainage network over

Table 1. The information of Elder Creek watershed for USGS DEM

Cell size (m)	Mean elevation (m)	Perimeter (km)	Area (km^2)	Total length of drainage network (km)	Number of sub-basins
10	849.8869	27.1000	16.8774	23.8416	33
30	849.0160	26.5800	16.8696	24.0451	32
50	849.0160	25.8000	16.8025	24.2359	31
70	850.2207	25.3400	16.8462	23.9094	34
90	848.2297	25.3800	16.5402	23.1565	32

Table 2. The information of Elder Creek watershed for LiDAR DEM

Cell size (m)	Mean elevation (m)	Perimeter (km)	Area (km^2)	Total length of drainage network (km)	Number of sub-basins
1	819.9754	29.5720	16.9737	27.2641	35
10	819.0834	27.7400	16.9932	25.7459	36
30	820.2189	26.6400	16.9218	24.6556	37
50	819.7681	26.0000	16.8000	25.7527	39
70	821.0421	25.4800	16.7531	24.1774	33
90	834.6452	23.0400	16.0785	25.5454	42

the watershed is shown on Figs 9, 10. The comparison of spatial distribution of TI over the resolution decrease in Figs. 9 and 10 individually, demonstrates redundant information in high resolution DEMs which leads to underlying noise in distribution of TI. Moreover, the comparison of figs 9 and 10 proves more noise in spatial distribution of TI in LiDAR DEM than USGS DEM in the corresponding resolution. It is noted that the reason that make the high resolution LiDAR DEM too noisy is originated from consideration of artificial manmade structures as the surface digital elevation in each point of the ground. Considering LiDAR DEM with 90 meter resolution (Figure 3-d), it is obvious that the extracted shape for watershed of Elder Creek is altered from its main shape through the aggregation of DEMs. In this way, it could be concluded that destruction of basic information in DEMs occur in advance at high resolution DEMs (e.g. LiDAR DEM) in comparison to low resolution DEMs (e.g. USGS DEM) while DEM resolution is decreased.

Hydro-Geomorphologic Parameter Extraction

In order to study the hydro-geomorphologic parameters of Elder Creek watershed (i.e., concentration time, drainage density, SCA, slope and TI) firstly the parameters

Figure 9. (a) The spatial distribution of topographic index of USGS DEM with 10 meters resolution, (b) The spatial distribution of topographic index of USGS DEM with 30 meters resolution, (c) The spatial distribution of topographic index of USGS DEM with 90 meters resolution

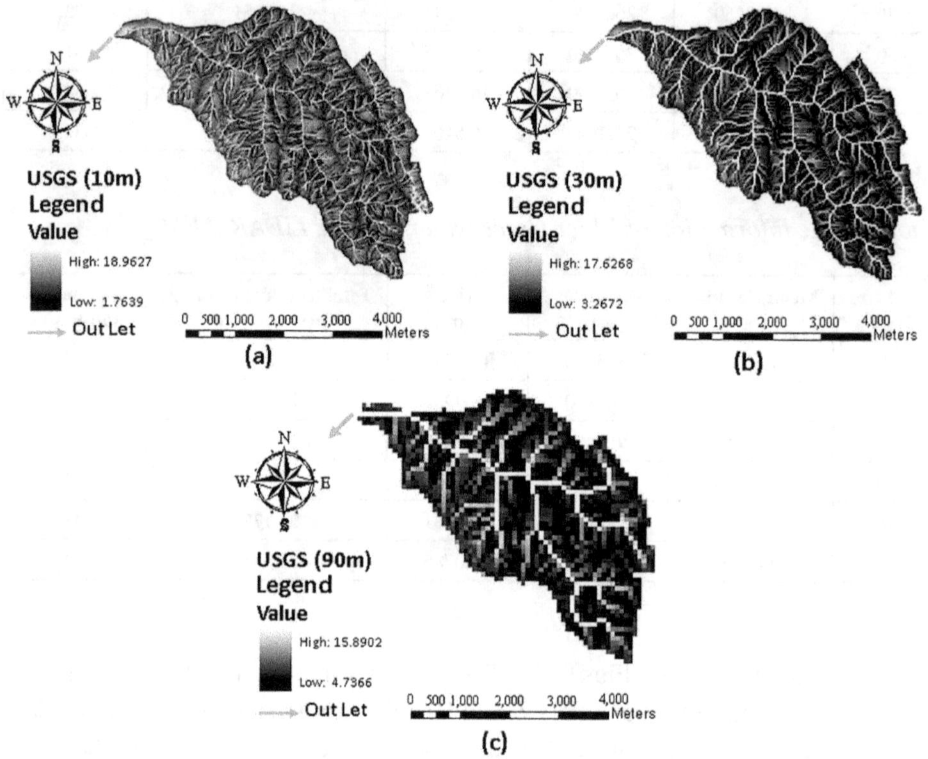

were calculated according the mentioned tips in the methodology section for each DEM type and resolution and then the results were compared according to Tables 3 and 4.

Tables 3 and 4 individually demonstrates that, since TC and drainage density are developed based on flow direction which alters through the various resolution inconstantly, TC and drainage density do not follow an especial subtractive and incremental trend. However the reminder three parameters of Tables 3 and 4 follow a constant increase or decrease pattern. The comparison of Tables 3 and 4 indicates that the values of TC and drainage density in LiDAR DEM are greater than USGS DEM in the same resolution (except for 90 meter LiDAR DEM due to the extraction of different watershed shape), which referred to high resolution of LiDAR DEM. In fact, it could be concluded that, in a watershed the high resolution DEMs (e.g. LiDAR DEM) shows faster reaction against rainfall than low resolution DEMs (e.g.

Figure 10. (a) The spatial distribution of topographic index of LiDAR DEM with 1 meter resolution, (b) The spatial distribution of topographic index of LiDAR DEM with 10 meters resolution, (c) The spatial distribution of topographic index of LiDAR DEM with 30 meters resolution, (d) The spatial distribution of topographic index of LiDAR DEM with 90 meters resolution

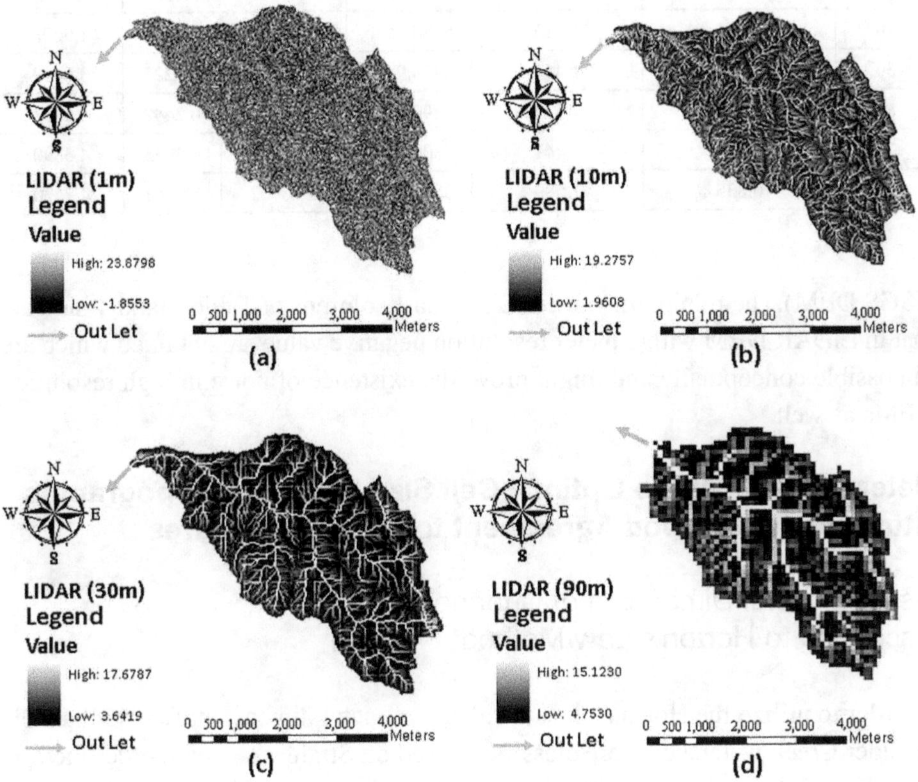

Table 3. The extracted hydro-geomorphologic parameters of Elder Creek watershed for USGS DEM

Cell size (m)	Time of concentration (TC) (hr)	Drainage density (km/km²)	Mean of SCA (m)	Slope (degree)	Mean of TI	Minimum of TI
10	0.969	1.413	4490.8841	25.4037	5.0128	1.7639
30	0.961	1.425	4786.3391	24.0073	5.6775	3.2672
50	0.975	1.442	4896.3068	22.361	6.1615	3.9100
70	0.947	1.419	4622.1262	20.9350	6.4658	4.4290
90	0.919	1.400	4551.9304	19.7024	6.7781	4.7366

Table 4. The extracted hydro-geomorphologic parameters of Elder Creek watershed for LiDAR DEM

Cell size (m)	Time of concentration (TC) (hr)	Drainage density (km/km²)	Mean of SCA (m)	Slope (degree)	Mean of TI	Minimum of TI
1	1.145	1.606	6182.2560	28.0312	3.2855	-1.8553
10	1.051	1.515	5280.4536	26.6748	4.7305	1.9608
30	0.978	1.457	4728.9556	24.1512	5.6227	3.6419
50	1.008	1.533	4890.4389	21.7784	6.2442	4.0343
70	0.971	1.441	4800.6547	20.7806	6.4933	4.3890
90	0.835	1.589	3863.9803	18.8002	6.9161	4.7530

USGS DEM).The minimum values of TI in last columns of Tables 3and 4 indicate that in LiDAR DEM with 1 meter resolution negative value are obtained which are impossible conceptually and might prove the existence of noise in high resolution DEMs as well.

Determination of the Optimal Cell Size to Extract Topographic Attributes With Good Agreement to the Real Features

Usage Fractal Dimension of Channels, According to Horton's Law Method

In order to utilize the Horton's law to calculate fractal dimension, firstly, channels of Elder Creek watershed were classified based on Strahler's stream order method via GIS and Arc Hydro tool. Then the length and number of each class as well as bifurcation ratio and length ratio of drainage network were determined. Finally, fractal dimension of drainage network was calculated based on Equation 4, in which the drainage networks were extracted from various resolutions of LiDAR and USGS DEMs as well as real image map of the Elder Creek watershed. The results were tabulated in Tables 5 and 6.

Owing to Table 5, it is obvious that, fractal dimension of 10 and 30 meters resolutions in USGS DEM are similar and approximately their fractal dimension are near to the fractal dimension of drainage network obtained from real image map of the Elder Creek watershed. The Table 5 shows that the fractal dimension of DEMs with 10, 30 and 90 meter resolutions are approximately equal, here a question arises: why 10 and 30 meter resolutions are selected as the most optimal resolutions based on fractal dimension similarity to the reality for USGS DEM while

Table 5. Extracted value of fractal dimension via GIS software for USGS DEM and comparison against real fractal dimension (Elder Creek watershed)

Cell size (m)	Software results			Real results		
	R_b	R_L	D	R_b	R_L	D
10	4.333	2.410	1.6669			
30	4.333	2.401	1.6740	-	-	-
50	4.167	2.312	1.7029	3.500	2.184	1.6037
70	4.500	2.381	1.7338	-	-	-
90	4.333	2.409	1.6677			

the fractal dimension of 90 meter resolution is roughly equal to the 10 and 30 meter resolution? The answer can be found in the adaptation of extracted channels with actual channels, which is figured out by observation and comparison of extracted drainage networks with reality (Figure 11). Owing to Figure 11 it is evident that matching rate of the extracted channel with actual channels in DEMs with 10 and 30 meter resolution is more than DEM with 90 meter resolution in USGS.

According to Table 6, the fractal dimension in DEM with 10 meter resolution has the most akin value to the fractal dimension of actual drainage network of Elder Creek watershed. As it was stated, one of the applications of fractal dimension is the measurement of geometric complexity in phenomena. When two phenomena have similar fractal dimensions, their geometric behavior is also similar together.

The collating of optimal results from Tables 5 and 6 revealed that the extracted drainage network from LiDAR DEM with 10 meter resolution could have the best

Table 6. Extracted value of fractal dimension via GIS software for LiDAR DEM and comparison against real fractal dimension (Elder Creek watershed)

Cell size (m)	Software results			Real results		
	R_b	R_L	D	R_b	R_L	D
1	4.5000	2.5229	1.6253	-	-	-
10	4.6667	2.6069	1.6077			
30	4.3750	3.1395	1.2900	-	-	-
50	3.1667	2.5084	1.2534	3.500	2.184	1.6037
70	4.3333	2.3892	1.6835	-	-	-
90	3.0333	1.8464	1.8100			

Figure 11. (a) Matching the extracted channel with actual channels for USGS DEM with 10 meter resolution, (b) Matching the extracted channel with actual channels for USGS DEM with 30 meter resolution, (c) Matching the extracted channel with actual channels for USGS DEM with 90 meter

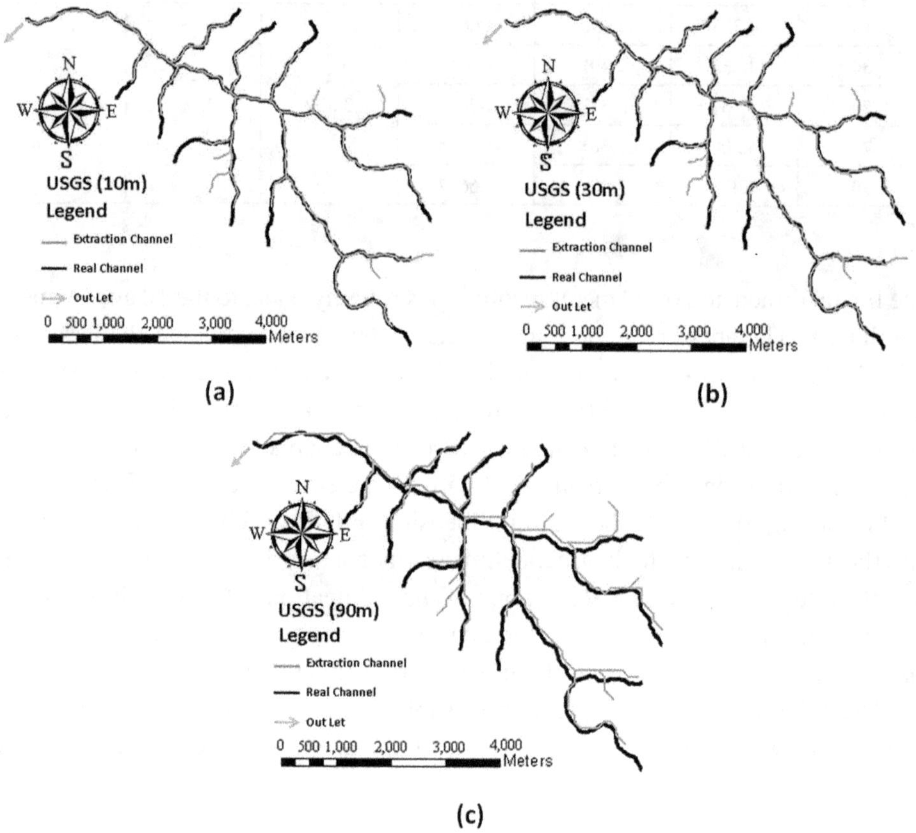

matching with the actual channel of Elder Creek watershed in comparison with 10 and 30 meter cell sizes of USGS DEM.

According to the obtained results 10 meter LiDAR DEM was selected as the most appropriate DEM in extraction of hydro-geomorphologic parameters. The high resolution LiDAR DEM encompasses extra information so that leads to expose of noise in extracting hydro-geomorphologic parameters. In this way, an efficient method which helps to reduce the involved noise in high resolution DEM with redundant information might be effective in hydrological models. To this end wavelet transform as a de-noising

Investigation the Effect of Wavelet-Based De-Noising Method on Extracted Hydro-Geomorphologic Parameters

The scale dependence of the TI is a known issue in geomorphic and hydrological modeling (Nourani & Mano, 2007; Saulnier et al., 1997). In particular Sorensen and Seibert (2007), in a study on the effect of the DEM resolution on the TI distribution, explicitly suggested how an interesting approach to deal with the problem could be to combine the TI computed for DEMs of different resolutions. The procedure presented in the current study follows this suggestion through a robust, multi-resolution analysis in the wavelet domain. In particular, we filter the DEM noise and redundant information by removing a certain fraction of the wavelet coefficients, determined through the definition of an appropriate (or "optimal") threshold. One novel aspect of our work is that the regularization process is carried out with the explicit purpose of producing a TI distribution which is hydrologically meaningful. To this end we use the Shannon entropy of the TI as a criterion to define a suitable threshold. Interestingly, the application to a relevant case study shows how the TI entropy, as a function of the threshold considered, has a characteristic pattern, which allows for an automatic identification of a suitable threshold for the regularization procedure. It is indeed clear in the entropy trend how after a certain threshold, the process starts removing significant information from the original DEM.

Wavelet-based soft threshold de-noising method was used in this study for de-noising approach. In this regard, DEM pre-processing and de-noising procedure started with various thresholds using wavelet transform, then the pre-processed DEM was used in ArcGIS to extract and compute the hydro-geomorphologic parameters such as slope, TI, TI entropy, etc. The appropriate mother wavelet and decomposition level of wavelet-based soft threshold de-noising method was selected according to the outcomes of previous study of the authors (i.e., Nourani & Zanardo, 2013), in this way, the 1-m resolution LiDAR DEM was de-noised with the db6 (Daubechies function with 6 vanishing moments) mother wavelet and decomposition level 3. The de-noising was performed according to the aforementioned procedures in methodology section. After de-noising procedure, wavelet reconstruction could derive the de-noised data from the approximation at resolution level 3 and processed detailed signals at all resolution levels. DEM reconstruction should be in a way that proportionally coincide the original DEM, thus beside de-noising and subsequently smoothing of the DEM the main properties of DEM are preserved. Results reveal that hydro-geomorphological parameters will alter parallel with soft threshold value (Table 7). The rejection of noise using wavelet transform not only eliminates noise in the topographic data, but also reduces the slope of the basin (Figure 12). In fact, by eliminating the noise in the DEM, the holes will significantly fill and the average slope will reduce.

Table 7. Behavior of hydro-geomorphologic parameters of Skunk Creek sub-basin with different threshold value (after the de-noising procedure)

Threshold value (m)	Mean of Slope (degree)	Mean of TI Ln((m))	Normalized value of Entropy TI
0	23.4643	3.3554	1
0.0001	23.4661	3.3559	0.999
0.0005	23.4635	3.3572	0.990
0.001	23.4627	3.3595	0.977
0.007	23.4534	3.3944	0.935
0.04	23.4199	3.4939	0.781
0.06	23.4013	3.5343	0.734
0.08	23.3883	3.5615	0.677
0.1	23.3771	3.5851	0.637
0.22	23.3312	3.6842	0.470
0.33	23.3045	3.7357	0.328
0.42	23.2878	3.7675	0.258
0.64	23.2595	3.8100	0.119
0.72	23.2520	3.8234	0.079
0.83	23.2433	3.8316	0.036
0.9	23.2386	3.8372	0.013
0.94	23.2360	3.8398	0

Figure 12. Changes of slope by threshold increasing in the Skunk Creek sub-basin

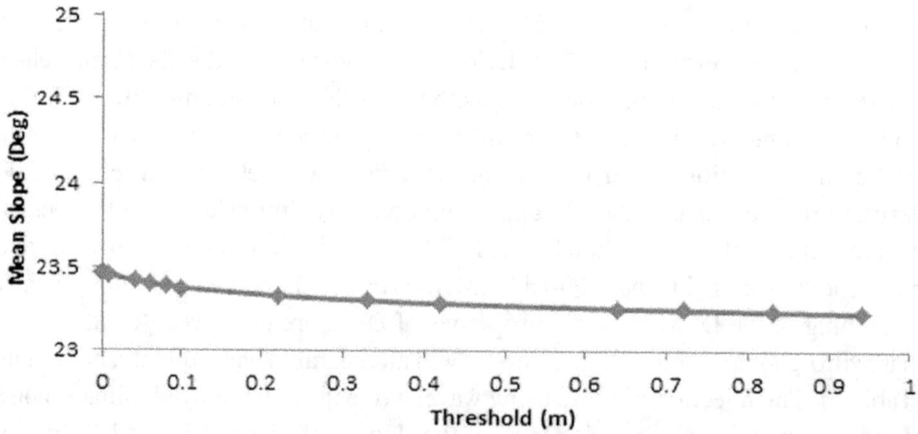

Figure 13 shows the change of the wavelet coefficients in study area. As expected, most of the wavelet coefficients have a relatively low value and therefore even the application of a small threshold will produce the removal of a high fraction of them. According to Figures 13-a, 13-b, 13-c and 13-d which show respectively the spatial distributions of TI within the basin after the de-noising procedure with different threshold value, for the finer threshold the structure of the TI distribution is much complex and irregular (Figure 13 exhibit the threshold decrease from 0.99 *m* (Figure 13-d) to 0 *m* (Figure 13-a)). A large number of TIs with low and even negative values can be seen in the computed distribution by 1-m resolution DEM.

According to TI formula (Equation 8(regions with high value of TI have the greatest potential to be saturated because they are the parts that constitute the river network, thus, such regions have the lightest slope with the most cumulative flow in comparison to the neighbor regions.

Figure 13. The spatial distribution of Topographic index (Skunk Creek basin)

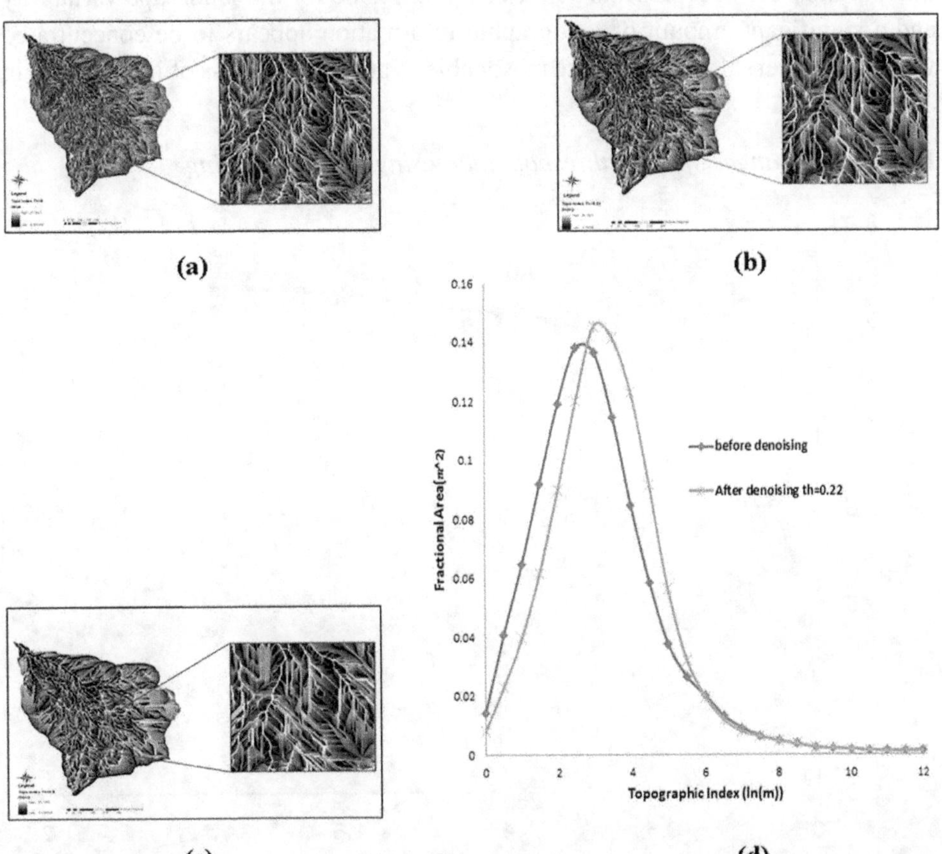

Increasing threshold value removes the noise of the DEM but after a specific value, it also removes main information of DEM and finally creates smoothed signal that topographical details and an artificial effect disappears altogether. As seen in Figure 14, after a threshold value of about one, average TI remains almost constant.

The suitable selection of threshold value is an important task in de-noising processes. A small threshold may yield a result close to the input, but may still be noisy. A large threshold on the other hand, produces a signal with a large number of zero coefficients. Through this study the selection of the threshold by entropy is focused on. The entropy of the TI for different thresholds for Skunk Creek River basin was computed. Figure15 shows how *I* vary as functions of the threshold adopted. It is interesting to observe the entropy plot shows three distinct regions; in the first and third regions a slight, linear decrease of the information content is observed, whereas in the second region, approximately in the interval of thresholds 0.04-0.1 *m*, a significant loss of information content is shown and the decay rate increases with the threshold. Since *I* is a measure of topographic variability, the threshold interval 0.04-0.1 *m* represents a sort of critical zone for the landscape variability and a significant amount of topographic information appears to be concentrated within that interval. It is therefore reasonable to attribute the loss of information in

Figure 14. Changes of TI by threshold increasing after de-noising

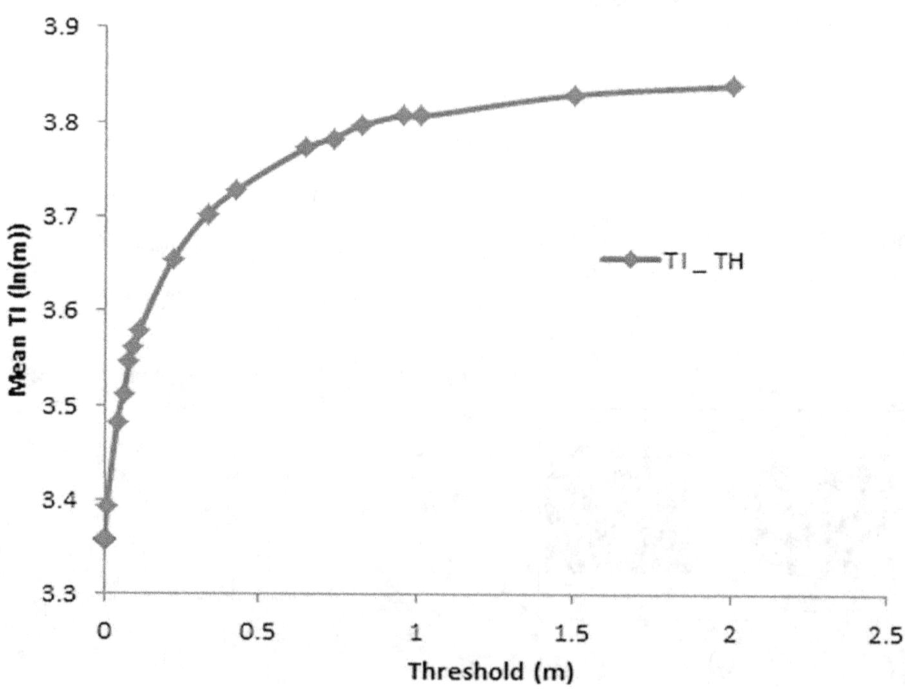

the first region of the plot to the elimination of noise and discontinuities, while, as soon as the threshold $\tau \approx 0.04\ m$ is exceeded, main topographic features start to be removed. Eventually, when the threshold $\tau \approx 0.1\ m$ is also exceeded the landscape seems to have lost its main characteristics as further regularization (by increasing the threshold) introduces variation in the entropy with a much slower rate. For these reasons the value $\tau \approx 0.04\ m$, represents a suitable threshold to be used in our regularization procedure.

Increase in wavelet-based threshold value cause to more filling of the sinks and subsequently more flattening of the DEM, which ultimately leads to reduce of the slope and TI entropy value (Figure 15). This result was expected since the minimum entropy is in accordance with smoother surfaces.

CONCLUSION

This book chapter dealt with the influence of DEM types and resolutions on hydro-geomorphologic parameters extraction in the watershed. To this end firstly, the sub-basins, drainage networks, and spatial distribution of TI as well as hydro-

Figure 15. Cumulative distribution of the wavelet coefficients after 3 level of decomposition

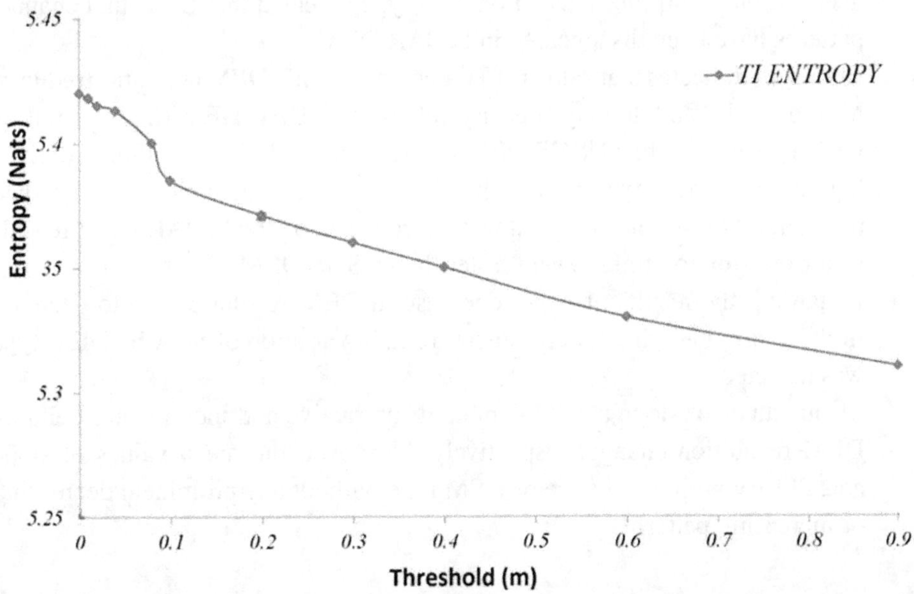

geomorphologic parameters (i.e., TC, drainage density, specific catchment area, slope, TI) were extracted and then compared through DEM type (i.e., LiDAR and USGS) and resolutions. The below results obtained;

- Extraction of sub-basins and drainage networks revealed that through various DEM types and resolutions the extracted pattern of sub-basins and drainage networks differed which led to different area, perimeter, and length of channel in sub-basins.
- The investigations on TI spatial distribution by alteration on DEM type and resolution demonstrated that; firstly the change of resolution on a specific DEM type showed that the decrease of resolution in each DEM (i.e., from 10 meter to 90 meter in USGS DEM and 1meter to 90 meter in LiDAR DEM) increased the transparent visualization of the drainage networks due to the exhibition of main channels, however in lower cell sizes (i.e., high resolutions) the redundant information led to generation of secondary and unimportant channels which might be figured out as noise. Secondly, the comparison of spatial distribution of TI in similar resolutions of USGS and LiDAR DEMs disclosed that the involved noise in LiDAR DEM is greater than USGS DEM. It is noted that the extra increase of cell size in DEMs led to elimination of basic information from DEMs. Considering this fact it is observed that, the elimination of basic information in LiDAR DEM by increasing the cell size occurred faster than USGS DEM, thereafter, the comparison of TI spatial distribution on 90 meter DEM on both type revealed that the actual channel patterns have been disappeared in LiDAR DEM.
- The result indicates that values of TC decreases while DEM resolution reduces. Moreover, TC and drainage density values for LiDAR DEM (high resolution DEM) are higher than USGS DEM in a similar resolution. Assuming that the high drainage density in a watershed leads to faster reaction of the watershed to a rainfall event, the analyzed watershed based on the LiDAR DEM reveals faster reaction to rainfall event faster than USGS DEM.
- Regarding the results of paper decrease in DEM resolution led to decrease in SCA, provided that no prominent trend in variation of SCA by DEM type was detected.
- Mean values of slope and TI exhibited subtractive and incremental trend by DEM resolution change, respectively. Moreover, the mean values of slope and TI vary with the change of DEM type without any prominent decreasing or increasing pattern.

- The minimum TI value in 1 meter resolution LiDAR DEM was negative and proved the existence of noise in it. This consequence reconfirmed the existence of noise in LiDAR DEM which approved by studying the spatial distribution of TI. The origin of noise in LiDAR DEM might be related to this fact that manmade structures (e.g. bridges, roads, Buildings and etc.) are modeled as terrain surface in high resolution LiDAR DEM.

The explained hydro-geomorphologic parameters in the current study are the necessities and prerequisites of many hydrological models and water resources management tools in a watershed area. Therefore, the application of hydro-geomorphologic parameters derived from low quality DEM data utilized in hydro-environmental models can impose irreparable costs from financial and human resources and environmental viewpoints in the catchment area. In this light, selection of an optimal cell size in extracting hydro-geomorphologic parameters has prime importance in hydrological processes modeling and was investigated in this study via fractal dimension effect. In this regard, geographic complexity of drainage networks in a watershed was figured out through fractal dimension values and the optimal fractal dimension values for both type of DEMs were attained according to the affinity of DEM-based extracted fractal dimension to actual fractal dimension as well as channel shape adoption with actual channel shape. The results demonstrated that;

- Although, USGS DEM with 10 and 30 meter resolutions performed sufficiently in conformity of fractal dimension to actual fractal dimension, the LiDAR DEM with 10 meter resolution is the most appropriate DEM to extract the hydro-geomorphologic parameters since best matches the actual properties of drainage network in Elder Creek watershed.

Noise can be a major cause of quality reduction of high resolution DEM. Prior to any use of the DEM it is essential to perform a primary pre-processing and noise removal in order to create a DEM with better quality. In this study, the noise of DEM was removed by using soft threshold wavelet transform and resulted in;

- The filtering and de-noising procedure of the wavelet coefficients by a soft thresholding scheme, not only discarded the coefficients lower than the selected threshold, but also reduced the remaining coefficients of a quantity equal to the threshold value. As a consequence, not only did this scheme smooth the topography by removing noise and discontinuities, but also it reduced the overall slope of the landscape.

REFERENCES

Addison, P.S. (2002). *The illustrated wavelet transform Hand-Book*. London: Institute of Physics Publishing.

Beven, K., & Kirkby, M. J. (1979). A physically based, variable contributing area model of basin hydrology. *Hydrological Sciences Bulletin*, *24*(1), 43–69. doi:10.1080/02626667909491834

Bian, L., & Walsh, S. J. (1993). Scale dependencies of vegetation and topography in a mountainous environment of Montana. *The Professional Geographer*, *45*(1), 1–11. doi:10.1111/j.0033-0124.1993.00001.x

Biswas, A., Cresswell, H., Viscarra Rossel, R., & Si, C. B. (2012, April). *Application of curvelet transform in denoising digital elevation model*. In the 5th Global Workshop on Digital Soil Mapping, Sydney, Australia.

Bjorke, J. T., & Nilsen, S. (2003). Wavelets applied to simplification of digital terrain models. *International Journal of Geographical Information Science*, *17*(7), 601–621. doi:10.1080/1365881031000135500

Brasington, J., & Richards, K. (1998). Interactions between model predictions, parameters, and DTM Scales for TOPMODEL. *Computers & Geosciences*, *24*(4), 299–314. doi:10.1016/S0098-3004(97)00081-2

Burrough, P. A. (1986). *Principles of geographic information systems for earth resources assessment*. Oxford, UK: Clarendon Press.

California Culvert Practice. (1955). Department of Public Works, Division of Highways, Sacramento, CA.

Chang, S. G., Yu, B., & Vetterli, M. (2000). Adaptive wavelet thresholding for image denoising and ccompression. *IEEE Transactions on Image Processing*, *9*(9), 1532–1546. doi:10.1109/83.862633 PMID:18262991

Chen, Y., Zhu, D., & Zhao, J. (2004, September). Small basin flash flood simulation with TOPMODEL. In *International Conference of GIS and Remote Sensing in Hydrology, Water Resources and Environment (ICGRSHWE), Three Gorges Dam* (pp. 41-49). Academic Press.

Cracknell, A. P., & Hayes, L. W. B. (1991). *Introduction to remote sensing*. London: Taylor and Francis.

Crawford, N. H., & Linsley, R. K. (1966). *Digital simulation in hydrology: Stanford Watershed Model IV.* Department of Civil Engineering Technical Report 39. Palo Alto, CA: Stanford University.

De Cola, L. (1989). Fractal analysis of a classified landsat scene. *Photogrammetric Engineering and Remote Sensing*, *55*(5), 601–610.

Donoho, D. L. (1995). De-Noising by soft-thresholding. *IEEE Transactions on Information Theory*, *41*(3), 613–627. doi:10.1109/18.382009

Donoho, D. L., & Johnstone, I. M. (1994). Ideal spatial adaptation via wavelet shrinkage. *Biometrika*, *81*(3), 425–455. doi:10.1093/biomet/81.3.425

Falorni, G., Teles, V., Vivoni, E. R., Bras, R. L., & Amaratunga, K. S. (2005). Analysis and characterization of the vertical accuracy of digital elevation models from the Shuttle Radar Topography Mission. *Journal of Geophysical Research*, *110*(F2), F02005. doi:10.1029/2003JF000113

Farajalla, N. S., & Vieux, B. E. (1995). Capturing the essential spatial variability in distributed hydrological modeling: Infiltration parameters. *Hydrological Processes*, *8*(1), 55–68. doi:10.1002/hyp.3360090106

Foufoula-Georgiou, E., & Kumar, P. (1995). *Wavelet in geophysics*. New York: Academic Press.

Haan, C. T., Barfield, B. J., & Hayes, J. C. (1994). *Design hydrology and sedimentology for small catchments*. Academic Press.

Horton, R. E. (1945). Erosional Development of Streams and their Drainage Basins: Hydro-Physical Approach to Quantitative Morphology. *Geological Society of America Bulletin*, *56*(3), 275–370. doi:10.1130/0016-7606(1945)56[275:EDOSAT]2.0.CO;2

Ibbitt, R., & Woods, R. (2004). Re-Scaling the topographic index to improve the representation of physical processes in catchment models. *Journal of Hydrology (Amsterdam)*, *293*(1-4), 205–218. doi:10.1016/j.jhydrol.2004.01.016

Ichoku, C., Karnieli, A., & Verchovsky, I. (1996). Application of fractal techniques to the comparative evaluation of two methods of extracting channel networks from digital elevation models. *Water Resources Research*, *32*(2), 389–399. doi:10.1029/95WR02762

James, S.M. (1983). *South Fork Eel Watershed erosion investigation.* Northern District, California Department of Water Resources.

Jansen, M. (2001). *Noise reduction by wavelet thresholding. 161 of Lecture Notes in Statistics*. New York: Springer-Verlag. doi:10.1007/978-1-4613-0145-5

Jensen, S., & Domingue, J. O. (1988). Extracting topographic structure from digital elevation model data for geographic information system analysis. *Photogrammetric Engineering and Remote Sensing*, *54*(11), 1593–1600.

Jiang, L., & Wong, W. S. (2010). Effects of DEM Source on Hydrologic applications. *Computers, Environment and Urban Systems*, *34*(3), 251–261. doi:10.1016/j.compenvurbsys.2009.11.002

Jones, K. H. (1998). A Comparison of Two Approaches to Ranking Algorithms Used to Compute Hill Slopes. *GeoInformatica*, *2*(3), 235–256. doi:10.1023/A:1026472421094

Kagan, A. M., Linnik Yu, V., & Rao, C. R. (1973). *Characterization problems in mathematical statistics*. New York: Wiley.

Kienzle, S. (2004). The effect of DEM raster resolution on first order, second order and compound terrain derivatives. *Transactions in GIS*, *8*(1), 83–111. doi:10.1111/j.1467-9671.2004.00169.x

Koutsoyiannis, D. (2005). Uncertainty, entropy, scaling and hydrological statistics. 1. Marginal distributional properties of hydrological processes and state scaling. *Hydrological Sciences Journal*, *50*(3), 381–404. doi:10.1623/hysj.50.3.381.65031

Kuchment, L. S., & Gelfan, A. N. (2007). Statistical self-similarity of spatial variations of snow cover and its application for modelling snowmelt runoff generation in basins with a sparse snow measurement network. *Predictions in ungauged basins: PUB Kick-off, 309*, 156-163.

Kuo, W. L., Steenhuis, T. S., McCulloch, C. E., Mohler, C. L., Weinstei, D. A., DeGloria, S. T., & Swaney, D. (1999). Effect of grid size on runoff and soil moisture for a variable-source-area hydrology model. *Water Resources Research*, *35*(11), 3419–3428. doi:10.1029/1999WR900183

Labat, D. (2005). Recent advances in wavelet analyses: Part 1–A Review of Concepts. *Journal of Hydrology (Amsterdam)*, *314*(1-4), 275–288. doi:10.1016/j.jhydrol.2005.04.003

Lam, N. S. (1990). Description and measurement of landsat TM images using fractals. *Photogrammetric Engineering and Remote Sensing*, *56*(2), 187–195.

Lashermes, B., Foufoula-Georgiou, E., & Dietrich, W. E. (2007). Channel network extraction from high resolution topography using wavelets. *Geophysical Research Letters*, *34*(23), n/a. doi:10.1029/2007GL031140

Lee, J. S. (1981). Speckle analysis and smoothing of synthetic aperture radar images. *Computer Graphics and Image Processing, 17*(1), 24–32. doi:10.1016/S0146-664X(81)80005-6

Li, Z., Zhu, Q., & Gold, C. (2005). *Digital terrain modeling: Principles and methodology*. Boca Raton, FL: CRC Press.

Maidment, D. R. (1993). GIS and hydrologic modeling, in environmental modeling with GIS. In M. F. Goodchild, B. O. Parks, & L. T. Steyaert (Eds.), (pp. 147–167). New York: Oxford University Press.

Mallat, S., & Hwang, W. L. (1992). Singularity detection and processing with wavelets. *IEEE Transactions on Information Theory, 38*(2), 617–643. doi:10.1109/18.119727

Mallat, S. G. (1998). *A wavelet tour of signal processing* (2nd ed.). San Diego, CA: Academic Press.

Mandelbrot, B. B. (1967). How long is the coast of Britain? Statistical self-Similarity and fractional dimension. *Science. New Series, 156*(3775), 636–638. PMID:17837158

Mandelbrot, B. B. (1982). The fractal geometry of nature. W. H. Freeman and Company.

Maruyama, T., Kawachi, T., & Singh, V. P. (2005). Entropy-Based assessment and clustering of potential water resources availability. *Journal of Hydrology (Amsterdam), 309*(1–4), 104–113. doi:10.1016/j.jhydrol.2004.11.020

Melton, M. A. (1957). *An analysis of the relations among elements of climate, surface properties, and geomorphology* (Tech. Rep. No. 3). Office of Naval Research Project NR389-042. New York: Columbia University, Department of Geology.

Mendicino, G., & Sole, A. (1997). The information content theory for the estimation of the topographic index distribution used in TOPMODEL. *Hydrological Processes, 11*(9), 1099–1114. doi:

Moore, I. D. (1988). A contour-Based terrain analysis program for the environmental sciences (TAPES). *Transactions - American Geophysical Union, 69*(16), 345.

Moore, I. D., Gessler, P. E., & Nielsen, G. A. (1993). Terrain attributes: estimation methods and scale effects. In A. J. Jakeman, M. B. Beck, & M. McAleer (Eds.), *Topics in Modeling Change in Environmental Systems* (pp. 189–214). London: Wiley.

Moore, I. D., Grayson, R. B., & Ladson, A. R. (1991). Digital terrain modeling: A review of hydrological, geomorphological and biological applications. *Hydrological Processes, 5*(1), 3–30. doi:10.1002/hyp.3360050103

Niedda, M. (2004). Up scaling hydraulic conductivity by means of entropy of terrain curvature representation. *Water Resources Research*, *40*(4). doi:10.1029/2003WR002721

Nourani, V. (2011). Application of continues wavelet transform to investigate the resolution effect of digital elevation model on topographic index distribution. *International Journal of Soft Computing and Engineering*, *1*(5), 405–410.

Nourani, V., Komasi, M., & Mano, A. (2009). A multivariate ANN-Wavelet approach for rainfall-runoff modeling. *Water Resources Research*, *23*(14), 2877–2894.

Nourani, V., & Mano, A. (2007). Semi-distributed flood runoff model at the subcontinental scale for southwestern Iran. *Hydrological Processes*, *21*(23), 3173–3180. doi:10.1002/hyp.6549

Nourani, V., & Zanardo, S. (2013). Wavelet-Based regularization of the extracted topographic index from high-resolution topography for hydro-geomorphic applications. *Hydrological Processes*, *28*(3), 1345–1357. doi:10.1002/hyp.9665

O'Callaghan, J. F., & Mark, D. M. (1984). The extraction of drainage networks from digital elevation data. *Computer Vision Graph*, *28*(3), 323–344. doi:10.1016/S0734-189X(84)80011-0

O'Loughlin, E. M. (1986). Predictions of surface saturation zones in natural catchments by topographic analysis. *Water Resources Research*, *22*(5), 794–804. doi:10.1029/WR022i005p00794

O'Neill, R. V., Krummel, J. R., Gardner, R. H., Sugihara, G., Jackson, B., DeAngelis, D. L., ... Graham, R. L. (1988). Indices of landscape pattern. *Landscape Ecology*, *1*(3), 153–162. doi:10.1007/BF00162741

Pan, F., & King, A. W. (2012). Downscaling 1-km topographic index distributions to a finer resolution for the TOPMODEL-Based GCM hydrological modeling. *Journal of Hydrologic Engineering*, *17*(2), 243–251. doi:10.1061/(ASCE)HE.1943-5584.0000438

Panuska, J. C., Moore, I. D., & Kramer, L. A. (1991). Terrain analysis: Integration into the agriculture nonpoint source (AGNPS) pollution model. *Journal of Soil and Water Conservation*, *46*(1), 59–64.

Passalacqua, P., Tarolli, P., & Foufoula-Georgiou, E. (2010). Testing space-scale methodologies for automatic geomorphic feature extraction from LIDAR in a complex mountainous landscape. *Water Resources Research*, *46*(11). doi:10.1029/2009WR008812

Quinn, P. F., Beven, K. J., & Lamb, R. (1995). The ln (a/tanb) index: How to calculate it and how to use it within the TOPMODEL framework. *Hydrological Processes, 9*(2), 161–182. doi:10.1002/hyp.3360090204

Rodriguez-Iturbe, I., Rinaldo, A., Rigon, R., Bras, R. L., Ijjasz-Vasquez, E., & Marani, A. (1992). Fractal structures as least energy patterns: The case of river networks. *Geophysical Research Letters, 19*(9), 889–892. doi:10.1029/92GL00938

Sang, Y. F., Wang, D., Wu, J. C., Zhu, Q. P., & Wang, L. (2009). Entropy-based wavelet de-Noising method for time series analysis. *Entropy (Basel, Switzerland), 11*(4), 1123–1147. doi:10.3390/e11041123

Saulnier, G., Obled, C., & Beven, K. (1997). Analytical compensation between DTM grid resolution and effective values of saturated hydraulic conductivity within the TOPMODEL framework. *Hydrological Processes, 11*(9), 1331–1346. doi:

Shannon, C. E. (1948). A mathematical theory of communications I and II. *The Bell System Technical Journal, 27*(3), 379–443. doi:10.1002/j.1538-7305.1948.tb01338.x

Shapiro, J. M. (1993). Embedded image coding using zero-trees of wavelet coefficients. *Transactions on Signal Processing, 11*, 3115–3162.

Simoncelli, E. P., & Adelson, H. E. (1996). Noise removal via bayesian wavelet coring. *Proceedings - International Conference on Image Processing, 1*, 379–382.

Singh, V. P. (2011). Hydrologic synthesis using entropy theory. *Journal of Hydrologic Engineering, 16*(5), 421–433. doi:10.1061/(ASCE)HE.1943-5584.0000332

Sorensen, R., & Seibert, J. (2007). Effects of DEM resolution on the calculation of topographical indices: TWI and its components. *Journal of Hydrology (Amsterdam), 347*(1-2), 79–89. doi:10.1016/j.jhydrol.2007.09.001

Tarboton, D. G., Bras, R. L., & Rodriguez-Iturb, I. (1988). The fractal nature of river networks. *Water Resources Research, 24*(8), 1317–1322. doi:10.1029/WR024i008p01317

Todini, E. (1996). The ARNO rainfall-runoff model. *Journal of Hydrology (Amsterdam), 175*(1-4), 339–382. doi:10.1016/S0022-1694(96)80016-3

Tokunaga, E. (2003). Tiling properties of drainage basins and their physical bases. In I. S. Evans, R. Dikau, E. Tokunaga, H. Ohmori, & M. Hirano (Eds.), *Topics in Concepts and Modelling in Geomorphology: International Perspectives* (pp. 147–166). Tokyo, Japan: Terrapub.

Vieux, B. E. (1993). DEM Aggregation and smoothing effects on surface runoff modeling. *Journal of Computing in Civil Engineering*, *7*(3), 310–338. doi:10.1061/(ASCE)0887-3801(1993)7:3(310)

Wang, D., & Cai, X. (2006). Scale and resolution effects of topographic index by 2-D continuous wavelet transform. In *Proceedings of Chinese PUB International Symposium* (vol. 322, pp. 108-114). Beijing: IAHS Publication.

Wang, Z., Batelaan, O., & De Smedt, F. (1996). *Distributed model for water and energy transfer* between soil, plants and atmosphere (WetSpa). *Physics and Chemistry of the Earth*, *21*(3), 189–193. doi:10.1016/S0079-1946(97)85583-8

Wilson, J. P., & Gallant, J. C. (2000). *Digital terrain analysis. In terrain analysis: principles and applications*. John Wiley & Sons.

Wolock, D. M., & Price, C. V. (1994). Effects of digital elevation model map scale and data resolution on a topography-based watershed model. *Water Resources Research*, *30*(11), 3041–3052.

Wood, E. F., Roundy, J. K., Troy, T. J., van Beek, L. P. H., Bierkens, M. F. P., Blyth, E., ... Whitehead, P. (2011). Hyperresolution global land surface modeling: Meeting a grand challenge for monitoring Earth's terrestrial water. *Water Resources Research*, *47*(5). doi:10.1029/2010WR010090

Zhang, W., & Montgomery, D. R. (1994). Digital elevation model grid size, landscape representation, and hydrologic simulations. *Water Resources Research*, *30*(4), 1019–1028. doi:10.1029/93WR03553

KEY TERMS AND DEFINITIONS

Digital Elevation Model (DEM): A digital model or 3D representation of a terrain's surface created from terrain elevation data.

Entropy: A measure of the amount of chaos or of the lack of information about the system.

Fractal: A fractal is an object or quantity that displays self-similarity, in a somewhat technical sense, on all scales. The object need not exhibit exactly the same structure at all scales, but the same "type" of structures must appear on all scales.

Terrain Analysis: The collection, analysis, evaluation, and interpretation of geographic information on the natural and manmade features of the terrain, combined with other relevant factors.

Thresholding: The minimum intensity or value of a signal, etc. that will produce a response or specified effect.

Topographic Index (TI): A measure of the extent of flow accumulation at the given point of the topographic surface.

Wavelet De-Noising: To reject noise by damping or thresholding in the wavelet domain.

Chapter 7
Keyword Search Mechanisms in Geo-Spatial Databases

Priya M.
Bharathiyar College of Engineering and Technology, India

Kalpana R.
Pondicherry Engineering College, India

ABSTRACT

Most web and mobile applications are based on searching the location-based objects called spatial objects. In spatial database systems, searching such objects is a challenging task since it deals with geo-spatial capabilities. Sometimes, the spatial queries are associated with text information in order to obtain the most relevant answers nearest to the given location. Such queries are called spatial textual query. Conventional spatial indexes and text indexes are not suitable for resolving such queries. Since these indexes use various approaches to perform searching, they can cause performance degradation. Effective processing of the query mainly depends on the index structure, searching algorithms, and location-based ranking. This chapter reviews the different hybrid index structures and search mechanisms to extract the spatial objects, the different ranking model it supports, and the performance characteristics.

INTRODUCTION

Spatial query is the special type of database query supported by the geospatial databases or spatial databases. Nowadays, most of the research areas are focused on web applications and mobile based applications. Often these applications are in need to search the answer for the queries and find out the most relevant answer

DOI: 10.4018/978-1-5225-5039-6.ch007

Figure 1. Spatial objects

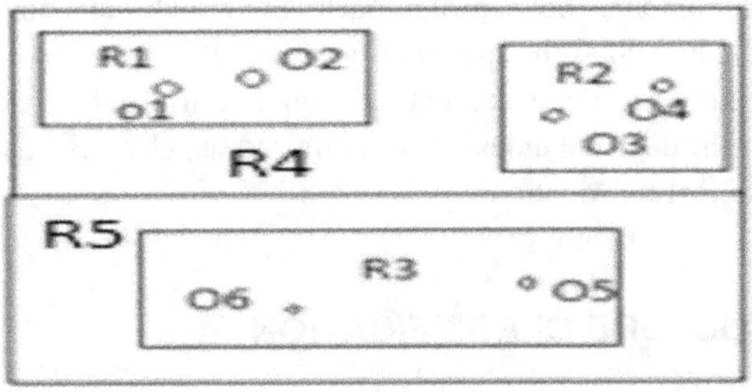

to the query. The query may result in finding the spatial objects shown in Figure 1 that are nearest to the given locations called location based objects.

Sometimes the spatial query can also be enriched with some textual information called keywords as shown in Table 1. Each object has associated text information and the number of times its occurrence within the document. This type of query is referred as Spatial-Textual query (ST). For example, the query may be in the form "Find the list of hotels within 2km with the facility of pool". ST query can be resolved by resolving two predicates called spatial predicates and string predicates. In order to resolve these predicates, different types of index structures which optimizes the search space are constructed

Index structure for spatial predicates falls under three categories. They are tree based, grid based and Quad tree based Index structures. Inverted file index, bitmap index and signature file are used to resolve the string predicates. Initially, two approaches such as Geospatial approach and textual approach are used to resolve the ST query. In Geospatial approach, the spatial predicates are resolved first and then the textual information is searched. The textual approach, first output the query answer that satisfies the textual information, and then the results are checked against

Table 1. Objects associated with text information

Object	Text information Associated
O1	(AC,5) (Box,5) (Expensive,3)
O2	(Café,5) (Box,4) (Expensive,2)
O3	(AC,3) (Balcony, 3) (Expensive,1)
O4	(Box,4) (Balcony,2) (Expensive,2)
O5	(AC, 3) (Online reservation,4)
O6	(Cafe,3)(Online reservation,2)

the spatial coordinates. In order to reduce the search space, a new hybrid index structure which combines both the approach is required. Each node is augmented with both textual as well as the spatial information. It is necessary to consider both the spatial indexing as well as the text indexing to retrieve the spatial object. This paper discusses the different indexing structure and search mechanisms that support both spatial and text information.

INDEX STRUCTURE CLASSIFICATION

Queries with spatial condition can be answered by a tree based index like R tree and its variants (Bechmann, Kriegel, Schneider & Seeger, 1990; Guttman, 1984), KD tree, Quad tree, etc. R tree is used as the most effective basic index shown in Figure 2 for representing the spatial condition.

Most commonly used indexes for answering the query keyword are trie based index and inverted-index. In Trie based indexes, the string collections are stored as a trie. Most of the trie-based indexes are specific to edit distance and its variants, and do not support other similarity measures such as Jaccard. An inverted-index approach supports a family of similarity metrics, like edit distance, Jaccard, etc (Li, Wang & Yang, 2007).

The two methods differ in their performance characteristics, but the inverted index approach getting importance to answer the query keywords. Grams based string collections are used to build the inverted index. Several searching algorithms and fast indexing are discussed in Hariharan, Hore, Li and Mehrotra (2007), Li, Wang, and Yang (2007), Papadias, Mamoulis, and Theodoris (1993). Search techniques can be categorized into exact search, approximate search which are used to search the textual information. Different types of hybrid index structure and search mechanisms to handle the spatial and a string predicates to extract the spatial object are shown in Figure 3.

Figure 2. R tree index

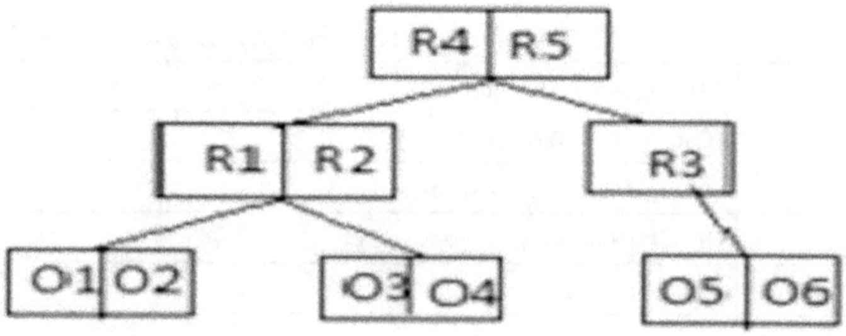

Figure 3. Classifications of indices

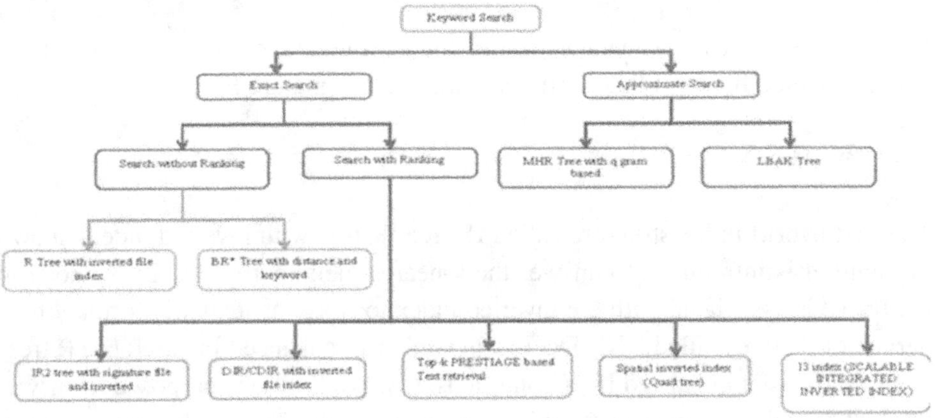

EXACT SEARCH MECHANISM

Generally spatial query processing can be classified into range query, nearest neighbor query and join query (Papadias, Mamoulis, & Theodoris, 1993; Sarawagi & Kirpal, 2004). Different types of queries are introduced in the literature to extract the spatial objects that are associated with the textual information (Chen, Cong, Christian & Wu, n.d.). mCk query is used to extract the object that consists of m keyword out of the k keywords (Zahng, Chee, Mondal, Tung & Kitsuregawa, 2009). The queries such as LkT (Cong, Jensen & Wu, 2009; Zhou, Xie, Wang & Gong, 2005) (Location aware Top-k Text retrieval), LkPT (Cao, Cong & Jensen, 2010) (Location aware Top-k Prestige based Text retrieval) are based on location proximity and document similarity. Distance first top-k Query introduced by Felipe et al (2008) to extract top most k spatial objects is based on distance. The query k-SB (top-k Spatial Boolean query) is allowed to introduce the Boolean operators such as AND, OR, NOT into the set of keywords which are associated with spatial query (Cary, Wolfson & Rishe, 2010). A special type of range query called SAS (Spatial Approximate String) is introduced to augment the spatial queries with approximate strings (Yao, Tang & Hadjieleftheriou, 2013). Approximate String Query (Alsubaiee, Behm & Li, 2010) is the query which appends a string with the spatial objects with some threshold limits. Most of the applications today associates a string with relational operators like <,>, = with the spatial information. The performance of these queries depends on three factors such as index structure used for spatial and string predicates, the pruning technology used to restrict the search space and the ranking model used to find the most relevant answer to the query.

Indices for Exact Search

Indices for exact search can be organized under three categories based on the index structure called Inverted Index, Bitmap index and Signature File.

Inverted Index

The first hybrid Index structure called IR tree (R tree with inverted index) shown in Figure 4 is introduced to answer the k-nearest neighbor query. Every node in the R tree is associated with the inverted index to store the textual information. A variant of IR-tree, called DIR (Document similarity enhanced Inverted file R tree) and CIR (Cluster enhanced IR) (Cong, Jensen & Wu, 2009) is proposed to answer the LkT based on location proximity, document similarity and the text relevance (Cong, Jensen & Wu, 2009). Cao, Cong, and Jensen (2010) introduced an index structure called TkPIR (Top-k Prestige based IR tree) to process LkPT query which considers both location proximity and relevance of the nearest objects (Berkhin, 2011; Cao, Cong, & Jensen, 2010).

Rocha-Junior et al. (2010, 2011) proposed an index called S2I (Spatial Inverted Index) based on R tree and Inverted file. It employs a tree structure, called aggregated R tree (aR tree), for each keyword in the query that maps into a distinct R tree and store the objects with the given keyword. The terms used in the spatial documents can be categorized into frequent and infrequent terms according to Zipf's law. Document similarity is considered to construct the inverted index. The aR tree imposed two different data structures for handling the frequent and infrequent terms. Terms can be partitioned and can be executed by traditional distributed search engines.

A new index structure I3 (Scalable Integrated Inverted Index) is introduced by Zhang et al (2013) which uses the concept of textual first partition. The spatial document is split into tuples and each tuple is associated with one keyword. The inverted file is built for each partition and the tuples are inserted into it based on their weight. The quad tree index is used for spatial pruning.

Figure 4. IR tree

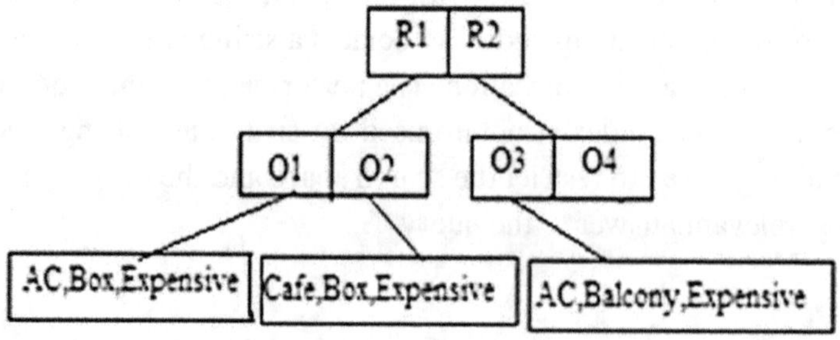

Bitmap Index

A new hybrid index bR* tree shown in the Figure 5, is an extension of the R* tree with bitmap index (Zhang, Chee, Mondal, Tung & Kitsuregawa, 2009), is proposed to answer the mCk spatial-textual query. For each keyword in the database, one bit is associated with bitmap index. If there are N keywords in the database, then each node is associated with the N bit keyword bitmap.

Cary et al (2010), proposed an index structure called spatial keyword index (SKI) to process k-SB query. At each node, the inverted file maintains a list of terms, document frequencies, and the spatial information from R tree. Each super node maintains a sorted order of terms bitmap.

Signature File

Felipe et al (2008) introduced an efficient index structure to answer the spatial-textual query called IR2 tree (Information Retrieval R tree) shown in Figure 6. It is a combination of R tree with signature file index. IR2 tree based on signature length are well suited to retrieve the top-k objects based on the distance.

Figure 5. bR tree*

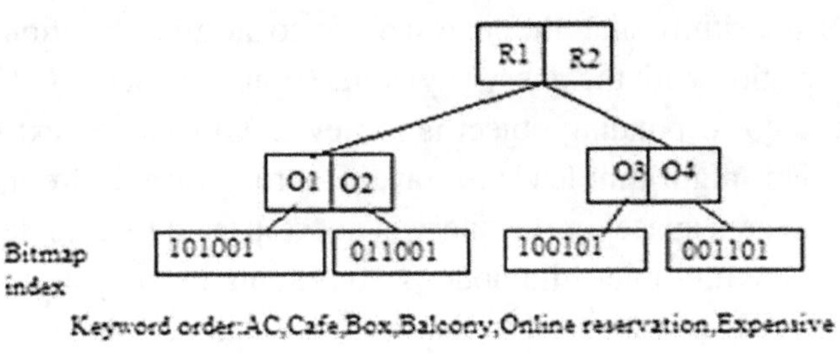

Figure 6. IR2 tree

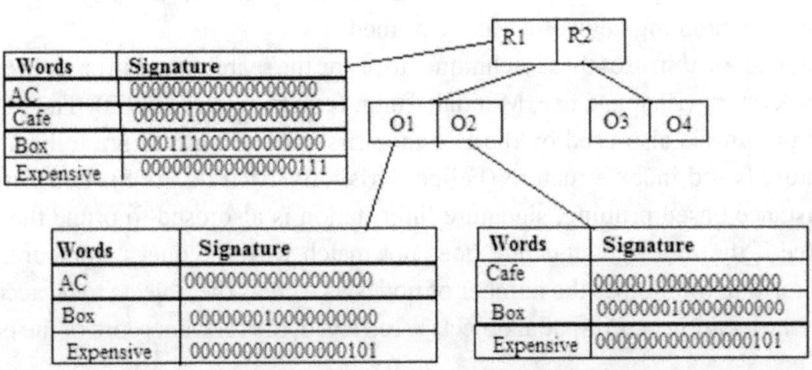

Each node is augmented with signature information. Each leaf node is augmented with the signature file of the object. Each non leaf node is enhanced with the signature of the node and the sub tree underlying it. The signature of the node is calculated by applying OR operations over all of its entries. A variant of the IR2 tree is MIR2 tree (Multi level IR2 tree) uses the concept of varying the signature file length.

Search Space Pruning

The basic algorithm used to resolve the spatial predicates is Nearest Neighbor search algorithm (Roussopoulos, Kelly & Wincent, 1995a, 1995b). The depth first, breadth first and branch and bound techniques are used to prune the search space in this algorithm. Initially, R tree base line algorithm and Inverted Index Only algorithm are used to answer the Spatial-textual query. R tree base line algorithm finds all objects nearest to the given location and then compares the textual description with the query keyword. Inverted Index only algorithm (IIO) first finds all the documents whose textual descriptions match with the keywords, and then the spatial predicates are checked. These two step approaches leads to the wastage of computational cost and IO cost. Bin yao (2013) reviews that the pruning power of these algorithms are based on either spatial or string predicates. So the pruning power based on both spatial and string predicates is required to optimize the search space.

An improvement over the base line algorithm called Incremental Nearest Neighbor algorithm (NN algorithm) finds the nearest object to the given location and compares its textual description with the query keywords (Finkel & Bentley, 1974). If match occurs, then the corresponding object is retrieved. Otherwise next nearest object is considered. This algorithm leads to traverse through the entire index structure. Sometimes the search space for the query is not restricted to a single child node. It can span across the multiple child node. Two pruning techniques such as distance based and keyword based are used to handle this situation. If the minimum distance between two node is larger than the given threshold δ*(smallest distance between two nodes), then the nodes can be pruned. Similarly, the minimum distance between two keywords are greater than δ*(smallest distance between two keywords), then the nodes containing keywords can be pruned.

Priori algorithm uses these techniques to prune the search space in order to answer the mCk query (Zhang, Chee, Mondal, Tung, & Kitsuregawa, 2009). The distance based pruning is also used by the Distance-first-IR2 tree algorithm to handle the Signature based index structure (Felipe, Hristidis & Rishe, 2008). In addition to the distance based pruning, signature information is also used to prune the entire sub tree if the root tree signature does not match with the query signature. Thus the algorithm minimizes the number of nodes as well as the objects to be accessed. When the location based spatial objects is retrieved, the relevant score of the nearest objects are used as the metric to restrict the search space. For example if there is a shoe shop, then there will be a socks shop nearest. Figure 6 shows the different index structure for exact search mechanisms.

Ranking Model

The objective of the ranking function is to improve the accuracy of the search results by retrieving the relevant objects based on their relevance score. In Inverted File Only algorithm, the text relevance score of the objects is considered and the objects are ranked based on that score. The list is then checked against the spatial proximity to return the top-k objects. R tree based Inverted File (RIF) algorithm finds the nearest neighbors from the given location and considers the text relevance score of the objects to return the top-k spatial objects (Ilyas, Beskales & Soliman, 2008). Most of the index structure uses the ranking function that combines both the distance proximity and the text relevance score. The text relevance score can be calculated using the probabilistic based ranking function or linear ranking function to answer the top-k results (Cao, Cong & Jensen, 2010). Terms frequency and the document length are also considered to calculate the text relevance score.

APPROXIMATE SEARCH MECHANISMS

Searching the strings that are similar to a query string is called approximate string search (Gravano et al., 2001). Approximate string search is the most suitable search mechanism to handle the fuzzy search conditions (Alsubajee & Li, 2010; Gand & Widom, 2003) such as spell checking, natural language processing, and web search. The search is based on similarity measures. Similarity between two strings is measured using edit distance, jaccard efficient, cosine similarity and n-gram based measurement (Li, Lu & Lu, 2008). The edit distance and n-gram based similarity measure get attractiveness to find the similarity. The edit distance between two strings s1 and s2 is the minimum number of operations required to transform one string to another. Several algorithms have focused on approximate string queries using the edit distance function (Arasu, Ganti & Kaushik, 2006; Levenshtein, 1965). In n-grams based similarity measure, each string is represented as a token set and the distance between two strings is calculated by finding the similarity measure between two token sets. Hence, Similarity measure between two token sets is considered as the set-similarity problem (Acharya, Poosala & Ramaswamy, 1999; Alsubajee & Li, 2010; Arasu, Ganti & Kaushik, 2006; Hadjieleftheriou, Chandal, Koudas & Srivastava, 2008a).

Indices for String Search

Different index structures are proposed to perform the string search. Inverted index based trees (Hadjieleftherious, Chandal, Koudas & Srivastava, 2008b) such as BK-tree (Burkhard–Keller), FQ-tree (Fixed-Queries) and FQH-tree (FQ for fixed Height) are used to perform approximate searching based on edit distance. Inverted index is considered as the most effective indexing scheme for set-similarity problem (Zobel & Moffat, 2006). An n-gram based inverted list with different filtering techniques such as length, position and prefix, is proposed for set-similarity problem.

The performance of the inverted list is limited due to its size. The size of the inverted list can be reduced by introducing various compression techniques. The inverted list is divided into fixed segment, each segment is compressed and assigned with an integer encoding scheme. Further, the size of the inverted list can be reduced by introducing a technique called VGRAM (Li, Wang & Yang, 2007; Alsubajee, Behm & Li, 2010) (variable length gram). The n-gram based signature file with various filters is also used to search the set-similarity. But it does not get attractiveness due to its signature hash functions. Bed tree is the best tree based indexing scheme for string processing based on the edit distance measure which is built on B+ tree (Jagadish et al., 2005). It transforms string domain into integer space to answer the selection and join queries. String order, comparability and lower bound are the concept used for string pruning

Indices for Approximate Search

Yao et al (2013) proposed an approximate index to answer the range query called MHR tree (Min wise signature hash R tree) shown in Figure 7 (Pagel, Six, Tobin & Widmayer, 1993). The spatial predicates are resolved using R tree or its variants R* tree in which each node is associated with the minimum bounding rectangle under

Figure 7. MHR tree

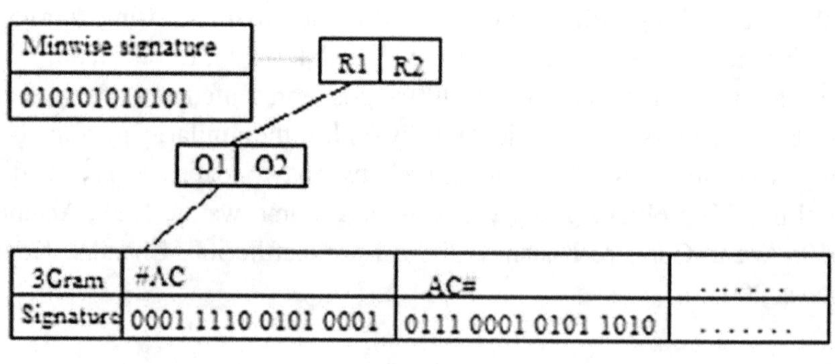

it. Each minimum bound rectangle contains all the points under that region. The n-gram based signature index is used to store all the strings under each node. Instead of storing the n-gram of the node and its sub tree, each non-leaf node calculates the min wise signature of the node set using min wise permutation function and embeds this into the R tree (Broder, Charikar, Frieze & Mitzenmacher, 1998). Thereby, MHR tree restricts the search space.

Sattam Alsubaiee (2010) proposed an index structure called LBAK tree, to answer location-based approximate-keyword queries (Chaudhuri, Ganti & Kaushik, 2006). It combines gram-based inverted index to perform approximate string search augmented to each node in the spatial distance tree shown in Figure 8.

Nodes in spatial tree are categorized into three types, S-Node, SA-Node and SK-Node. Each SA-Node is used to the store the union of the keywords and its approximate indexes. Each S-node is augmented with the n-gram of the texts. Each SK-node is used to resolve the spatial predicates.

Search Space Pruning

Two types of pruning power have to be considered for approximate search. Distance based pruning is used to restrict the search space in spatial domain and the n-gram based edit distance pruning is used to restrict the search space in inverted lists. In order to avoid the scanning of the entire inverted list, different filtering techniques based on length, position and prefix is used. The inverted list pruning power can be improved by dividing the inverted list into separate list containing relevant portion and irrelevant portion and discarding the inverted list which is not relevant.

Several algorithms such as MergeOpt, Heap, ScanCount and DivideSkip are used to process n-gram based inverted list (li, Lu & Lu, 2008). An improvement over the n-gram based pruning is made by introducing two level index or variable length n-gram. In n-gram based signature, the pruning power is based on the min

Figure 8. LBAK tree

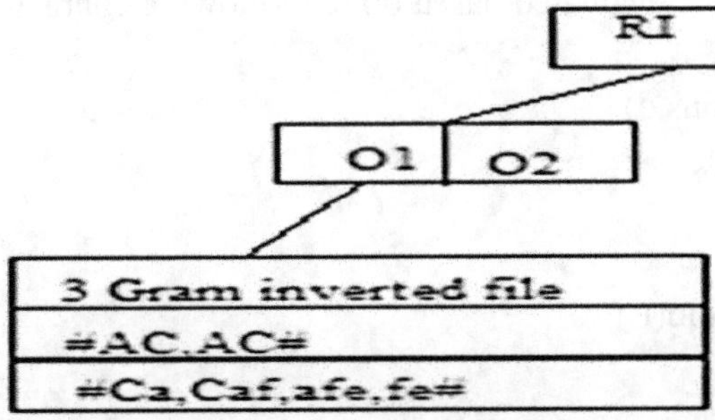

wise signature embeds in R tree. The signature of the query string is compared with the min wise value of the non leaf node .If the intersection value is greater than the threshold limit, and then the corresponding sub tree will be searched. Otherwise the entire tree will prune.

PERFORMANCE CHARACTERISTICS

The basic parameters to consider the efficiency of the hybrid index schemes are

- Size of the dataset(DS)
- The number of keywords associated with each objects(TK)
- The total number of keywords used

Exact Search

The Table 2 shows the characteristics of different index structure of exact search mechanism and its performance. Efficiency of the Geo-spatial keyword search is improved due to the hybrid indexing scheme which is tightly bounded. Space complexity depends on the number of disk access since the pages are disk resident. The performance parameters of the exact searching mechanism are

- The k in top-k (k)
- The query preference parameter (α/β) to decide the preference between spatial proximity and text relevancy
- Variance Threshold (ξ)
- Distance Parameter (λ).

Approximate Search

The Table 3 list the characteristics of different index structure of approximate search and the performance can be evaluated based on the following characteristics

- Number of dimensions d)
- Signature Length (l)
- Query area (Θ)
- Data Size (N)
- Edit distance threshold(Γ)

Table 2. Comparisons of exact searching mechanisms

Tree	Index Structure	Type of query processed	Algorithm	Similarity	Space	Response Time	Performance characteristics
bR *	R* tree with Bitmap Index	mCk	Priori Algorithm		Incur fixed storage cost	Good	Performance decrease when TK and DS is less
IR	R tree with inverted file	LkT	Best first traversal algorithm with priority queue	Cosine similarity	More space than R tree algorithm	Good	Performance depends on the distance parameter
DIR	R tree with inverted file	LkT	Best first traversal algorithm with priority queue	Document similarity	Less than IR tree	Better than IR tree	Affected when the number of documents increases
CIR	R tree with inverted file	LkT	Best first traversal algorithm with priority queue	Cluster enhanced document similarity	More than IR/DIR	Better than IR/DIR	Response time decreases when the number of cluster increases.
TkPIR	IR tree with inverted file	LkPR	ES-EBC, S-EBC algorithm	Vector space model	More space for S-EBC	Good	A Runtime decrease when high PR score (α) is required.
IR^2	R tree with signature index	Distance first top-k spatial query	IR^2 tree algorithm		Good	Good	Affected due to the fixed size Signature file
MIR^2	R tree with signature index	Distance first top-k spatial query	IR^2 tree algorithm		Better than IR^2	Better than IR^2	Affected due to maintenance operations of multilevel.
SKI	R tree with Bitmap index	k-SB	Best first traversal algorithm	Document frequency	Less space	Good	Performance limited due to the size of bitmap index
S2I	aR* -aggregate R tree with Inverted Index.	Top-k Spatial Keyword	SKA,MKA algorithm	Cosine similarity	Space cost is more due to the more number of trees.	Access less disk pages to answer, so that the query cost is less	Depends on size of the index structure
I^3	Quad Tree with inverted file	Top-k Spatial Keyword query	AND-OR algorithm	Textual relevance score	Space Efficient	Least time to access and build the tree	Depends on number of keywords and the varying dataset

COMPARISON OF KEYWORD SEARCH

bR* tree is good for keyword search, but it has the restriction due to the fixed size bitmap index. The performance of bR*-tree degrades slightly with the increase of data size and the total number of keywords. It integrates all the data points in one tree, leading to higher access cost. DIR/CIR tree based on inverted file is used to answer the location based query, but it does not consider the relevance of the nearest

Table 3. Comparisons of approximate searching mechanisms

Tree	Index Structure	Type of query processed	Algorithm	Similarity	Space	Response Time	Performance characteristics
MHR	R* tree with signature Indexes	SAS	Greedy principle	Edit distance based	Less space cost	Query time increases with the index size.	Recall is constantly 70% -80% for most of the applications
LBAK	R* tree with approximate indexes	LBAK	Knapsack Problem	q-gram based	Space cost depends on the approximate index size which is more.	Average query time reduced	Better recall percentage due to the average query time

objects. The family of IR tree has the drawback of spending more processing time to search the nodes that are relevant to the query keyword. TkPIR tree are based on Inverted File which is the best to answer the location based query, since it considers the nearest relevance objects. It has the drawback of propagating the PR score to the nearest objects which will take more running time.IR2 tree based on signature length are well suited to retrieve the top-k objects based on the distance. It has the drawback of fixed size Signature file. MIR2 tree introduced the signature file with varying size, but it has the drawback of increasing the complexity of maintenance operations and identifies more false positives.

S2I uses the concept of different data structure to support frequent and infrequent word. S2I reduce the number of I/O access since only the relevant inverted file is accessed. When the keywords are frequent, it creates the partial aggregation of different R trees. It results in higher cost and large number of random access. It has higher overhead of transferring the flat file of infrequent keyword into a new R tree when the keyword becomes frequent.I3 uses the concept of quad tree which replace the hybrid index R tree. It has the advantage of less building time index, less update cost and storage cost. Quad tree based index structure can be used in multilayer, but it supports only spatial search mechanism and it is purely based on main memory (Kanth, Ravada & Abugov, 2002). The performance of the index structure also depends on the number of keywords used, but there is a possibility of reducing the search time if relationship among the keywords is known.

The index tree structure and searching mechanisms discussed above supports only exact search solution and not approximation.MHR tree and LBAK tree support approximation solution and are based on Inverted index. MHR-tree uses probabilistic signature scheme to represent textual information in tree nodes. LBAK tree uses approximate indexes to store textual information. It chooses nodes from the tree to store approximate indexes, such that the average query time of a given workload is minimized. LBAK approach guarantees to find all true answers while MHR tree miss some answers due to its probabilistic nature. MHR tree produces the constant performance for most of the applications, while LBAK tree increases the performance characteristics. LBAK tree has some drawbacks such as the space cost and lookup time which depends on the approximate index size.

CONCLUSION AND RESEARCH ISSUES

This paper discusses the different index structures, ranking model and its performance characteristic based on exact searching mechanism and approximate searching mechanism. The approximate solution gets more importance over the exact solution since it provides better performance in most of the fuzzy search condition. But approximate solution depends upon predefined gram dictionary and threshold limits. The lookup time of gram dictionary depends on the dense of the spatial objects. There is a possibility of missing some answers due to the threshold limits. An increase in the number of disk access and process time would result if the threshold limit raise. A search mechanism based on learning strategies can be devised to overcome the issues associated with the approximate solution. A probabilistic based search mechanism can be introduced to perform string search.

REFERENCES

Acharya, S., Poosala, V., & Ramaswamy, S. (1999). Selectivity estimation in spatial databases. *SIGMOD '99 Proceedings of the 1999 ACM SIGMOD international conference on Management of data*, 13–24. 10.1145/304182.304184

Alsubaiee, S., Behm, A., & Li, C. (2010) Supporting Location-Based Approximate-Keyword Queries. *Proceeding GIS '10 Proceedings of the 18th SIGSPATIAL International Conference on Advances in Geographic Information Systems*, 61-70.

Alsubaiee, S., & Li, C. (2010). Fuzzy keyword search on spatial data. In *International Conference on Database Systems for Advanced Applications. DASFAA*. Springer. 10.1007/978-3-642-12098-5_51

Arasu, A., Ganti, V., & Kaushik, R. (2006) Efficient exact set-similarity joins. *Proceeding VLDB '06 Proceedings of the 32nd international conference on Very large data bases*, 918–929.

Beckmann, N., Kriegel, H. P., Schneider, R., & Seeger, B. (1990). The R*-tree: An efficient and robust access method for points and rectangles. *SIGMOD '90 Proceedings of the 1990 ACM SIGMOD international conference on Management of data*, 322-331.

Berkhin, P. (2011). Bookmark-coloring algorithm for personalized page rank computing. *Internet Mathematics*, *3*(1), 41–62. doi:10.1080/15427951.2006.10129116

Broder, A. Z., Charikar, M., Frieze, A. M., & Mitzenmacher, M. (1998). *Min wise independent permutations*. STOC. Retrieved from http://citeseerx.ist.psu.edu/viewdoc/download?doi=10.1.1.121.8215&rep=rep1&type=pdf

Cao, X., Cong, G., & Jensen, C. S. (2010). Retrieving top-k prestige-based relevant spatial web objects. *Proceedings of the VLDB Endowment International Conference on Very Large Data Bases*, *3*(1).

Cary, A., Wolfson, O., & Rishe, N. (2010). Efficient and scalable method for processing top-k spatial boolean queries. *22nd International Conference on Scientific and Statistical Database Management, SSDBM 2010*, 87-95. 10.1007/978-3-642-13818-8_8

Chaudhuri, S., Ganti, V., & Kaushik, R. (2006). A primitive operator for similarity joins in data cleaning. *22nd International Conference on Data Engineering (ICDE'06)*, 5–16. 10.1109/ICDE.2006.9

Cong, G., Jensen, C. D., & Wu, D. (2009). Efficient retrieval of the top-k most relevant spatial web objects. *Proceedings of the VLDB Endowment International Conference on Very Large Data Bases*, *2*(1), 337–348. doi:10.14778/1687627.1687666

Felipe, I. D., Hristidis, V., & Rishe, N. (2008). *Keyword search on spatial databases*. ICDE.

Finkel, R. A., & Bentley, J. L. (1974). Quad trees: A data structure for retrieval on composite keys. *Acta Informatica*, *4*(1), 1–9. doi:10.1007/BF00288933

Gravano, L., Ipeirotis, P. G., Jagadish, H. V., Koudas, N., Muthukrishnan, S., & Srivastava, D. (2001). Approximate string joins in a database. *Proceedings of the 27th VLDB Conference*, 491–500.

Guttman, A. (1984). R-trees: A dynamic index structure for spatial searching. *SIGMOD '84 Proceedings of the 1984 ACM SIGMOD international conference on Management of data*, 47–57.

Hadjieleftheriou, M., Chandel, A., Koudas, N., & Srivastava, D. (2008a). *Set similarity selection queries at interactive speeds*. ICDE.

Hadjieleftheriou, M., Chandel, A., Koudas, N., & Srivastava, D. (2008b). Fast indexes and algorithms for set similarity selection queries. *Proceeding ICDE '08 Proceedings of the 2008 IEEE 24th International Conference on Data Engineering*, 267-276. 10.1109/ICDE.2008.4497435

Hariharan, R., Hore, B., Li, C., & Mehrotra, S. (2007). Processing spatial keyword queries in geographic information retrieval systems. *19th International conference on scientific and statistical database management. SSDBM.*

Ilyas, I.F., Beskales, G., & Soliman, M.A. (2008). A survey of top-k query processing techniques in relational database systems. *ACM Computing Surveys, 40*(4), 1–58.

Jagadish, H. V., Ooi, B. C., Tan, K., Yu, C., & Zhang, R. (2005). iDistance: An adaptive B+ tree based indexing method for nearest neighbor search. *ACM Transactions on Database Systems, 30*(2), 364–397. doi:10.1145/1071610.1071612

Jeh, G., & Widom, J. (2003). *Scaling personalized web search*. Retrieved from: http://infolab.stanford.edu/~glenj/spws.pdf

Joachims, T. (2001). A statistical learning model of text classification for support vector machines. *Proc. of ACM*, 128–136. 10.1145/383952.383974

Kanth, K. V. R., Ravada, S., & Abugov, D. (2002). Quadtree and r-tree indexes in oracle spatial: a comparison using gisdata. *SIGMOD '02 Proceedings of the 2002 ACM SIGMOD international conference on Management of data*, 546-557.

Khan, S. & Mustafa, J. (2014). Effective Semantic search using thematic similarity. *Journal of King Saud University – Computer and Information Sciences, 26*(2), 161-169.

Khodaei, A., Shahabi, C., & Li, C. (2010). *Hybrid indexing and seamless ranking of spatial and textual features of web documents*. DEXA. doi:10.1007/978-3-642-15364-8_37

Levenshtein, V. (1965). Binary codes capable of correcting spurious insertions and deletions of ones. *Soviet Physics, 10*(8).

Li, C., Lu, J., & Lu, Y. (2008). Efficient merging and filtering algorithms for approximate string searches. *2008 IEEE 24th International Conference on Data Engineering*.

Li, C., Wang, B., & Yang, X. (2007). VGRAM: improving performance of approximate queries on string collections using variable length grams. *Proceeding VLDB '07 Proceedings of the 33rd international conference on Very large data bases*, 303–314.

Mazeika, A., Bohlen, A. M., Koudas, N., & Srivastava, D. (2007). Estimating the selectivity of approximate string queries. *ACM TODS*, *32*(2), 12–52, es. doi:10.1145/1242524.1242529

Navarro, G. (2001). A guided tour to approximate string matching. *ACM Computing Surveys*, *33*(1), 31–88. doi:10.1145/375360.375365

Pagel, B. U., Six, H. U., Toben, H., & Widmayer, P. (1993). Towards an analysis of range query performance in spatial data structures. *PODS '93 Proceedings of the twelfth ACM SIGACT-SIGMOD-SIGART symposium on Principles of database systems*, 214-221. 10.1145/153850.153878

Papadias, D., Mamoulis, N., & Theodoridis, Y. (1999). Processing and optimization of multi way spatial joins using R-trees. *Proceeding PODS '99 Proceedings of the eighteenth ACM SIGMOD-SIGACT-SIGART symposium on Principles of database systems*, 44-55. 10.1145/303976.303981

Ponte, J. M., & Croft, W. B. (1998). *A language modeling approach to information retrieval*. SIGIR. doi:10.1145/290941.291008

Rocha-Junior, J. B., Gkorgkas, O., Jonassen, S., & Nørvag, K. (2011). Efficient processing of top-k spatial keyword queries. *International Symposium on Spatial and Temporal Databases*, 205–222. 10.1007/978-3-642-22922-0_13

Rocha-Junior, J. B., Vlachou, A., Doulkeridis, C., & Nørvag, K. (2010). Efficient processing of top-k spatial preference queries. *Proceedings of the VLDB Endowment International Conference on Very Large Data Bases*, *4*(2), 93–104. doi:10.14778/1921071.1921076

Roussopoulos, N., Kelley, S., & Vincent, F. C. (1995a). *Nearest Neighbor Queries*. San Jose, CA: SIGMOD ACM Intl. Conf. on Management of Data. doi:10.1145/223784.223794

Roussopoulos, N., Kelley, S., & Vincent, F. C. (1995b). Nearest neighbour queries. *Proc. SIGMOD*, 71–79.

Sarawagi, S., & Kirpal, A. (2004) Efficient set joins on similarity predicate. *Proceeding SIGMOD '04 Proceedings of the 2004 ACM SIGMOD international conference on Management of data*, 743-754. 10.1145/1007568.1007652

Yang, X., Wang, B., & Li, C. (2008). *Cost-based variable-length gram selection for string collections to support approximate queries efficiently*. SIGMOD. doi:10.1145/1376616.1376655

Yao, B., Tang, M., & Hadjieleftheriou, M. (2013). Spatial approximate string search. *IEEE Transactions on Knowledge and Data Engineering*.

Zhang, D., Chee, Y. M., Mondal, A., Tung, A. K. H., & Kitsuregawa, H. (2009). Keyword search spatial databases: Towards searching by document. *2009 IEEE 25th International Conference on Data Engineering*, 688–699.

Zhang, D. X., Tan, K.-L., & Tung, A. K. H. (2013). *Scalable top-k spatial keyword search*. ACM.

Zhou, Y., Xie, X., Wang, C., Gong, Y., & Ma, W. Y. (2005). *Hybrid index structures for location-based web search* (pp. 155–162). CIKM. doi:10.1145/1099554.1099584

Zobel, J., & Moffat, A. (2006). Inverted files for text search engines. *ACM Computing Surveys*, *8*(2), 1–56.

KEY TERMS AND DEFINITIONS

Edit Distance: Edit distance is a way of quantifying how dissimilar two strings are to one another by counting the minimum number of operations required to transform one string into the other.

Inverted Index: An inverted index is an index data structure storing a mapping from content, such as words or numbers, to its locations in a database file, or in a document or a set of documents.

N-Gram Model: An *n*-gram model is a type of probabilistic language model for predicting the next item in the sequence and it uses only N−1 words of prior context.

Quad Tree: A quad tree is a tree data structure in which each internal node has exactly four children. Quad trees are most often used to partition a two-dimensional space by recursively subdividing it into four quadrants or regions.

R Tree: R-trees are tree data structures used for spatial access methods, for indexing multi-dimensional information such as geographical coordinates, rectangles, or polygons.

Signature File: A signature file is a short text file created for use as a standard appendage at the end of the e-mail notes or Usenet messages.

Spatial Database: A spatial database or geo database is a database that is optimized to store and query data that represents objects defined in a geometric space. Most spatial databases allow representing simple geometric objects such as points, lines, and polygons.

Spatial Index: A spatial index is a type of extended index that allows you to index a spatial column. A spatial column is a table column that contains data of a spatial data type, such as geometry or geography.

Spatial Query: A spatial query is a special type of database query supported by geo databases and spatial databases. The queries differ from non-spatial SQL queries in several important ways.

Chapter 8
Role of Free and Open Source GIS in River Rejuvenation

Smart Kundassery
Cochin University of Science and Technology, India

Babu C. A.
Cochin University of Science and Technology, India

ABSTRACT

Rivers represent one of the sources of freshwater. The existence and survival of humanity is intricately connected with the availability of freshwater resources. It is valuable to provide comprehensive information of all water resources through online geo-visual interfaces. Specific web slices can be of use, such as the illustration of headwater sources along with details of streams of various orders and tributaries up to the point of merging with the sea downstream. Since the local self-government departments, Grama Panchayat, can play a crucial role, they too are eager to know the details of the drainage network flowing within each administration unit. The possibilities emerging out of integration of free and open source GIS is enormous and can eventually succeed in bringing forth a ray of hope to the forlorn riverine ecosystem.

INTRODUCTION

The water resources perform a pivotal role in sustaining life on the Planet Earth. Myriad forms of flora and fauna flourish and nourish through the intake of water. Human race too is not an exception. For the humanity to sustain and thrive on the Planet Earth, the indispensable prerequisite is the availability of freshwater. All

DOI: 10.4018/978-1-5225-5039-6.ch008

cross-sections of the society, primitive and present, have depended on freshwater for subsistence. In the terrestrial landscape, the major sources of freshwater are the rivers. Ample supply of freshwater from the rivers well ensured and nurtured the initial vibes of the civilizations to sprout along the valleys of rivers, be it the Tigris-Euphrates river valley (4400 B.C.-2000 B.C.) along the Tigris and Euphrates rivers in Mesopotamia (Jason, 2010), Harappan civilization (2300 B.C.-1700 B.C.) along the Indus river valley (Fairservis, 1967), Nile river valley (3000 B.C.-30 B.C.) along the Nile river in Egypt and Yellow river valley (2100 B.C.-1 A.D.) along the Yellow river in China (Gernet, 1996). These civilizations are hence aptly known ever since as the river valley civilizations. Manifold ecological functions are infused by the rivers and extended numerous invaluable services to the ecosystem as well. The river continuum is thus regarded not just a flow of water over terrestrial landscapes but a flux of energy and matter of immense ecological concern.

Ever since the dawn of the civilizations, the rivers have espoused a momentous role in many religious ceremonies too. Most of the rivers of India are depicted as the avatars of Goddesses. Consequently, the rivers upheld a divine status and waters hold significant role in the holy rituals. The largest of the certain massive human congregation (Figure 1) occur alternately in the banks of the rivers, viz. the Ganges at Haridwar; the confluence of the Ganges and the Yamuna and the mythical Saraswati at Allahabad; the Godawari at Nashik; and the Shipra at Ujjain. On reminiscence, the rivers are thus indeed a living and sacred entity in holistic insights and not just a commercial commodity awaiting exploitation. In recognition of the many facets rivers enrich human life and endeavour, United Nations General Assembly has declared 2003 as the 'International Year of Freshwater [55/196]'; 2005-2015 as the International Decade for action, with the focal theme "Water for Life" (United Nations, 2004). The year 2013 is observed as the 'International Year of Water Cooperation', United Nations, 2011). 'World Rivers Day' is being celebrated on the last Sunday of September. Apart from these, annually, March 14 is observed as the 'International Day of Action for Rivers'. 'World Water Day' is observed on March 22 every year.

In quintessence, the vitality and vibrancy of the Nature is exalted by numerous rivers gushing across the planetary terrain. However, world over, the rivers are facing interferences from various natural as well as human induced processes. The natural processes are often related with the variations associated with the global warming and climatic changes. The human induced processes have multifarious roots. For long, the creation of dams and reservoirs were a compelling necessity for the prevention of recurring floods, generation of electricity and irrigating large tracts of lands. Later, in the process, the structures caused physical alterations to the river flow, caused irrecoverable damage and disturbance to the riverine system and even blocked the migration routes (Marmulla, 2001) of fishes (Figure 2). Eventually, the dams fragmented the natural floodplains of rivers to the utmost destructive extent (Figure 3/4). Unabated deforestation in the catchment area enhances the risk of catastrophic floods to a considerable extent. The natural forests are highly efficient in storing the precipitation and reduce the direct runoff. The loss of the forests, wetlands, grasslands and floodplains exasperated the flood risk factor. The best illustration of this intricately correlated scenario is the summer floods in the Yangtze River in China during 1998 (Zhou, et al., 2002). Excessive water abstraction for agricultural irrigation, urban consumption, industrial use and pollution cause immense threat to the riverine system. Threats to rivers are on the rise.

Consequently, the rivers are undergoing a fast-paced shrinkage of the headwater sources. The emerging scenario of the scarcity of water trends the water resources as a precarious entity. Globally, critical scenarios have already evolved to adopt fair water usage. The incurring compulsion calls for the effective management of the available freshwater resources. All the freshwater resources with dim prospects have to be identified for revival and rejuvenation. The free and Open Source Geographical Information System (GIS) has a significant role in aiding river revival and rejuvenation processes.

WATER RESOURCES ON THE PLANET EARTH

In the Solar System, the human race resides in a habitat known as the Planet Earth. The Planet Earth is one among eight planets in the Solar System. As of now, these eight planets together are endowed with 178 natural satellites. Remarkably, nine among these natural satellites have already expounded clear evidences for the presence of water bodies. These water bodies are now known as 'planetary oceans'. The ocean on the Earth, which we all are familiar, harbours on a planet. Curiously, the newly discovered abodes of oceans are found to exist in some of the natural satellites of the planets of the Solar System! These planetary objects are the natural satellites of the planets viz., the Jupiter, the Saturn, the Uranus and the Neptune. Presently, subsurface oceans exist on the Europa, the Ganymede, the Callisto; the Mimas, the Dione; the Ariel; and the Triton. A subsurface ocean means the water body lies sandwiched between an icy crust and a rocky interior. In contrast, on the Enceladus, an open ocean; and on the Titan, instead of a globe encircling ocean, numerous random and unrelated seas are found to dot the northern hemisphere (Kundassery and Babu, 2015).

All these pristine environments favour air-sea interaction processes and do contribute toward the modulation of the planetary climatology. These planetary objects are also regarded as the ideal environments where the hope and probability of finding signatures of alien life does exist. In essence, the Solar System abounds with remarkable deposits of water. Still incredible is the scenario that among these diverse planetary and satellite landscapes, the Planet Earth is the one and only one habitat where life as we sense does exist. Thus, the Earth holds a unique and distinct position in the spatial and temporal perspectives of the Solar System. The uniqueness relies in the presence of a congenial environment ardently supporting the evolution of life. The elixir nurturing the evolution, sustenance, nourishing and flourishing of life is nothing but the ubiquitous presence of water resources.

The formation scenario of water on the surface of the Earth is noteworthy at this moment. It is now widely accepted that the Planet Earth was formed around 4.54 billion years ago (Tera, 1980). Recent scientific evidences (Sarafian et al., 2014) ascertain that the water (H_2O) appeared primitively on the surface of the Earth during the formation of the planet itself. On the Planet Earth, ~70.900% of the surface area is covered with the water (H_2O). The compositions of the planetary oceans' elsewhere are as diverse and varied as ethane (C_2H_6) and methane (CH_4). Over the terrestrial landscape, water occurs naturally in ice caps, glaciers, snow and permafrost and as soil moisture. Much of the surface water is contained in rivers and lakes; while beneath the surface as groundwater. The solar irradiance causes evaporation of the molecules from the ocean. The molecules, on condensation, form clouds and subsequently fall as rain or snow and ultimately reach back to the ocean. Thus, the main source of water over the terrestrial landscapes is the ocean.

Of the water (H_2O) content available on the Planet Earth, the oceans, seas and bays hold 96.540% of the total water content. The water composition of the oceans, seas and bays in the terrestrial landscape is inherently saline. Saline groundwater holds 0.930% and saline lakes holds 0.006%. The freshwater resources on the Planet Earth thus hold a meagre 2.524%. The gist is, of all the water repository on the Planet Earth, saline water makes up about 97.476% and only 2.524% is fresh water! Of the total freshwater content of the Planet Earth, about 68.600% exists as ice caps, glaciers and permanent snow. 30.100% of the freshwater resides as groundwater content. The surface water and other freshwater sources hold the remaining 1.200%.

We have every reason to be more concerned and conscious about the remaining 1.200% of the surface and other freshwater resources. Because, within this 1.200%, ice caps, glaciers and permanent snow hold 68.600%. Freshwater lakes hold 0.260%, fresh groundwater holds 30.100% and swamps hold 0.030%. Soil moisture holds 0.050%. The atmosphere contains 0.040%, whereas biological water holds 0.003%. The rest of the fresh water is contained in rivers. Interestingly, rivers thus hold just 0.006% of the total freshwater content (Shiklomanov, 1993) of the Planet Earth! In other words, the critical scenario slowly unveiling and clearly emerging is that the fresh water available for the human consumption from rivers is just 0.0002% of the total water contained in the Planet Earth (Figure 5)!

WATER ECONOMY OF INDIA

The statistics enumerated through the recent 15th census of India during 2011 (http://censusindia.gov.in) reveals a total population of 1,210,569,573 people (over 1.2 billion) inhabiting India's geographical area of 329 million hectares. Agrarian economy, sanitation requirements, moreover, survival of the ever exploding population of the nation is rigorously met from the precipitation process. On an average, the geographical extent of India receives about 3840 Billion Cubic Meters (BCM) of precipitation (UNICEF, FAO and SaciWATERs, 2013), rainfall and snowfall together, every year. However, this precipitation too is not uniformly distributed over the entire land area. It varies from less than 100 millimetres in Rajasthan to more than 3,000 millimetres in Kerala and over 11,000 millimetres in Meghalaya. Moreover, the precipitation is seasonal. Precipitation is thus uneven in distribution over space and time. Of all the rain that falls on the terrestrial landscape, some evaporates back in to the atmosphere, some percolates in the ground and some is used by the forests. The remaining flows into the rivers.

The total surface water resource runoff potential of the country is estimated as 1,869.37 BCM (Billion Cubic Meters). Another 431.42 BCM can be used from the ground water recharge potential of the country. Thus, the available water potential of the nation is 2300.79 BCM. However, out of the available surface water resource runoff potential, the estimated utilisable freshwater potential to support ever exploding population of India is just 654 BCM (FAO UINICEF, 2013). An alarming situation arising in due course is the fact that the current water usage of the nation is marked as 634 BCM (FAO UINICEF, 2013). With a population of over 1.2 billion, the 'average per capita water availability' of the nation during 2011 was 1545 m^3/year, and projected towards 2050 is 1140 m^3/year. Almost 29% of the rural population and 23% of the urban population would still lack access to adequate safe, sustainable and potable water. A per capita availability of less than 1700 m^3 is termed as a 'water-stressed' condition while per capita availability below 1000 m^3 is termed as a 'water-scarcity' condition (Chaturvedi, 2012). In gist, the nation is marching towards a water-stressed condition (Figure 6). Under the spell of all these complex environmental aspects, the need of the hour is the implementation of a clear 'River Literacy' programme.

Necessity for the Implementation of River Literacy

In the regional scenario, Kerala, the southernmost state in the west coast of India, is endowed with 44 rivers. Most of them originate from the Western Ghats and drains into the Arabian Sea. Among these, there are 3 rivers, viz., Kabbini, Bhavani and Pambar which also originate from the Western Ghats, but they flow eastwards into the neighbouring States of Karnataka and Tamil Nadu. The societal requirements for livelihood are met from these water resources. During the past few years, most of the riverine ecosystems in Kerala, as elsewhere, have been heading towards a continuous degradation stage. Kerala is already included in moderate category of desertification process (Chouhan, 2005). Alarmingly, the signs of the desertification are observed in the region under the bounty of River Bharathapuzha, the second longest river in the state. Almost 175 villages are dependent on the River Bharathapuzha for subsistence. As elsewhere, the climatic factors, deforestation, clay mining, illicit human activities such as river sand mining, among a plethora of other factors, have contributed and played a cumulative role to aggravate the riverine degradation processes (Figure 7). According to the Planning Commission (2007), Government of India, in the national scenario, non-availability of potable water is posed to be a major crisis in terms of spread and severity. In the global context also, mighty rivers too are drying up at an alarming rate. Worldwide, the emerging situations have serious consequences not only for humankind, but also for the flora and fauna and on the future of the Planet Earth. The existence and survival of human race is thus intricately connected with the presence of freshwater resources, better within the immediate reach.

In this context, 'River Literacy' is defined as the 'ability to appreciate the manifold functions lotic ecosystems bestow on inhabitants of Planet Earth so as to initiate and embark on a prolonged process of river conservation, restoration, revival and rejuvenation process in volition' (Kundassery & Babu, 2013). Implementation of 'River Literacy' is envisaged to enhance community perception and awareness of the diverse and crucial aspects water resources behold to nurture and sustain life on Planet Earth. It is envisaged that 'River Literacy' will eventually lead to adopt optimum conservation, management and utilization principles of water resources to ensure prolonged access to freshwater resources. Moments when potable water dips almost to the level of being scarce, it is commendable and valuable to provide comprehensive information of the detailed status of all freshwater resources through online geo-visual Interfaces. The availability of the pertinent critical information at the fingertips will invoke participation of the enlightened mass in the conservation, revival and rejuvenation of the freshwater resources. RiverineGiS, a geo-visual illustration showcasing the rivers of a southern state along the west coast of India, is envisaged and implemented in this perspective.

RiverineGiS Geo-Visuals

Geo-visual illustrations of water resources herald manifold purposes. As explained elsewhere, down through the millennium, all drainage networks are under continuous mode of adapting itself to the climatic and natural changes in the terrestrial landscapes. As the world progressively heads to the 21st century, the Indus in Asia, the Nile in north-eastern Africa, the Colorado in United States, notable among several others, now struggles to merge with the water mass. There are instances of tributaries of rivers, such as Noyyal (Figure 8) of River Kaveri, in Tamil Nadu, completely dried up beyond interventions for probable rejuvenation. Rivers such as Uthrapalli, in Southern Kerala, is just extinct beyond revival even almost a century ago. Requirements for the water abstraction always arise still many more rivers are constantly under peril arising out of unwarranted human intentions and interventions. The perspectives emerging from the scrutiny of trends is very clear. Do preserve, conserve, inherit and adapt technologies for the optimum conservation and utilization of the natural resources and all sources of potable water for posterity too to live amicably on planet Earth! However, in the process of river revival, rejuvenation, moreover, in the restoration processes, few scenarios emerge at the global front as well as at grassroots. The foremost is the realization that various schemes envisaged so far, for river conservation does not bring the desired results. The schemes implemented for river restoration too could not bring anticipated results and fail miserably at grassroots.

A poignant point is becoming increasingly evident. Many well-documented resources of natural and environmental resources do exist. However, the realistic scenario existing at the micro-level, for the community working at the grassroots, is that utter scarcity of information pertaining to the natural drainage resources still do exist. It is the nearby community who knew the river health pulses well and not the occasional onlookers! The local community need to be enlightened, thoroughly lit with river literacy snippets, well-illustrated and included in all river restoration process to reap the envisaged benefits of river rejuvenation programs. Harsh reality of non-accessibility of crucial pertinent information can hamper constructive intervention of activists. Unless we address these crucial instances of relevant information retrieval at the grassroots, blown up to the full extent of severity, at later decades, the situation could seriously hamper the national progress outlined through various indicators of societal well-being. A paradigm shift is imminent and now proceeding with fast pace to include all stakeholders especially the inspired and motivated near the riverbanks in the rejuvenation process, be it the mighty Ganges or the River Bharathapuzha.

Rejuvenating a river requires participation of activists of different capacity from a wide spectrum of society, from policy makers, administrators and workers in the grassroots. All these inter-dependent work forces require detailed drainage maps. A public domain map, apart from the traditional paper-based maps, too is essential because there should not be a situation where lack of information derails the initiatives to conserve every rivers of the Planet Earth. The technologies available today are apt and mature enough to develop such an Interface in public domain. Wikimapia, Google Earth and Bhuvan have leapfrogged technology in this regard. However, region-specific, especially of the water resources of Kerala, is so far absent in public domain (Kundassery & Babu, 2013).

From the laymen perspectives, this scenario happens because some of the information is laid in documents having restricted access for content. Moreover, most of the literature remains inside bound volumes but not as geo-spatial digitized content, even if it exists, it is only in the closed camp of relevant professionals. Comprehensive aspects concerning the rivers in an easily accessible digital form are still less available. For example, scientific literature suggests River Periyar originates from the Sivagiri group of hills. River Periyar is the longest river in the state. However, it is very difficult to locate Sivagiri hills from existing web mapping domains. The situation is not much different, be it Google Maps (http://maps.google.co.in), (Wikimapia (http://wikimapia.org), or even ISRO Bhuvan (http://bhuvan.nrsc.gov.in). To know the geo-spatial locations of tributaries Mullayar and Periyar of River Mullaperiyar need much exercise; situations are not much different for other rivers elsewhere. In the process, there exist a few probable hurdles to overcome to initiate community participation in the conservation, revival and rejuvenation of the freshwater resources at the grass roots. Foremost is to mitigate the lack of digital access even to basic information pertaining to the riverine environment!

The community participation in the protection, conservation, revival, rejuvenation and management of the freshwater resources has to be ensured to make the programs viable. Therefore, the first step in this regard is the need and presence of a medium to present and access basic riverine information. Equally, the medium could also be conceptualized as a Web-GIS Interface to collate and present the riverine database. This Interface can then be designed as a dynamic user Interface through appropriate web technologies as a portal to deliver crucial scientific elements for societal empowerment through easily perceptual knowledge snippets. Recent technological advancements in web based GIS make it possible that the relevant riverine literature content can now be made available to larger cross sections of the society including student community, research scholars, planners and elite public alike. This will also be helpful for river activists. Being available, all relevant riverine information in public domain will increase rapid availability as well as accessibility to evoke a much wider movement and participation. Alternatively, enthusiasts have all the relevant knowledge at fingertips to initiate congress for river restoration process.

Ultimately, the success of river restoration process is intricately connected with the amalgamation of committed brave heart's spanning the entire ecosystem. There are serious researchers who do analysis of the changes in the land use pattern in different spatial and temporal scales and influencing parameters on its evolution. Science communication activists bring the results of hard-core scientists to grassroots in order to enlighten the masses and make them understand the relevance to appreciate subtleties of Grandma Nature. In addition, there are activists who need geo-spatial information at fingertips and appropriate technologies to initiate new vibes of restoration of the ecosystem. The utmost essential requirement for this to ensue in a constructive way is the availability of geo-spatial data framed with appropriate modes of information technology. Thus, it seems essential and meaningful for an attempt to generate a GIS based geo-visual framework of the water resources.

The deficit of sufficient regional information in the public domain motivated the proposition of RiverineGiS. RiverineGiS is the Web-GIS illustration of the water resources of Kerala. The Interface is conceived, developed and implemented for supporting the river literacy campaigns at the grass roots. Through RiverineGiS, geo-visuals of rivers and snippets of the critical problems faced by the riverine environment are presented to stress the need for immediate implementation of a comprehensive river conservation movement. The conservation and optimum usage of natural resources can occur only when community perception through river literacy is enhanced to ensure community participation. Free and Open Source Software are used to develop RiverineGiS.

FREE AND OPEN SOURCE WEB-GIS

Traditionally, map and atlas were used to illustrate geographical information. In print media, a map could be as simple as displaying the glimpses of a geographical area. Because of the static nature of these maps, the information portrayed had limited accessibility of content. Whoever possesses and procures only could use the map and atlas for the intended purposes. The emergence of geo-spatial techniques has resulted in a radical visual enrichment in the representation and analysis of the terrestrial landscapes. At the core of this enhanced functionality is geographically referenced data. Geographically referenced data outlines the criteria to help identify its location in the real world through geo-referenced coordinates such as latitude and longitude.

Web GIS refers to a variety of internet-based applications that rely on geographically referenced data. Web GIS is essentially a stripped Geographic Information System but distributed over a network using HTML protocols as its core mode of communication. Indeed the resources are generated through a Desktop GIS, which is then served through an online Web GIS application. In fact, the transformation from a Desktop GIS into a Web-based GIS has specific motivation. From the localhost restriction, Web GIS provides online connectivity and access of geographic information through various protocols to wider audience geographically apart.

The design and deliverance of geographically referenced data through a Web GIS application assumes a few indispensable features. Foremost, the focus is on what spatial data is intended to distribute through the geo-spatial application. The specific datasets and parameters are subsequently formulated to fulfil the user engagement and interaction to the portal. Later, based on the theme of the application, the spatial database is distinctively configured as base maps and overlays. This is achieved through generation and gathering of spatial data that can be shared and having implications for societal benefit. There could be one or more base maps, while there are certain restrictions in the number of operational layers rendered online. For example, OpenLayers, a front-end library, impose a restriction for 100 layers. Both raster and vector datasets are possible to serve through the application.

For individuals working at the grassroots for river rejuvenation, the most appropriate entity to intervene in this process is the access to appropriate geo-spatial software. Free and Open Source GIS software has their niche in this regard. As is well demonstrated, all proprietary software needs massive and periodical financial investment. For the activists working in the grassroots as well as for most Non-governmental Organizations (NGO), recurring software expenditure is a major hindrance. For wider acceptance and mass deployment, the natural implementation choices of Web-GIS stacks are preferably Free and Open Source Software (FOSS) than proprietary. Free and Open-source Software (FOSS) is the comprehensive term which embraces both domains of the free software and open-source software philosophies (Miller, et al., 2010). Apart from the financial liberation for prolong and continuous usage, free and open source also advocate the liberty to re-use and modify the source code according to the personal requirements of the individual and re-distribution of the improved content.

RiverineGiS: Geo-Visual Slices of Water Resources of Kerala

Nature has bestowed a bountiful of 44 rivers, most belonging to the category of medium rivers according to the national norms. Rivers has certain sources and flows down slopes carving a course until its end at the mouth. Beyond these generalizations, if somebody wants to know the geographical locations from where each of these rivers arises, then, matter gets worse. Well, the details are inside paperbacks, but not as interactive and geo-spatial content in the public domain. RiverineGiS depict selected natural resources and environmental parameters for the geo-visualization of the water resources of Kerala. The RiverineGiS Interface presents the entire drainage network draped over the Digital Elevation Model (DEM) of the terrain for the natural realization of the panorama. Initially, the water resource of Kerala, a southern state along the west coast of India is portrayed through RiverineGiS. Through the RiverineGiS Interface, each of the 44 rivers of Kerala and their tributaries are illustrated from headwater sources along with details of streams of various orders and tributaries up to the river mouth.

The critical aspects which makes imperative to make available online the relevant geo-spatial content are as follows. Analysed in minute details, an activist is confronted with certain critical circumstances. River Periyar, the longest river in the state of Kerala, has 283 tributaries. Being fed and enriched by the monsoon rains not all tributaries can be identified throughout the year. However, there exist perennial streams and tributaries of the river system. It is hard to figure out the names of those numerous tributaries as well. Neither Google Earth, Google Maps, Wikimapia nor ISRO Bhuvan fails miserably in providing the details. They do provide but of the datasets, they are of custodian and accumulated over time. Unless we identify and take constructive steps to conserve those precious rivulets in terms of the long-term sustenance oriented process, humanity will be slowly aiding the depletion of finite natural resources one after another.

In the system of public governance, there exist different entities entitled to perform specific function. Since the local self-government departments (LSGD) and grama panchayat can play a crucial role in the river rejuvenation process, they too are eager to know the details of the drainage network flowing within the capacity of each administration unit. A local community too exists who knew well the pulses of the ecosystem. Using Open Source GIS, they can contribute the local and traditional knowledge for the societal well-being. It is the apt tool to illustrate the burning problems in the neighbourhood. For those enthusiastic activists, the best way to unleash their creativity for aiding national progress is nothing but Open Source Geographical Information System. The Web GIS Interface can then be used to educate as well as to illustrate where lies and the extent of the imminent geographical problem. That thread is enough to lay spark for more people to contribute to the Interface and will eventually led to a conglomeration of expertise in the revival and rejuvenation process. Of course, wonders have happened even without the application of geo-spatial technologies as in Rajasthan. Nevertheless, the possibilities emerging out of integration of free and open source GIS is enormous and can eventually succeed in bringing forth a ray of hope to the forlorn riverine ecosystem.

Illustration of RiverineGiS Geo-Visual Interface

Through the RiverineGiS Interface (Figure 9), clients can pan, zoom and browse for information related to the water resources of Kerala. This effort has increased rapid availability as well as accessibility to evoke a much wider awareness and participation in the conservation efforts of rivers in the state. The interface can be understood through three options. 'Select a River' combo box allows easy auto panning to the originating geographical extent of each River. This tool is significant in understanding the drainage course of a single River, from origin up to the river mouth. 'Pan and Zoom' toolbar embedded within the map window allows flexible panning and viewing of random geographical extents. Different base layers are provided to tag streams, tributaries and rivers to different administration entities.

Geo-Visual Concepts

Geo-informatics is defined as "Science and Technology dealing with the structure and character of spatial information, its capture, its classification and qualification, its storage, processing, portrayal and dissemination, including the infrastructure necessary to secure optimal use of this information" (Raju, 2003). Geographic Information System (GIS) is one of the branches of Geo-informatics. GIS enables stacking of layers of information from varied sources into a unified software domain to analyse the relationships, if any, and correlate the physical and environmental entities within the geographical context to derive an optimum deduction. The database thus generated can be organized to distribute and enabled to access over a local or wide area network. The database can also be hosted in a server for easier access in public domain.

Data Sources

The RiverineGiS Interface is composed of raster and vector data. Raster data is the base layer; GEBCO (www.gebco.net) is used to depict bathymetry, and the data from the Shuttle Radar Topographic Mission (SRTM) for topography. In the present dataset, depth varies up to 8698 m, whereas elevation up to 2536 m and the variations are depicted through a colour index. Topographical sheets of entire Kerala from Survey of India at 1: 50,000 scale (http://www.surveyofindia.gov.in) are used to generate vector layers except administrative boundaries, airports, KSRTC and railway stations, mud banks and sand mining kadavu in Palakkad District. Administrative boundaries are generated from GADM Database of Global Administrative Areas (www.gadm.org). Airports, KSRTC and railway stations are generated from Google Maps. Mud banks and sand-mining Kadavu in Palakkad District (CESS, 2004) are generated from various scientific publications corroborated through field surveys. Kerala State Gazetteer (Ramachandran, 1986) and Water Resources of Kerala: A Compendium of Advance Report (Vaidyanathan, 1958) provided basic literature for rivers of Kerala.

Software

Free and Open Source Software is used in development. Geographic Resources Analysis Support System (GRASS GIS) and Quantum GIS Version 1.7.1 through 2.10.1 was used for basic GIS operations and for the generation of Vector layers and its attributes. OGC compliant Web Map Servers, viz., MapServer and Geoserver are used to develop the mapping application. MapServer is hosted in Apache Server whereas Geoserver requires either an Apache Tomcat or Jetty Server; we used

Apache Tomcat. PostgreSQL together with PostGIS 1.5 is used to generate the vector database. The database is then fed into Geoserver (2.1.3), and front-end is developed using OpenLayers 2.13. Geoserver map script was generated using java script. RiverineGiS MapServer Interface was generated using MapScript.

Thematic Layers

The following themes were generated: drainage of 44 rivers of Kerala along with ocean boundaries, maritime boundary, and bathymetry and administration units consisting of district, panchayat, and village and watershed boundary. Connectivity, viz., national highway, state highway, railway, airport, port and harbour, railway and KSRTC stations, place names, mountains and reserved forests, beaches, light houses, mud banks along Kerala coast, sand mining kadavu in Palakkad are also incorporated into the RiverineGiS Interface.

Applications of RiverineGiS Interface

The RiverineGiS Interface also demonstrates usefulness of the RiverineGiS geo-visuals as an innovative tool for interactive academic and scientific research under the backdrop of natural landscapes. The highlights of the RiverineGiS Interface are as follows. Through the RiverineGiS Interface, the geo-visual information pertaining to the entire rivers of Kerala is delivered into public domain. Through the Interface, clients can pan, zoom and browse for information related to the water resources of Kerala in the backdrop of natural landscapes (Figure 10). The clients can pan right from the headwater sources, each of the tributaries through river mouth and beyond the coastal oceans.

The ecologically sensitive areas in Western Ghats lies within the 500 m contour above M.S.L. (Gadgil, 2011). The significance of this zone is illustrated through the theme 'Ecologically Sensitive Areas' in the Western Ghats (Kerala). Out of the 44 rivers of Kerala, 34 rivers originate from the Western Ghats. Out of these 34 rivers, 17 rivers originate from ecologically sensitive villages in Western Ghats (Kerala). These 17 rivers have 33 origin locations spread across the 'Ecologically Sensitive Areas' in the Western Ghats (Kerala). These aspects alone vouch for the significance of the implementation of the RiverineGiS Interface. We cannot demarcate that a river should flow along a particular direction favouring population needs, rather we should constrain a zone ecologically favouring riverine environment, which then sustains population in its cradle.

Critical aspects of the riverine environment that require immediate attention, viz. illicit river sand mining are highlighted. Illegal sand mining poses serious environmental issues. Sand mining persists in almost all the rivers of Kerala. It is more rampant in Bharathapuzha (Figure 11). It is estimated that the sand extraction from River Bharathapuzha is 1652 truckloads per day (TLPD); whereas in River Bhavani, it is 40 TLPD, (CESS, 2004). We have also developed a database through which it is easier to identify the kadavu, panchayat and village details and are part of 'RiverineGiS'. The Interface presents the geographical locations where rampant illegal sand mining persists. The locations are identified and mapped. The Interface also supplements scientific literature explaining the environmental impacts of sand mining along with visual narratives from the riverbanks by various channels for enhanced awareness of the socio-economic impacts. Science communication and awareness information explaining the adverse impacts imposed is also provided.

The OpenWeatherMap (http://openweathermap.org) service provides free weather data and forecast API suitable for any cartographic services like web and smartphone applications. Weather data is received from global meteorological broadcast services. The service use data from two weather services - NOAA that use GFS model, and Environment Canada. Both models are global, and determine the total atmospheric dynamics on the planet as a whole; with a grid size of about 50 km and longer ranged forecast of 5-7 days. RiverineGiS Interface is over layered with OpenWeather WMS to yield weather conditions persisting in river basins. Currently, in the weather mashup (Figure 12), meteorological parameters such as clouds and precipitation forecasts, pressure, temperature and wind magnitude are supported. Mashup enables forecasting of meteorological conditions prevailing in the catchment areas of a river basin. This allows the possibility of monitoring formation of severe weather phenomena to take appropriate safety measures to avert human and property loss. Thus, RiverineGiS can be used simultaneously for educational and planning purposes. Inclusion of appropriate themes allows supplementing on-going river rejuvenation efforts of various user groups.

CONCLUSION

Rivers as a source of freshwater are the lifeline of societies and arteries of the habitable Planet Earth. Without drainage system, what waits is a barren desert. A dewdrop or a raindrop, each droplet beholds and nurture life on Planet Earth. We have to save and preserve it for posterity to survive and continue habitat on Planet Earth. RiverineGiS is the first step towards achieving this end.

ACKNOWLEDGMENT

The motivation for the development of the RiverineGiS Web-GIS Interface was inspired through numerous chance encounters with the community who ardently hoped to revive River Bharathapuzha. Heartfelt thanks to the grass root community for imparting traditional knowledge of the riverine habitat. The assistance extended by Mr. Praveen Balakrishnan and Mrs. Sajna Ansari, ORGIN, during various stages of the project is duly acknowledged.

REFERENCES

CESS. (2004). *Bharathapuzha and Its Problems with Special Preference to Sand Mining from the River Stretch between Chamravattom and Thirunavaya*. Trivandrum: Centre for Earth Science Studies.

Chaturvedi, M. C. (2010). *India's Waters: Advances in Development and Management*. CRC Press.

Chouhan, T. S. (2005). *Degree, Extent and Treatment of Desertification Hazards in India. Sociedade & Natureza, 1*(1), 901–919.

Fairservis, W. A. (1967). The origin, character, and decline of an early civilization. *American Museum Novitates*, 2302.

Gernet, J. (1996). *A History of Chinese Civilization*. Cambridge University Press.

Kundassery & Babu. (2013). RiverineGIS-A Geo-visual Framework for enhancing River Literacy. *Indian Cartographer, 33*, 453–458.

Kundassery & Babu. (2015). An Oceanographic Exploration of the Solar System. *PLANEX*, 7-11.

Madhav Gadgil Commission. (2011). *Report of the Western Ghats Ecology Expert Panel*. The Ministry of Environment and Forests, Government of India.

Marmulla, G. (Ed.). (2001). *Dams, fish and fisheries-Opportunities, challenges and conflict resolution. FAO Fisheries Technial Paper, Issue 419*. Food and Agriculture Organization of the United Nations.

Miller, K. W., Voas, J., & Costello, T. (2010). Free and Open Source Software. *IT Professional, 12*(6), 14–16. doi:10.1109/MITP.2010.147

Planning Commission. (2007). *Report of the Steering Committee on Water Resources for Eleventh Five Year Plan (2007-2012)*. Government of India.

Raju, P. L. N. (2003). Fundamentals of Geographic Information System. *Proc. of the Training Workshop, held at Dehra Dun*, 103-120.

Ramachandran Nair, K. K. (1986). *Kerala State Gazetteer* (Vol. 1). Thiruvananthapuram, Kerala: Kerala Council for Historical Research.

Sarafian, A. R., Nielsen, S. G., Marschall, H. R., McCubbin, F. M., & Monteleone, B. D. (2014). Early accretion of water in the inner solar system from a carbonaceous chondrite-like source. *Science, 346*(6209), 623–626. doi:10.1126cience.1256717 PMID:25359971

Shiklomanov, I. (1993). World fresh water resources. In Water in Crisis: A Guide to the World's Fresh Water Resources. Oxford University Press.

Tera, F. (1980). Reassessment of the Age of the Earth. *Carnegie Institution of Washington Year Book, 79*, 524–531.

UNICEF, FAO, & SaciWATERs. (2013). *Water in India: Situation and Prospects*. Author.

United Nations General Assembly Resolution. (2004). *A/RES/58/217, International Decade for Action, "Water for Life", 2005-2015*. Author.

United Nations General Assembly Resolution. (2011). *A/RES/65/154, International Year of Water Cooperation, 2011*. Author.

Ur, J. A. (2010). Cycles of Civilization in Northern Mesopotamia, 4400–2000 BC. *Journal of Archaeological Research, 18*(4), 387–431. doi:10.100710814-010-9041-y

Vaidyanathan, P. H. (1985). Water Resources of Kerala: A Compendium of Advance Report. Kerala, India: Public Works Department, Govt. of Kerala, Trivandrum.

Zhou, X., Lin, D., Yang, H., Chen, H., Sun, L., Yang, G., ... Brown, L. (2002). Use of landsat TM satellite surveillance data to measure the impact of the 1998 flood on snail intermediate host dispersal in the lower Yangtze River Basin. *Acta Tropica, 82*(2), 199–205. doi:10.1016/S0001-706X(02)00011-6 PMID:12020893

APPENDIX

Figures

Figure 1. Pilgrims of Kumbh Mela gather to take bath in the Ganges
(© 2010, Coupdoeil / Philipp Eyer. Distributed under a CC BY-SA 3.0 license).

Figure 2. Reduced flow downstream of dams affects fish abundance
(© 2008, Smart Kundassery).

Figure 3. Where is the river here?
(© 2013, Smart Kundassery).

Figure 4. Full reservoir level in the alternate side of dam
(© 2013, Smart Kundassery).

Figure 5. River bed is dry except these few drops!
(© 2008, Smart Kundassery).

Figure 6. Other side of 'Digital India'
(© 2015, Smart Kundassery.).

Figure 7. View of second largest river, Bharathapuzha, in Kerala
(© 2008, Smart Kundassery).

Figure 8. River Noyyal in Tamil Nadu
(©2013, Smart Kundassery).

Figure 9. RiverineGiS interface
(© 2015, Smart Kundassery).

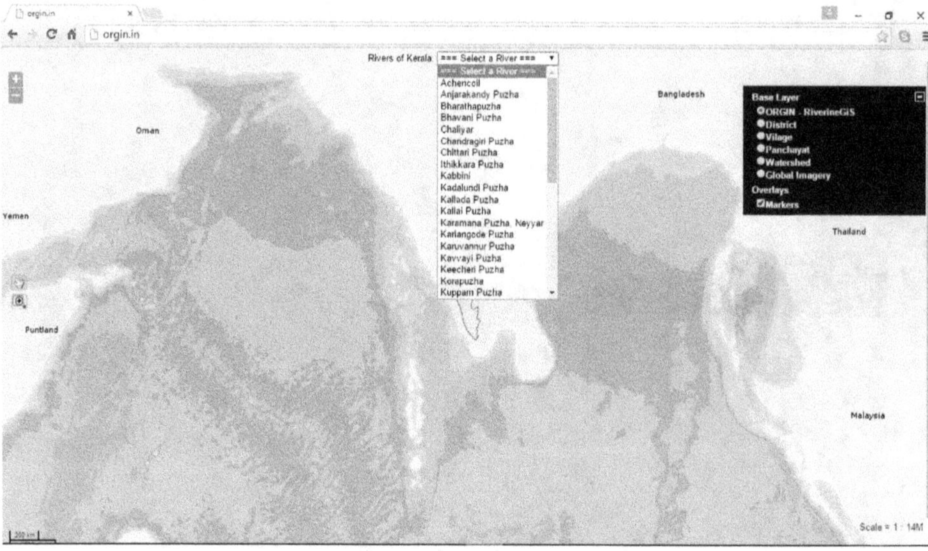

Figure 10. Geo-visuals of rivers
(©2015, Smart Kundassery).

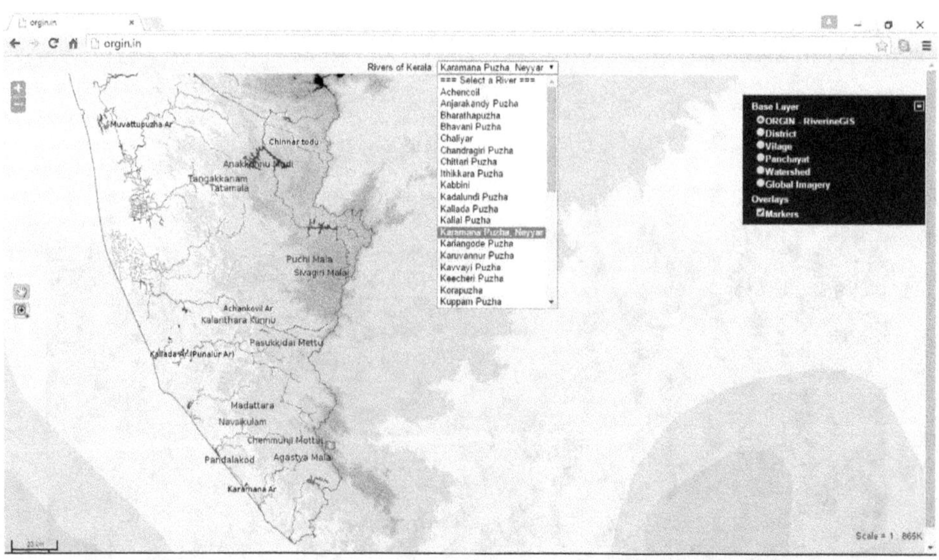

Figure 11. River sand mining
(©2013, Smart Kundassery).

Figure 12. RiverineGiS weather mashup
(©2015, Smart Kundassery).

Chapter 9
Use of Cloud, Multimedia, and QR Codes to Enhance Print Maps

Harpinder Singh
Punjab Remote Sensing Centre, India

Dheeraj Gambhir
Punjab Remote Sensing Centre, India

Sagar Taneja
Punjab Remote Sensing Centre, India

Amardeep Singh
Punjab Remote Sensing Centre, India

ABSTRACT

Quick response (QR) codes are two-dimensional codes. They are machine-readable optical labels which contain information of the object on which they are attached. Due to their large storage capacity and faster readability, they are becoming more popular than one-dimensional bar codes. They are widely used in creation of inventories, product tracking, document management, etc. With the widespread use of smart phones, QR codes are becoming more popular. QR code interpreting apps are freely available and all modern smartphones are equipped with a camera and internet to process and interpret these codes. The value of printed maps can also be enhanced by using QR codes. This chapter reviews how the value of printed or hard copy maps can be enhanced by adding QR codes that have multimedia content placed on the Cloud.

DOI: 10.4018/978-1-5225-5039-6.ch009

INTRODUCTION

Maps are tools with the help of which the situation of the particular area can be visualized topographically and thematically. Now day's web interactive maps are more popular than print or hard copy maps. Google Maps and Bing Maps are best examples of such websites, where a user can PAN, zoom and query the data. Multimedia data like images, video, music, animations, slide shows, documents, and website links etc. can be linked to the map features. One such example is the 2012 U.S. presidential campaign website (https://www.google.com/maps/d/viewer?mid=zNvDJoOATiic.kpXHGQXy0L58) where YouTube videos are linked to the map features. In Google Earth also we can see various examples of images being linked to the map features. These are examples of multimedia maps where the multimedia is attached or embedded in the maps. Multimedia helps the user to grasp information faster and it improves user experience also. These types of maps help to present geographical information in an intuitive manner. In such web portals the maps as well as the multimedia content is placed on the internet or a web server. As these maps are digital they can be viewed only on a computer or a mobile device.

Cloud computing is a style of computing where IT related capabilities (software, infrastructure, platform) are provided as a service through the internet. The important features of the cloud are its low cost, flexibility, fast deployment and scalability. Nowadays people can leverage the cloud facilities like getting storage, access to application software and also processing power. One of the best examples of cloud is the Google Docs application where you can create, manage and store your documents, spreadsheets and presentations online using a simple browser.

Quick response (QR) codes are two dimensional codes. They are machine readable optical labels which contain information of the object on which it is attached. Due to its large storage capacity and faster readability, it is becoming popular day by day than a one dimensional bar code. Fig.1 It consists of a square grid in which black dots are arranged on a white background. Information in the form of URL, text, numbers, email, SMS, visiting card, location and images, etc can be converted into a QR code and a QR code scanning software along with an imaging device like a camera is required to read and interpret the code.

QR codes are becoming popular today and we can see them on various products like magazines, books, medicines, rail tickets, household products etc and the advantage of using them is that they can be printed on paper, cloth, wood, plastic, cardboard and any type of opaque surface. A QR code on a product e.g. medicine bottle, can be scanned with the help of a QR code scanner app in the mobile phone and the details regarding the medicine like manufacturing, composition and expiry etc can be viewed by the consumer in a web page. It can also be used by the seller or the manufacturer for billing and inventory purpose.

QR codes are being used in diverse applications worldwide. Ashford (2010) has improved the library processes by using QR codes. Massis (2011) describes it as a motivating technology for libraries to enhance their own offerings to users with greater potential and possibilities. UC Irvine libraries has linked print and virtual resources using QR codes according to Kane and Schneidewind (2011). Chang et al. (2007) presents a wayfinding prototype system based geo-coded QR codes for individuals with cognitive impairments. The results show the prototype is user friendly and promising with high reliability. Conde-Lagoa et al. (2010) proposes a new model for secure eTickets based on QR-Codes with user-encrypted content. It is compatible with access control systems with minimal infrastructures. Kan et al. (2009) presents an augmented reality (AR) application based on the QR Code. The system can extract the information embedded in a QR Code and show the information in an extended 3D form with the QR Code being the traditional AR marker. An innovative method for television billing has been illustrated by Schaefer and Kennedy (2011). Barrera et al. (2013) describes the use of these codes for secure and noise free information retrieval.

PROPOSED FRAMEWORK

The proposed framework makes use of QR codes which are printed on the hardcopy or print maps and act as an intermediate object to link multimedia like images, video, music, animations, slide shows, documents etc which are placed on the cloud storage, to the maps.

A user needs a smart phone with a QR code scanner application and internet connection to scan the QR code and view the linked media or website. Cloud storage is used to store the data which is linked to the map. The proposed process is as shown in Figure 2.

An example map with QR codes is shown in Figure 3. It is a tourism map which can be printed on any opaque surface and placed in a tourist place. QR codes have been generated for various places of interest and placed in the district boundary map of Punjab. Slideshows, Videos, Images, Hyperlinks of website and even contact details have been linked to this map using QR codes for various tourist places of Punjab.

Some other advantages of using QR codes on print maps are:

- The route to a specific place from the current location can also be added to QR codes. Geocoded QR codes can also be generated.
- Paper can be saved by adding the QR link on printed maps for the detailed description (pdf, word format) or metadata of the maps.

Table 1. QR code scanning mobile apps

S. No.	Android	iOS	Windows
1	QR Code Reader	QR Reader	QR Code Reader
2	QR Droid Code Scanner	QR Code Reader and Scanner	QR Code Scanner

- Contact details like SMS, vCard, phone number, email address can also be added to QR codes.
- QR Codes can be printed on any opaque surface like a paper, cloth, wood, plastic, cardboard etc

Figure 1. QR code

Figure 2. Proposed process

A simple smart phone is required to scan and interpret a QR code. Table 1 describes some popular mobile apps which can be used to scan QR codes.

The following is the list of some popular websites to create QR Codes.

- https://app.qr-code-generator.com
- http://goqr.me/
- https://www.unitag.io/qrcode

Figure 3. Tourism map

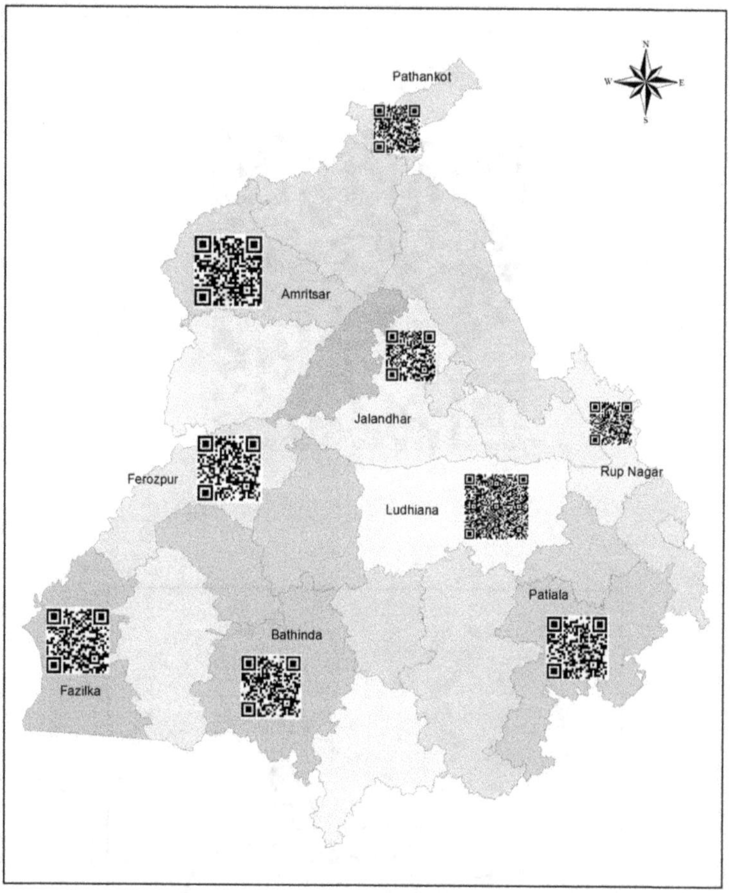

DISCUSSIONS

This paper amply demonstrates the use of Cloud storage, multimedia and QR codes to enhance print or hardcopy maps.

A new technology "Augmented Reality" (AR) can outshine QR codes in the near future and it is changing the way we view the world. It blurs the line between what's computer-generated what's real by enhancing what we see, feel, smell and hear. But the issue with this technology is that currently creation of AR content is complex. AR can revolutionize the way we view, understand and perceive maps.

REFERENCES

Ashford, R. (2010). QR codes and academic libraries Reaching mobile users. *College & Research Libraries News*, *71*(10), 526–530. doi:10.5860/crln.71.10.8454

Barrera, J. F., Mira, A., & Torroba, R. (2013). Optical encryption and QR codes: Secure and noise-free information retrieval. *Optics Express*, *21*(5), 5373–5378. doi:10.1364/OE.21.005373 PMID:23482108

Chang, Y.-J., Tsai, S.-K., Chang, Y.-S., & Wang, T.-Y. (2007). A novel wayfinding system based on geo-coded QR codes for individuals with cognitive impairments. *Proceedings of the 9th international ACM SIGACCESS conference on Computers and accessibility*. 10.1145/1296843.1296887

Conde-Lagoa, D., Costa-Montenegro, E., Gonzalez-Castao, F., & Gil-Castieira, F. (2010). *Secure eTickets based on QR-Codes with user-encrypted content.* Paper presented at the Consumer Electronics (ICCE), 2010 Digest of Technical Papers International Conference on. 10.1109/ICCE.2010.5418880

Kan, T.-W., Teng, C.-H., & Chou, W.-S. (2009). Applying QR code in augmented reality applications. *Proceedings of the 8th International Conference on Virtual Reality Continuum and its Applications in Industry*. 10.1145/1670252.1670305

Kane, D., & Schneidewind, J. (2011). QR codes as finding aides: Linking electronic and print library resources. *Public Services Quarterly*, *7*(3-4), 111–124. doi:10.1080/15228959.2011.623599

Massis, B. E. (2011). QR codes in the library. *New Library World*, *112*(9/10), 466–469. doi:10.1108/03074801111182058

Schaefer, G., & Kennedy, J. T. (2011). *Billing with QR Codes*. Google Patents.

Related References

To continue our tradition of advancing knowledge management and discovery research, we have compiled a list of recommended IGI Global readings. These references will provide additional information and guidance to further enrich your knowledge and assist you with your own research and future publications.

Abril, R. M. (2011). The quality attribution in data, information and knowledge. In D. Schwartz & D. Te'eni (Eds.), *Encyclopedia of knowledge management* (2nd ed.; pp. 1343–1354). Hershey, PA: IGI Global. doi:10.4018/978-1-59904-931-1.ch129

Abufardeh, S. (2013). KM and global software engineering (GSE). In S. Saeed & I. Alsmadi (Eds.), *Knowledge-based processes in software development* (pp. 12–34). Hershey, PA: IGI Global. doi:10.4018/978-1-4666-4229-4.ch002

Aggestam, L., Backlund, P., & Persson, A. (2010). Supporting knowledge evaluation to increase quality in electronic knowledge repositories. *International Journal of Knowledge Management, 6*(1), 23–43. doi:10.4018/jkm.2010103002

Aggestam, L., Backlund, P., & Persson, A. (2012). Supporting knowledge evaluation to increase quality in electronic knowledge repositories. In M. Jennex (Ed.), *Conceptual models and outcomes of advancing knowledge management: New technologies* (pp. 24–44). Hershey, PA: IGI Global. doi:10.4018/978-1-4666-0035-5.ch002

Aiken, P., Gillenson, M., Zhang, X., & Rafner, D. (2011). Data management and data administration: Assessing 25 years of practice. *Journal of Database Management, 22*(3), 24–45. doi:10.4018/jdm.2011070102

Akabawi, S., & Hodeeb, H. (2013). Implementing business intelligence in the dynamic beverages sales and distribution environment. In M. Khosrow-Pour (Ed.), *Cases on performance measurement and productivity improvement: Technology integration and maturity* (pp. 194–221). Hershey, PA: IGI Global. doi:10.4018/978-1-4666-2618-8.ch010

Al-Busaidi, K. A. (2011). A social and technical investigation of knowledge utilization from a repository knowledge management system. In M. Al-Shammari (Ed.), *Knowledge management in emerging economies: Social, organizational and cultural implementation* (pp. 122–139). Hershey, PA: IGI Global. doi:10.4018/978-1-61692-886-5.ch007

Al-Busaidi, K. A. (2012). The impact of supporting organizational knowledge management through a corporate portal on employees and business processes. In M. Jennex (Ed.), *Conceptual models and outcomes of advancing knowledge management: New technologies* (pp. 208–229). Hershey, PA: IGI Global. doi:10.4018/978-1-4666-0035-5.ch011

Alaraifi, A., Molla, A., & Deng, H. (2013). An empirical analysis of antecedents to the assimilation of sensor information systems in data centers. *International Journal of Information Technologies and Systems Approach*, *6*(1), 57–77. doi:10.4018/jitsa.2013010104

Alguezaui, S., & Filieri, R. (2010). Social capital: Knowledge and technological innovation. In P. López Sáez, G. Castro, J. Navas López, & M. Delgado Verde (Eds.), *Intellectual capital and technological innovation: Knowledge-based theory and practice* (pp. 271–296). Hershey, PA: IGI Global. doi:10.4018/978-1-61520-875-3.ch013

Alhashem, A., & Shaqrah, A. A. (2012). Exploring the relationship between organizational memory and business innovation. *International Journal of Knowledge-Based Organizations*, *2*(3), 32–46. doi:10.4018/ijkbo.2012070102

Allan, M. B., Korolis, A. A., & Griffith, T. L. (2011). Reaching for the moon: Expanding transactive memory's reach with wikis and tagging. In M. Jennex (Ed.), *Global aspects and cultural perspectives on knowledge management: Emerging dimensions* (pp. 144–156). Hershey, PA: IGI Global. doi:10.4018/978-1-60960-555-1.ch010

Alsmadi, I., & Alda, S. (2013). Knowledge management and semantic web services. In S. Saeed & I. Alsmadi (Eds.), *Knowledge-based processes in software development* (pp. 35–48). Hershey, PA: IGI Global. doi:10.4018/978-1-4666-4229-4.ch003

Alstete, J. W., & Meyer, J. P. (2011). Expanding the model of competitive business strategy for knowledge-based organizations. *International Journal of Knowledge-Based Organizations, 1*(4), 16–31. doi:10.4018/ijkbo.2011100102

Alstete, J. W., & Meyer, J. P. (2013). Expanding the model of competitive business strategy for knowledge-based organizations. In J. Wang (Ed.), *Intelligence methods and systems advancements for knowledge-based business* (pp. 132–148). Hershey, PA: IGI Global. doi:10.4018/978-1-4666-1873-2.ch008

Alves da Silva, N. S., Alvarez, I. M., & Rogerson, S. (2011). Glocality, diversity and ethics of distributed knowledge in higher education. In G. Morais da Costa (Ed.), *Ethical issues and social dilemmas in knowledge management: Organizational innovation* (pp. 131–159). Hershey, PA: IGI Global. doi:10.4018/978-1-61520-873-9.ch009

Anantatmula, V. S., & Kanungo, S. (2011). Strategies for successful implementation of KM in a university setting. In M. Jennex & S. Smolnik (Eds.), *Strategies for knowledge management success: Exploring organizational efficacy* (pp. 262–276). Hershey, PA: IGI Global. doi:10.4018/978-1-60566-709-6.ch014

Andreu, R., & Sieber, S. (2011). External and internal knowledge in organizations. In D. Schwartz & D. Te'eni (Eds.), *Encyclopedia of knowledge management* (2nd ed.; pp. 298–307). Hershey, PA: IGI Global. doi:10.4018/978-1-59904-931-1.ch029

Andriessen, D. (2011). Metaphor use in knowledge management. In D. Schwartz & D. Te'eni (Eds.), *Encyclopedia of knowledge management* (2nd ed.; pp. 1118–1124). Hershey, PA: IGI Global. doi:10.4018/978-1-59904-931-1.ch107

Angelopoulos, S., Kitsios, F., & Moustakis, V. (2012). Transformation of management in the public sector: Exploring the strategic frameworks of e-government. In T. Papadopoulos & P. Kanellis (Eds.), *Public sector reform using information technologies: Transforming policy into practice* (pp. 44–58). Hershey, PA: IGI Global. doi:10.4018/978-1-60960-839-2.ch003

Anselma, L., Bottrighi, A., Molino, G., Montani, S., Terenziani, P., & Torchio, M. (2013). Supporting knowledge-based decision making in the medical context: The GLARE approach. In J. Wang (Ed.), *Intelligence methods and systems advancements for knowledge-based business* (pp. 24–42). Hershey, PA: IGI Global. doi:10.4018/978-1-4666-1873-2.ch002

Arh, T., Dimovski, V., & Blažic, B. J. (2011). ICT and web 2.0 technologies as a determinant of business performance. In M. Al-Mutairi & L. Mohammed (Eds.), *Cases on ICT utilization, practice and solutions: Tools for managing day-to-day issues* (pp. 59–77). Hershey, PA: IGI Global. doi:10.4018/978-1-60960-015-0.ch005

Ariely, G. (2011). Operational knowledge management in the military. In D. Schwartz & D. Te'eni (Eds.), *Encyclopedia of knowledge management* (2nd ed.; pp. 1250–1260). Hershey, PA: IGI Global. doi:10.4018/978-1-59904-931-1.ch119

Assefa, T., Garfield, M., & Meshesha, M. (2014). Enabling factors for knowledge sharing among employees in the workplace. In Y. Al-Bastaki & A. Shajera (Eds.), *Building a competitive public sector with knowledge management strategy* (pp. 246–271). Hershey, PA: IGI Global. doi:10.4018/978-1-4666-4434-2.ch011

Assudani, R. H. (2011). Negotiating knowledge gaps in dispersed knowledge work. *International Journal of Knowledge-Based Organizations*, *1*(3), 1–21. doi:10.4018/ijkbo.2011070101

Assudani, R. H. (2013). Negotiating knowledge gaps in dispersed knowledge work. In J. Wang (Ed.), *Intelligence methods and systems advancements for knowledge-based business* (pp. 75–96). Hershey, PA: IGI Global. doi:10.4018/978-1-4666-1873-2.ch005

Atkins, R. (2011). Supply chain knowledge integration in emerging economies. In M. Al-Shammari (Ed.), *Knowledge management in emerging economies: Social, organizational and cultural implementation* (pp. 104–121). Hershey, PA: IGI Global. doi:10.4018/978-1-61692-886-5.ch006

Aung, Z., & Nyunt, K. K. (2014). Constructive knowledge management model and information retrieval methods for software engineering. In *Software design and development: Concepts, methodologies, tools, and applications* (pp. 253–269). Hershey, PA: IGI Global. doi:10.4018/978-1-4666-4301-7.ch014

Aziz, M. W., Mohamad, R., & Jawawi, D. N. (2013). Ontology-based service description, discovery, and matching in distributed embedded real-time systems. In M. Nazir Ahmad, R. Colomb, & M. Abdullah (Eds.), *Ontology-based applications for enterprise systems and knowledge management* (pp. 178–190). Hershey, PA: IGI Global. doi:10.4018/978-1-4666-1993-7.ch010

Badia, A. (2011). Knowledge management and intelligence work: A promising combination. In D. Schwartz & D. Te'eni (Eds.), *Encyclopedia of knowledge management* (2nd ed.; pp. 612–623). Hershey, PA: IGI Global. doi:10.4018/978-1-59904-931-1.ch059

Badr, K. B., & Ahmad, M. N. (2013). Managing lessons learned: A comparative study of lessons learned systems. In M. Nazir Ahmad, R. Colomb, & M. Abdullah (Eds.), *Ontology-based applications for enterprise systems and knowledge management* (pp. 224–245). Hershey, PA: IGI Global. doi:10.4018/978-1-4666-1993-7.ch013

Bakshi, K. (2014). Technologies for big data. In W. Hu & N. Kaabouch (Eds.), *Big data management, technologies, and applications* (pp. 1–22). Hershey, PA: IGI Global. doi:10.4018/978-1-4666-4699-5.ch001

Ballou, D. P., Belardo, S., & Pazer, H. L. (2012). A project staffing model to enhance the effectiveness of knowledge transfer in the requirements planning phase for multi-project environments. In M. Jennex (Ed.), *Conceptual models and outcomes of advancing knowledge management: New technologies* (pp. 77–98). Hershey, PA: IGI Global. doi:10.4018/978-1-4666-0035-5.ch005

Baporikar, N. (2011). Knowledge management and entrepreneurship cases in India. In M. Al-Shammari (Ed.), *Knowledge management in emerging economies: Social, organizational and cultural implementation* (pp. 325–346). Hershey, PA: IGI Global. doi:10.4018/978-1-61692-886-5.ch020

Baporikar, N. (2014). Knowledge management initiatives in Indian public sector. In Y. Al-Bastaki & A. Shajera (Eds.), *Building a competitive public sector with knowledge management strategy* (pp. 53–89). Hershey, PA: IGI Global. doi:10.4018/978-1-4666-4434-2.ch002

Baporikar, N. (2014). Organizational barriers and facilitators in embedding knowledge strategy. In M. Chilton & J. Bloodgood (Eds.), *Knowledge management and competitive advantage: Issues and potential solutions* (pp. 149–173). Hershey, PA: IGI Global. doi:10.4018/978-1-4666-4679-7.ch009

Barioni, M. C., Kaster, D. D., Razente, H. L., Traina, A. J., & Júnior, C. T. (2011). Querying multimedia data by similarity in relational DBMS. In L. Yan & Z. Ma (Eds.), *Advanced database query systems: Techniques, applications and technologies* (pp. 323–359). Hershey, PA: IGI Global. doi:10.4018/978-1-60960-475-2.ch014

Baroni de Carvalho, R., & Tavares Ferreira, M. A. (2011). Knowledge management software. In D. Schwartz & D. Te'eni (Eds.), *Encyclopedia of knowledge management* (2nd ed.; pp. 738–749). Hershey, PA: IGI Global. doi:10.4018/978-1-59904-931-1.ch072

Barroso, A. C., Ricciardi, R. I., & Junior, J. A. (2012). Web 2.0 and project management: Reviewing the change path and discussing a few cases. In I. Boughzala & A. Dudezert (Eds.), *Knowledge management 2.0: Organizational models and enterprise strategies* (pp. 164–189). Hershey, PA: IGI Global. doi:10.4018/978-1-61350-195-5.ch009

Baskaran, V., Naguib, R., Guergachi, A., Bali, R., & Arochen, H. (2011). Does knowledge management really work? A case study in the breast cancer screening domain. In A. Eardley & L. Uden (Eds.), *Innovative knowledge management: Concepts for organizational creativity and collaborative design* (pp. 177–189). Hershey, PA: IGI Global. doi:10.4018/978-1-60566-701-0.ch010

Bebensee, T., Helms, R., & Spruit, M. (2012). Exploring the impact of web 2.0 on knowledge management. In I. Boughzala & A. Dudezert (Eds.), *Knowledge management 2.0: Organizational models and enterprise strategies* (pp. 17–43). Hershey, PA: IGI Global. doi:10.4018/978-1-61350-195-5.ch002

Becerra-Fernandez, I., & Sabherwal, R. (2011). The role of information and communication technologies in knowledge management: A classification of knowledge management systems. In D. Schwartz & D. Te'eni (Eds.), *Encyclopedia of knowledge management* (2nd ed.; pp. 1410–1418). Hershey, PA: IGI Global. doi:10.4018/978-1-59904-931-1.ch134

Benbya, H. (2013). Valuing knowledge-based initiatives: What we know and what we don't know. In M. Jennex (Ed.), *Dynamic models for knowledge-driven organizations* (pp. 1–15). Hershey, PA: IGI Global. doi:10.4018/978-1-4666-2485-6.ch001

Berends, H., van der Bij, H., & Weggeman, M. (2011). Knowledge integration. In D. Schwartz & D. Te'eni (Eds.), *Encyclopedia of knowledge management* (2nd ed.; pp. 581–590). Hershey, PA: IGI Global. doi:10.4018/978-1-59904-931-1.ch056

Berger, H., & Beynon-Davies, P. (2011). Knowledge-based diffusion in practice: A case study experience. In A. Eardley & L. Uden (Eds.), *Innovative knowledge management: Concepts for organizational creativity and collaborative design* (pp. 40–55). Hershey, PA: IGI Global. doi:10.4018/978-1-60566-701-0.ch003

Berio, G., Di Leva, A., Harzallah, M., & Sacco, G. M. (2012). Competence management over social networks through dynamic taxonomies. In I. Boughzala & A. Dudezert (Eds.), *Knowledge management 2.0: Organizational models and enterprise strategies* (pp. 103–120). Hershey, PA: IGI Global. doi:10.4018/978-1-61350-195-5.ch006

Bhatt, S., Chaudhary, S., & Bhise, M. (2013). Migration of data between cloud and non-cloud datastores. In A. Ionita, M. Litoiu, & G. Lewis (Eds.), *Migrating legacy applications: Challenges in service oriented architecture and cloud computing environments* (pp. 206–225). Hershey, PA: IGI Global. doi:10.4018/978-1-4666-2488-7.ch009

Bloodgood, J. M., Chilton, M. A., & Bloodgood, T. C. (2014). The effect of knowledge transfer motivation, receiver capability, and motivation on organizational performance. In M. Chilton & J. Bloodgood (Eds.), *Knowledge management and competitive advantage: Issues and potential solutions* (pp. 232–242). Hershey, PA: IGI Global. doi:10.4018/978-1-4666-4679-7.ch013

Boersma, K., & Kingma, S. (2011). Organizational learning facilitation with intranet (2.0): A socio-cultural approach. In D. Schwartz & D. Te'eni (Eds.), *Encyclopedia of knowledge management* (2nd ed., pp. 1280–1289). Hershey, PA: IGI Global. doi:10.4018/978-1-59904-931-1.ch122

Bond, P. L. (2010). Toward a living systems framework for unifying technology and knowledge management, organizational, cultural and economic change. In D. Harorimana (Ed.), *Cultural implications of knowledge sharing, management and transfer: Identifying competitive advantage* (pp. 108–132). Hershey, PA: IGI Global. doi:10.4018/978-1-60566-790-4.ch006

Bordogna, G., Bucci, F., Carrara, P., Pepe, M., & Rampini, A. (2011). Flexible querying of imperfect temporal metadata in spatial data infrastructures. In L. Yan & Z. Ma (Eds.), *Advanced database query systems: Techniques, applications and technologies* (pp. 140–159). Hershey, PA: IGI Global. doi:10.4018/978-1-60960-475-2.ch006

Boughzala, I. (2012). Collaboration 2.0 through the new organization (2.0) transformation. In I. Boughzala & A. Dudezert (Eds.), *Knowledge management 2.0: Organizational models and enterprise strategies* (pp. 1–16). Hershey, PA: IGI Global. doi:10.4018/978-1-61350-195-5.ch001

Bratianu, C. (2011). A new perspective of the intellectual capital dynamics in organizations. In B. Vallejo-Alonso, A. Rodriguez-Castellanos, & G. Arregui-Ayastuy (Eds.), *Identifying, measuring, and valuing knowledge-based intangible assets: New perspectives* (pp. 1–21). Hershey, PA: IGI Global. doi:10.4018/978-1-60960-054-9.ch001

Bratianu, C. (2011). Universities as knowledge-intensive learning organizations. In A. Eardley & L. Uden (Eds.), *Innovative knowledge management: Concepts for organizational creativity and collaborative design* (pp. 1–17). Hershey, PA: IGI Global. doi:10.4018/978-1-60566-701-0.ch001

Breu, K., Ward, J., & Murray, P. (2000). Success factors in leveraging the corporate information and knowledge resource through intranets. In Y. Malhotra (Ed.), *Knowledge management and virtual organizations* (pp. 306–320). Hershey, PA: IGI Global. doi:10.4018/978-1-930708-65-5.ch016

Briones-Peñalver, A., & Poças-Rascão, J. (2014). Information technologies (ICT), network organizations, and information systems for business cooperation: A focus on organization and strategic knowledge management. In G. Jamil, A. Malheiro, & F. Ribeiro (Eds.), *Rethinking the conceptual base for new practical applications in information value and quality* (pp. 324–348). Hershey, PA: IGI Global. doi:10.4018/978-1-4666-4562-2.ch015

Brock, J. K., & Zhou, Y. J. (2011). MNE knowledge management across borders and ICT. In D. Schwartz & D. Te'eni (Eds.), *Encyclopedia of knowledge management* (2nd ed.; pp. 1136–1148). Hershey, PA: IGI Global. doi:10.4018/978-1-59904-931-1.ch109

Brunet-Thornton, R., & Bureš, V. (2011). Meeting Czech knowledge management challenges head-on: KM-Be.At-It. In M. Al-Shammari (Ed.), *Knowledge management in emerging economies: Social, organizational and cultural implementation* (pp. 20–46). Hershey, PA: IGI Global. doi:10.4018/978-1-61692-886-5.ch002

Bucher, T., & Dinter, B. (2012). Situational method engineering to support process-oriented information logistics: Identification of development situations. *Journal of Database Management, 23*(1), 31–48. doi:10.4018/jdm.2012010102

Burstein, F., & Linger, H. (2011). Task-based knowledge management approach. In D. Schwartz & D. Te'eni (Eds.), *Encyclopedia of knowledge management* (2nd ed.; pp. 1479–1489). Hershey, PA: IGI Global. doi:10.4018/978-1-59904-931-1.ch141

Butler, T. (2011). Anti-foundational knowledge management. In D. Schwartz & D. Te'eni (Eds.), *Encyclopedia of knowledge management* (2nd ed.; pp. 1–11). Hershey, PA: IGI Global. doi:10.4018/978-1-59904-931-1.ch001

Butler, T., & Murphy, C. (2011). Work and knowledge. In D. Schwartz & D. Te'eni (Eds.), *Encyclopedia of knowledge management* (2nd ed.; pp. 1556–1566). Hershey, PA: IGI Global. doi:10.4018/978-1-59904-931-1.ch148

Cabrilo, S., & Grubic-Nesic, L. (2013). The role of creativity, innovation, and invention in knowledge management. In S. Buckley & M. Jakovljevic (Eds.), *Knowledge management innovations for interdisciplinary education: Organizational applications* (pp. 207–232). Hershey, PA: IGI Global. doi:10.4018/978-1-4666-1969-2.ch011

Cabrita, M. D., Machado, V. C., & Grilo, A. (2010). Intellectual capital: How knowledge creates value. In E. O'Brien, S. Clifford, & M. Southern (Eds.), *Knowledge management for process, organizational and marketing innovation: Tools and methods* (pp. 237–252). Hershey, PA: IGI Global. doi:10.4018/978-1-61520-829-6.ch015

Cagliero, L., & Fiori, A. (2013). Knowledge discovery from online communities. In *Data mining: Concepts, methodologies, tools, and applications* (pp. 1230–1252). Hershey, PA: IGI Global. doi:10.4018/978-1-4666-2455-9.ch063

Camisón-Zornoza, C., & Boronat-Navarro, M. (2010). Linking exploration and exploitation capabilities with the process of knowledge development and with organizational facilitators. In M. Russ (Ed.), *Knowledge management strategies for business development* (pp. 159–179). Hershey, PA: IGI Global. doi:10.4018/978-1-60566-348-7.ch008

Carneiro, A. (2010). Change knowledge management: Transforming a ghost community into a real asset. In E. O'Brien, S. Clifford, & M. Southern (Eds.), *Knowledge management for process, organizational and marketing innovation: Tools and methods* (pp. 120–132). Hershey, PA: IGI Global. doi:10.4018/978-1-61520-829-6.ch007

Carrillo, F. J. (2010). Knowledge-based value generation. In K. Metaxiotis, F. Carrillo, & T. Yigitcanlar (Eds.), *Knowledge-based development for cities and societies: Integrated multi-level approaches* (pp. 1–16). Hershey, PA: IGI Global. doi:10.4018/978-1-61520-721-3.ch001

Cartelli, A. (2012). Frameworks for the benchmarking of digital and knowledge management best practice in SME and organizations. In A. Cartelli (Ed.), *Current trends and future practices for digital literacy and competence* (pp. 166–175). Hershey, PA: IGI Global. doi:10.4018/978-1-4666-0903-7.ch015

Castellano, G., Fanelli, A. M., & Torsello, M. A. (2010). Soft computing techniques in content-based multimedia information retrieval. In K. Anbumani & R. Nedunchezhian (Eds.), *Soft computing applications for database technologies: Techniques and issues* (pp. 170–192). Hershey, PA: IGI Global. doi:10.4018/978-1-60566-814-7.ch010

Chalkiti, K., & Carson, D. (2010). Knowledge cultures, competitive advantage and staff turnover in hospitality in Australia's northern territory. In D. Harorimana (Ed.), *Cultural implications of knowledge sharing, management and transfer: Identifying competitive advantage* (pp. 203–229). Hershey, PA: IGI Global. doi:10.4018/978-1-60566-790-4.ch010

Chang, W., & Li, S. (2011). Deploying knowledge management in R&D workspaces. In A. Eardley & L. Uden (Eds.), *Innovative knowledge management: Concepts for organizational creativity and collaborative design* (pp. 56–76). Hershey, PA: IGI Global. doi:10.4018/978-1-60566-701-0.ch004

Chawla, D., & Joshi, H. (2013). Impact of knowledge management dimensions on learning organization: Comparison across business excellence awarded and non-awarded indian organizations. In M. Jennex (Ed.), *Dynamic models for knowledge-driven organizations* (pp. 145–162). Hershey, PA: IGI Global. doi:10.4018/978-1-4666-2485-6.ch008

Chen, E. (2012). Web 2.0 social networking technologies and strategies for knowledge management. In I. Boughzala & A. Dudezert (Eds.), *Knowledge management 2.0: Organizational models and enterprise strategies* (pp. 84–102). Hershey, PA: IGI Global. doi:10.4018/978-1-61350-195-5.ch005

Chihara, K., & Nakamori, Y. (2013). Clarification of abilities and qualities of knowledge coordinators: The case of regional revitalization projects. In G. Yang (Ed.), *Multidisciplinary studies in knowledge and systems science* (pp. 1–17). Hershey, PA: IGI Global. doi:10.4018/978-1-4666-3998-0.ch001

Christidis, K., Papailiou, N., Apostolou, D., & Mentzas, G. (2011). Semantic interfaces for personal and social knowledge work. *International Journal of Knowledge-Based Organizations*, *1*(1), 61–77. doi:10.4018/ijkbo.2011010104

Christidis, K., Papailiou, N., Apostolou, D., & Mentzas, G. (2013). Semantic interfaces for personal and social knowledge work. In J. Wang (Ed.), *Intelligence methods and systems advancements for knowledge-based business* (pp. 213–230). Hershey, PA: IGI Global. doi:10.4018/978-1-4666-1873-2.ch012

Chua, C. E., Storey, V. C., & Chiang, R. H. (2012). Knowledge representation: A conceptual modeling approach. *Journal of Database Management*, *23*(1), 1–30. doi:10.4018/jdm.2012010101

Clinton, M. S., Merritt, K. L., & Murray, S. R. (2011). Facilitating knowledge transfer and the achievement of competitive advantage with corporate universities: An exploratory model based on media richness and type of knowledge to be transferred. In M. Jennex (Ed.), *Global aspects and cultural perspectives on knowledge management: Emerging dimensions* (pp. 329–345). Hershey, PA: IGI Global. doi:10.4018/978-1-60960-555-1.ch020

Colomb, R. M. (2013). Representation of action is a primary requirement in ontologies for interoperating information systems. In M. Nazir Ahmad, R. Colomb, & M. Abdullah (Eds.), *Ontology-based applications for enterprise systems and knowledge management* (pp. 68–76). Hershey, PA: IGI Global. doi:10.4018/978-1-4666-1993-7.ch004

Colucci, S., Di Noia, T., Di Sciascio, E., Donini, F. M., & Mongiello, M. (2011). Description logic-based resource retrieval. In D. Schwartz & D. Te'eni (Eds.), *Encyclopedia of knowledge management* (2nd ed.; pp. 185–197). Hershey, PA: IGI Global. doi:10.4018/978-1-59904-931-1.ch018

Connell, N. A. (2011). Organisational storytelling. In D. Schwartz & D. Te'eni (Eds.), *Encyclopedia of knowledge management* (2nd ed.; pp. 1261–1269). Hershey, PA: IGI Global. doi:10.4018/978-1-59904-931-1.ch120

Cooper, L. P., & Rober, M. B. (2012). Moving wikis behind the firewall: Intrapedias and work-wikis. In I. Boughzala & A. Dudezert (Eds.), *Knowledge Management 2.0: Organizational Models and Enterprise Strategies* (pp. 44–63). Hershey, PA: IGI Global. doi:10.4018/978-1-61350-195-5.ch003

Corallo, A., De Maggio, M., & Margherita, A. (2011). Knowledge democracy as the new mantra in product innovation: A framework of processes and competencies. In A. Eardley & L. Uden (Eds.), *Innovative knowledge management: Concepts for organizational creativity and collaborative design* (pp. 141–156). Hershey, PA: IGI Global. doi:10.4018/978-1-60566-701-0.ch008

Costa, G. (2011). Knowledge worker fair compensation: Ethical issues and social dilemmas. In G. Morais da Costa (Ed.), *Ethical issues and social dilemmas in knowledge management: Organizational innovation* (pp. 215–231). Hershey, PA: IGI Global. doi:10.4018/978-1-61520-873-9.ch013

Costello, R. (2014). Evaluating e-learning from an end user perspective. In M. Pańkowska (Ed.), *Frameworks of IT prosumption for business development* (pp. 259–283). Hershey, PA: IGI Global. doi:10.4018/978-1-4666-4313-0.ch017

Crasso, M., Zunino, A., & Campo, M. (2013). A survey of approaches to web service discovery in service-oriented architectures. In K. Siau (Ed.), *Innovations in database design, web applications, and information systems management* (pp. 107–138). Hershey, PA: IGI Global. doi:10.4018/978-1-4666-2044-5.ch005

Croasdell, D., & Wang, Y. K. (2011). Virtue-nets. In D. Schwartz & D. Te'eni (Eds.), *Encyclopedia of knowledge management* (2nd ed.; pp. 1545–1555). Hershey, PA: IGI Global. doi:10.4018/978-1-59904-931-1.ch147

Cucchiara, S., Ligorio, M. B., & Fujita, N. (2014). Understanding online discourse strategies for knowledge building through social network analysis. In H. Lim & F. Sudweeks (Eds.), *Innovative methods and technologies for electronic discourse analysis* (pp. 42–62). Hershey, PA: IGI Global. doi:10.4018/978-1-4666-4426-7.ch003

Cudanov, M., & Kirchner, K. (2011). Knowledge management in high-growth companies: A case study in Serbia. In M. Al-Shammari (Ed.), *Knowledge management in emerging economies: Social, organizational and cultural implementation* (pp. 227–248). Hershey, PA: IGI Global. doi:10.4018/978-1-61692-886-5.ch014

Cuel, R., Bouquet, P., & Bonifacio, M. (2011). Distributed knowledge management. In D. Schwartz & D. Te'eni (Eds.), *Encyclopedia of knowledge management* (2nd ed.; pp. 198–208). Hershey, PA: IGI Global. doi:10.4018/978-1-59904-931-1.ch019

Daidj, N. (2012). The evolution of KM practices: The case of the Renault-Nissan international strategic alliance. In I. Boughzala & A. Dudezert (Eds.), *Knowledge management 2.0: Organizational models and enterprise strategies* (pp. 190–213). Hershey, PA: IGI Global. doi:10.4018/978-1-61350-195-5.ch010

Daniel, B. K., Zapata-Rivera, J., & McCalla, G. I. (2010). A Bayesian belief network methodology for modeling social systems in virtual communities: Opportunities for database technologies. In K. Anbumani & R. Nedunchezhian (Eds.), *Soft computing applications for database technologies: Techniques and issues* (pp. 125–152). Hershey, PA: IGI Global. doi:10.4018/978-1-60566-814-7.ch008

Darchen, S., & Tremblay, D. (2010). Attracting and retaining knowledge workers: The impact of quality of place in the case of Montreal. In K. Metaxiotis, F. Carrillo, & T. Yigitcanlar (Eds.), *Knowledge-based development for cities and societies: Integrated multi-level approaches* (pp. 42–58). Hershey, PA: IGI Global. doi:10.4018/978-1-61520-721-3.ch003

Davenport, D. L., & Hosapple, C. W. (2011). Knowledge organizations. In D. Schwartz & D. Te'eni (Eds.), *Encyclopedia of knowledge management* (2nd ed.; pp. 822–832). Hershey, PA: IGI Global. doi:10.4018/978-1-59904-931-1.ch079

Davenport, D. L., & Hosapple, C. W. (2011). Social capital knowledge. In D. Schwartz & D. Te'eni (Eds.), *Encyclopedia of knowledge management* (2nd ed.; pp. 1448–1459). Hershey, PA: IGI Global. doi:10.4018/978-1-59904-931-1.ch138

De Maggio, M., Del Vecchio, P., Elia, G., & Grippa, F. (2011). An ICT-based network of competence centres for developing intellectual capital in the Mediterranean area. In A. Al Ajeeli & Y. Al-Bastaki (Eds.), *Handbook of research on e-services in the public sector: E-government strategies and advancements* (pp. 164–181). Hershey, PA: IGI Global. doi:10.4018/978-1-61520-789-3.ch014

Delbaere, M., Di Zhang, D., Bruning, E. R., & Sivaramakrishnan, S. (2014). Knowledge management and the roles it plays in achieving superior performance. In M. Chilton & J. Bloodgood (Eds.), *Knowledge management and competitive advantage: Issues and potential solutions* (pp. 90–108). Hershey, PA: IGI Global. doi:10.4018/978-1-4666-4679-7.ch006

Delgado-Verde, M., & Cruz-González, J. (2010). An intellectual capital-based view of technological innovation. In P. López Sáez, G. Castro, J. Navas López, & M. Delgado Verde (Eds.), *Intellectual capital and technological innovation: Knowledge-based theory and practice* (pp. 166–193). Hershey, PA: IGI Global. doi:10.4018/978-1-61520-875-3.ch008

Deltour, F., Plé, L., & Roussel, C. S. (2012). Knowledge sharing in the age of web 2.0: A social capital perspective. In I. Boughzala & A. Dudezert (Eds.), *Knowledge management 2.0: Organizational models and enterprise strategies* (pp. 122–141). Hershey, PA: IGI Global. doi:10.4018/978-1-61350-195-5.ch007

Derballa, V., & Pousttchi, K. (2011). Mobile technology for knowledge management. In D. Schwartz & D. Te'eni (Eds.), *Encyclopedia of knowledge management* (2nd ed.; pp. 1158–1166). Hershey, PA: IGI Global. doi:10.4018/978-1-59904-931-1.ch111

Dieng-Kuntz, R. (2011). Corporate semantic webs. In D. Schwartz & D. Te'eni (Eds.), *Encyclopedia of knowledge management* (2nd ed.; pp. 131–149). Hershey, PA: IGI Global. doi:10.4018/978-1-59904-931-1.ch014

Diosteanu, A., Stellato, A., & Turbati, A. (2012). SODA: A service oriented data acquisition framework. In M. Pazienza & A. Stellato (Eds.), *Semi-automatic ontology development: Processes and resources* (pp. 48–77). Hershey, PA: IGI Global. doi:10.4018/978-1-4666-0188-8.ch003

Donate-Manzanares, M. J., Guadamillas-Gómez, F., & Sánchez de Pablo, J. D. (2010). Strategic alliances and knowledge management strategies: A case study. In M. Russ (Ed.), *Knowledge management strategies for business development* (pp. 240–260). Hershey, PA: IGI Global. doi:10.4018/978-1-60566-348-7.ch011

Donnet, T., Keast, R., & Pickernell, D. (2010). Up the junction? Exploiting knowledge-based development through supply chain and SME cluster interactions. In K. Metaxiotis, F. Carrillo, & T. Yigitcanlar (Eds.), *Knowledge-based development for cities and societies: Integrated multi-level approaches* (pp. 179–195). Hershey, PA: IGI Global. doi:10.4018/978-1-61520-721-3.ch011

Douglas, I. (2011). Organizational needs analysis and knowledge management. In D. Schwartz & D. Te'eni (Eds.), *Encyclopedia of knowledge management* (2nd ed.; pp. 1290–1297). Hershey, PA: IGI Global. doi:10.4018/978-1-59904-931-1.ch123

Edvardsson, I. R., & Oskarsson, G. K. (2013). Outsourcing in knowledge-based service firms. In J. Wang (Ed.), *Intelligence methods and systems advancements for knowledge-based business* (pp. 97–113). Hershey, PA: IGI Global. doi:10.4018/978-1-4666-1873-2.ch006

Elenurm, T. (2013). Knowledge management and innovative learning. In S. Buckley & M. Jakovljevic (Eds.), *Knowledge management innovations for interdisciplinary education: Organizational applications* (pp. 108–131). Hershey, PA: IGI Global. doi:10.4018/978-1-4666-1969-2.ch006

Eppler, M. J., & Burkhard, R. A. (2011). Knowledge visualization. In D. Schwartz & D. Te'eni (Eds.), *Encyclopedia of knowledge management* (2nd ed.; pp. 987–999). Hershey, PA: IGI Global. doi:10.4018/978-1-59904-931-1.ch094

Ergazakis, K., Metaxiotis, K., & Ergazakis, E. (2011). Exploring paths towards knowledge cities developments: A research agenda. In D. Schwartz & D. Te'eni (Eds.), *Encyclopedia of knowledge management* (2nd ed.; pp. 288–297). Hershey, PA: IGI Global. doi:10.4018/978-1-59904-931-1.ch028

Eri, Z. D., Abdullah, R., Jabar, M. A., Murad, M. A., & Talib, A. M. (2013). Ontology-based virtual communities model for the knowledge management system environment: Ontology design. In M. Nazir Ahmad, R. Colomb, & M. Abdullah (Eds.), *Ontology-based applications for enterprise systems and knowledge management* (pp. 343–360). Hershey, PA: IGI Global. doi:10.4018/978-1-4666-1993-7.ch019

Erickson, G. S. (2014). Government as a partner in knowledge management: Lessons from the US freedom of information act. In Y. Al-Bastaki & A. Shajera (Eds.), *Building a competitive public sector with knowledge management strategy* (pp. 90–103). Hershey, PA: IGI Global. doi:10.4018/978-1-4666-4434-2.ch003

Erickson, G. S., & Rothberg, H. N. (2011). Assessing knowledge management needs: A strategic approach to developing knowledge. *International Journal of Knowledge Management*, *7*(3), 1–10. doi:10.4018/jkm.2011070101

Erickson, G. S., & Rothberg, H. N. (2011). Protecting knowledge assets. In D. Schwartz & D. Te'eni (Eds.), *Encyclopedia of knowledge management* (2nd ed.; pp. 1336–1342). Hershey, PA: IGI Global. doi:10.4018/978-1-59904-931-1.ch128

Erickson, G. S., & Rothberg, H. N. (2013). Assessing knowledge management needs: A strategic approach to developing knowledge. In M. Jennex (Ed.), *Dynamic models for knowledge-driven organizations* (pp. 180–189). Hershey, PA: IGI Global. doi:10.4018/978-1-4666-2485-6.ch010

Evermann, J., & Wand, Y. (2011). Ontology based object-oriented domain modeling: Representing behavior. In K. Siau (Ed.), *Theoretical and practical advances in information systems development: Emerging trends and approaches* (pp. 37–60). Hershey, PA: IGI Global. doi:10.4018/978-1-60960-521-6.ch003

Fadel, K. J., Durcikova, A., & Cha, H. S. (2011). An experiment of information elaboration in mediated knowledge transfer. In M. Jennex (Ed.), *Global aspects and cultural perspectives on knowledge management: Emerging dimensions* (pp. 311–328). Hershey, PA: IGI Global. doi:10.4018/978-1-60960-555-1.ch019

Fazel-Zarandi, M., Fox, M. S., & Yu, E. (2013). Ontologies in expertise finding systems: Modeling, analysis, and design. In M. Nazir Ahmad, R. Colomb, & M. Abdullah (Eds.), *Ontology-based applications for enterprise systems and knowledge management* (pp. 158–177). Hershey, PA: IGI Global. doi:10.4018/978-1-4666-1993-7.ch009

Ferri, F., & Grifoni, P. (2011). Sketching in knowledge creation and management. In D. Schwartz & D. Te'eni (Eds.), *Encyclopedia of knowledge management* (2nd ed.; pp. 1438–1447). Hershey, PA: IGI Global. doi:10.4018/978-1-59904-931-1.ch137

Filho, C. G., Baroni de Carvalho, R., & Jamil, G. L. (2011). Market knowledge management, innovation and product performance: Survey in medium and large Brazilian industrial firms. In M. Jennex & S. Smolnik (Eds.), *Strategies for knowledge management success: Exploring organizational efficacy* (pp. 32–50). Hershey, PA: IGI Global. doi:10.4018/978-1-60566-709-6.ch003

Fink, D., & Disterer, G. (2011). Knowledge management in professional service firms. In D. Schwartz & D. Te'eni (Eds.), *Encyclopedia of knowledge management* (2nd ed.; pp. 650–659). Hershey, PA: IGI Global. doi:10.4018/978-1-59904-931-1.ch063

Fink, K. (2011). Process model for knowledge potential measurement in SMEs. In M. Jennex & S. Smolnik (Eds.), *Strategies for knowledge management success: Exploring organizational efficacy* (pp. 91–105). Hershey, PA: IGI Global. doi:10.4018/978-1-60566-709-6.ch006

Fink, K., & Ploder, C. (2011). Knowledge management toolkit for SMEs. In M. Jennex (Ed.), *Global aspects and cultural perspectives on knowledge management: Emerging dimensions* (pp. 49–63). Hershey, PA: IGI Global. doi:10.4018/978-1-60960-555-1.ch004

Flynn, R., & Marshall, V. (2014). The four levers for change in knowledge management implementation. In Y. Al-Bastaki & A. Shajera (Eds.), *Building a competitive public sector with knowledge management strategy* (pp. 227–245). Hershey, PA: IGI Global. doi:10.4018/978-1-4666-4434-2.ch010

Fortier, J., & Kassel, G. (2011). Organizational semantic webs. In D. Schwartz & D. Te'eni (Eds.), *Encyclopedia of knowledge management* (2nd ed.; pp. 1298–1307). Hershey, PA: IGI Global. doi:10.4018/978-1-59904-931-1.ch124

Framinan, J. M., & Molina, J. M. (2010). An overview of enterprise resource planning for intelligent enterprises. In *Business information systems: Concepts, methodologies, tools and applications* (pp. 60–68). Hershey, PA: IGI Global. doi:10.4018/978-1-61520-969-9.ch005

Franco, M., Di Virgilio, F., & Di Pietro, L. (2014). Management of group knowledge and the role of E-WOM for business organizations. In M. Chilton & J. Bloodgood (Eds.), *Knowledge management and competitive advantage: Issues and potential solutions* (pp. 71–89). Hershey, PA: IGI Global. doi:10.4018/978-1-4666-4679-7.ch005

Franke, U. J. (2000). The knowledge-based view (KBV) of the virtual web, the virtual corporation and the net-broker. In Y. Malhotra (Ed.), *Knowledge management and virtual organizations* (pp. 20–42). Hershey, PA: IGI Global. doi:10.4018/978-1-930708-65-5.ch002

Freivalds, D., & Lush, B. (2012). Thinking inside the grid: Selecting a discovery system through the RFP process. In M. Popp & D. Dallis (Eds.), *Planning and implementing resource discovery tools in academic libraries* (pp. 104–121). Hershey, PA: IGI Global. doi:10.4018/978-1-4666-1821-3.ch007

Frieß, M. R., Groh, G., Reinhardt, M., Forster, F., & Schlichter, J. (2012). Context-aware creativity support for corporate open innovation. *International Journal of Knowledge-Based Organizations*, *2*(1), 38–55. doi:10.4018/ijkbo.2012010103

Fuller, C. M., & Wilson, R. L. (2011). Extracting knowledge from neural networks. In D. Schwartz & D. Te'eni (Eds.), *Encyclopedia of knowledge management* (2nd ed.; pp. 320–330). Hershey, PA: IGI Global. doi:10.4018/978-1-59904-931-1.ch031

Furquim, T. D., & do Amaral, S. A. (2011). Knowledge management practices in brazilian software organizations: The case of SERPRO. In M. Al-Shammari (Ed.), *Knowledge management in emerging economies: Social, organizational and cultural implementation* (pp. 213–226). Hershey, PA: IGI Global. doi:10.4018/978-1-61692-886-5.ch013

Gaál, Z., Szabó, L., Obermayer-Kovács, N., Kovács, Z., & Csepregi, A. (2011). Knowledge management profile: An innovative approach to map knowledge management practice. In A. Eardley & L. Uden (Eds.), *Innovative knowledge management: Concepts for organizational creativity and collaborative design* (pp. 253–263). Hershey, PA: IGI Global. doi:10.4018/978-1-60566-701-0.ch016

Ganguly, A., Mostashari, A., & Mansouri, M. (2013). Measuring knowledge management/knowledge sharing (KM/KS) efficiency and effectiveness in enterprise networks. In M. Jennex (Ed.), *Dynamic models for knowledge-driven organizations* (pp. 318–336). Hershey, PA: IGI Global. doi:10.4018/978-1-4666-2485-6.ch019

Gaumand, C., Chapdaniel, A., & Dudezert, A. (2012). Strategic knowledge management system framework for supply chain at an intra-organizational level. In I. Boughzala & A. Dudezert (Eds.), *Knowledge management 2.0: Organizational models and enterprise strategies* (pp. 142–163). Hershey, PA: IGI Global. doi:10.4018/978-1-61350-195-5.ch008

Ghazali, R., & Zakaria, N. H. (2013). Knowledge management processes in enterprise systems: A systematic literature review. In M. Nazir Ahmad, R. Colomb, & M. Abdullah (Eds.), *Ontology-based applications for enterprise systems and knowledge management* (pp. 1–24). Hershey, PA: IGI Global. doi:10.4018/978-1-4666-1993-7.ch001

Ghosh, B. (2011). Cross-cultural knowledge management practices to support offshore outsourcing. In M. Al-Shammari (Ed.), *Knowledge management in emerging economies: Social, organizational and cultural implementation* (pp. 249–260). Hershey, PA: IGI Global. doi:10.4018/978-1-61692-886-5.ch015

Gohil, U., Carrillo, P., Ruikar, K., & Anumba, C. (2013). Development of a business process model for a project-based service organisation. *International Journal of Knowledge-Based Organizations, 3*(1), 37–56. doi:10.4018/ijkbo.2013010103

Goldsmith, R. E., & Pillai, K. G. (2011). Knowledge calibration and knowledge management. In D. Schwartz & D. Te'eni (Eds.), *Encyclopedia of knowledge management* (2nd ed.; pp. 497–505). Hershey, PA: IGI Global. doi:10.4018/978-1-59904-931-1.ch048

Gomes de Andrade, F., & Baptista, C. D. (2013). An ontology-based approach to support information discovery in spatial data infrastructures. In C. Rückemann (Ed.), *Integrated information and computing systems for natural, spatial, and social sciences* (pp. 369–387). Hershey, PA: IGI Global. doi:10.4018/978-1-4666-2190-9.ch018

Gonçalo, C. R., & Jacques, E. J. (2010). Best practices of knowledge strategy in hospitals: A contextual perspective based on the implementation of medical protocols. In D. Harorimana (Ed.), *Cultural implications of knowledge sharing, management and transfer: Identifying competitive advantage* (pp. 180–202). Hershey, PA: IGI Global. doi:10.4018/978-1-60566-790-4.ch009

Górniak-Kocikowska, K. (2011). Knowledge management and democracy: A critical review of some moral issues and social dilemmas. In G. Morais da Costa (Ed.), *Ethical issues and social dilemmas in knowledge management: Organizational innovation* (pp. 28–44). Hershey, PA: IGI Global. doi:10.4018/978-1-61520-873-9.ch003

Gottschalk, P. (2014). Police knowledge management strategy. In M. Chilton & J. Bloodgood (Eds.), *Knowledge management and competitive advantage: Issues and potential solutions* (pp. 202–220). Hershey, PA: IGI Global. doi:10.4018/978-1-4666-4679-7.ch011

Goudos, S. K., Peristeras, V., & Tarabanis, K. (2010). Application of semantic web technology in e-business: Case studies in public domain data knowledge representation. In *Business information systems: Concepts, methodologies, tools and applications* (pp. 1223–1233). Hershey, PA: IGI Global. doi:10.4018/978-1-61520-969-9.ch075

Govindarajan, M., & Chandrasekaran, R. (2012). A hybrid multilayer perceptron neural network for direct marketing. *International Journal of Knowledge-Based Organizations*, *2*(3), 63–73. doi:10.4018/ijkbo.2012070104

Grant, J., & Minker, J. (2011). Logic and knowledge bases. In D. Schwartz & D. Te'eni (Eds.), *Encyclopedia of knowledge management* (2nd ed.; pp. 1022–1033). Hershey, PA: IGI Global. doi:10.4018/978-1-59904-931-1.ch097

Green, A. (2011). Engineering business reasoning, analytics and intelligence network (E-BRAIN): A new approach to intangible asset valuation based on Einstein's perspective. In B. Vallejo-Alonso, A. Rodriguez-Castellanos, & G. Arregui-Ayastuy (Eds.), *Identifying, measuring, and valuing knowledge-based intangible assets: New perspectives* (pp. 232–253). Hershey, PA: IGI Global. doi:10.4018/978-1-60960-054-9.ch011

Greenaway, K. E., & Vuong, D. C. (2010). Taking charities seriously: A call for focused knowledge management research. *International Journal of Knowledge Management*, *6*(4), 87–97. doi:10.4018/jkm.2010100105

Greenaway, K. E., & Vuong, D. C. (2012). Taking charities seriously: A call for focused knowledge management research. In M. Jennex (Ed.), *Conceptual models and outcomes of advancing knowledge management: New technologies* (pp. 333–344). Hershey, PA: IGI Global. doi:10.4018/978-1-4666-0035-5.ch017

Gunjal, B., Gaitanou, P., & Yasin, S. (2012). Social networks and knowledge management: An explorative study in library systems. In I. Boughzala & A. Dudezert (Eds.), *Knowledge management 2.0: Organizational models and enterprise strategies* (pp. 64–83). Hershey, PA: IGI Global. doi:10.4018/978-1-61350-195-5.ch004

Habhab-Rave, S. (2010). Knowledge management in SMEs: A mixture of innovation, marketing and ICT: Analysis of two case studies. In E. O'Brien, S. Clifford, & M. Southern (Eds.), *Knowledge management for process, organizational and marketing innovation: Tools and methods* (pp. 183–194). Hershey, PA: IGI Global. doi:10.4018/978-1-61520-829-6.ch011

Habicht, H., Möslein, K. M., & Reichwald, R. (2012). Open innovation maturity. *International Journal of Knowledge-Based Organizations*, *2*(1), 92–111. doi:10.4018/ijkbo.2012010106

Hamburg, I., & Hall, T. (2010). Readiness for knowledge management, methods and environments for innovation. In E. O'Brien, S. Clifford, & M. Southern (Eds.), *Knowledge management for process, organizational and marketing innovation: Tools and methods* (pp. 1–15). Hershey, PA: IGI Global. doi:10.4018/978-1-61520-829-6.ch001

Hamza, S. E. (2011). Capturing tacit knowledge from transient workers: Improving the organizational competitiveness. In M. Jennex (Ed.), *Global aspects and cultural perspectives on knowledge management: Emerging dimensions* (pp. 172–188). Hershey, PA: IGI Global. doi:10.4018/978-1-60960-555-1.ch012

Harorimana, D. (2010). Knowledge, culture, and cultural impact on knowledge management: Some lessons for researchers and practitioners. In D. Harorimana (Ed.), *Cultural implications of knowledge sharing, management and transfer: Identifying competitive advantage* (pp. 48–59). Hershey, PA: IGI Global. doi:10.4018/978-1-60566-790-4.ch003

Hasan, H. (2011). Formal and emergent standards in KM. In D. Schwartz & D. Te'eni (Eds.), *Encyclopedia of knowledge management* (2nd ed.; pp. 331–342). Hershey, PA: IGI Global. doi:10.4018/978-1-59904-931-1.ch032

He, G., Xue, G., Yu, K., & Yao, S. (2013). Business process modeling: Analysis and evaluation. In Z. Lu (Ed.), *Design, performance, and analysis of innovative information retrieval* (pp. 382–393). Hershey, PA: IGI Global. doi:10.4018/978-1-4666-1975-3.ch027

Heiman, B. A., & Hurmelinna-Laukkanen, P. (2010). Problem finding and solving: A knowledge-based view of managing innovation. In P. López Sáez, G. Castro, J. Navas López, & M. Delgado Verde (Eds.), *Intellectual capital and technological innovation: Knowledge-based theory and practice* (pp. 105–130). Hershey, PA: IGI Global. doi:10.4018/978-1-61520-875-3.ch005

Hendriks, P. H. (2011). Organizational structure. In D. Schwartz & D. Te'eni (Eds.), *Encyclopedia of knowledge management* (2nd ed.; pp. 1308–1318). Hershey, PA: IGI Global. doi:10.4018/978-1-59904-931-1.ch125

Hercheui, M. D. (2012). KMS for fostering behavior change: A case study on Microsoft Hohm. In I. Boughzala & A. Dudezert (Eds.), *Knowledge management 2.0: Organizational models and enterprise strategies* (pp. 214–232). Hershey, PA: IGI Global. doi:10.4018/978-1-61350-195-5.ch011

Hipkin, I. (2011). Perceptions of factors influencing knowledge-based technology management in conflict areas. In M. Al-Shammari (Ed.), *Knowledge management in emerging economies: Social, organizational and cultural implementation* (pp. 294–307). Hershey, PA: IGI Global. doi:10.4018/978-1-61692-886-5.ch018

Hofer, F. (2011). Knowledge transfer between academia and industry. In D. Schwartz & D. Te'eni (Eds.), *Encyclopedia of knowledge management* (2nd ed.; pp. 977–986). Hershey, PA: IGI Global. doi:10.4018/978-1-59904-931-1.ch093

Holjevac, I. A., Crnjar, K., & Hrgovic, A. V. (2013). Knowledge management and quality in Croatian tourism. In S. Buckley & M. Jakovljevic (Eds.), *Knowledge management innovations for interdisciplinary education: Organizational applications* (pp. 178–192). Hershey, PA: IGI Global. doi:10.4018/978-1-4666-1969-2.ch009

Holsapple, C. W., & Joshi, K. D. (2011). Knowledge management ontology. In D. Schwartz & D. Te'eni (Eds.), *Encyclopedia of knowledge management* (2nd ed.; pp. 704–711). Hershey, PA: IGI Global. doi:10.4018/978-1-59904-931-1.ch068

Holsapple, C. W., & Oh, J. (2014). Reactive and proactive dynamic capabilities: Using the knowledge chain theory of competitiveness. In M. Chilton & J. Bloodgood (Eds.), *Knowledge management and competitive advantage: Issues and potential solutions* (pp. 1–19). Hershey, PA: IGI Global. doi:10.4018/978-1-4666-4679-7.ch001

Huang, A., Xiao, J., & Wang, S. (2013). A combined forecast method integrating contextual knowledge. In G. Yang (Ed.), *Multidisciplinary studies in knowledge and systems science* (pp. 274–290). Hershey, PA: IGI Global. doi:10.4018/978-1-4666-3998-0.ch019

Huff, C. (2011). What does knowledge have to do with ethics? In G. Morais da Costa (Ed.), *Ethical issues and social dilemmas in knowledge management: Organizational innovation* (pp. 17–27). Hershey, PA: IGI Global. doi:10.4018/978-1-61520-873-9.ch002

Hürster, W., Wilbois, T., & Chaves, F. (2010). An integrated systems approach for early warning and risk management systems. *International Journal of Information Technologies and Systems Approach, 3*(2), 46–56. doi:10.4018/jitsa.2010070104

Iyer, S. R., Sharda, R., Biros, D., Lucca, J., & Shimp, U. (2011). Organization of lessons learned knowledge: A taxonomy and implementation. In M. Jennex (Ed.), *Global aspects and cultural perspectives on knowledge management: Emerging dimensions* (pp. 190–209). Hershey, PA: IGI Global. doi:10.4018/978-1-60960-555-1.ch013

Jacobson, C. M. (2011). Knowledge sharing between individuals. In D. Schwartz & D. Te'eni (Eds.), *Encyclopedia of knowledge management* (2nd ed.; pp. 924–934). Hershey, PA: IGI Global. doi:10.4018/978-1-59904-931-1.ch088

Jakovljevic, M. (2013). A conceptual model of creativity, invention, and innovation (MCII) for entrepreneurial engineers. In S. Buckley & M. Jakovljevic (Eds.), *Knowledge management innovations for interdisciplinary education: Organizational applications* (pp. 66–87). Hershey, PA: IGI Global. doi:10.4018/978-1-4666-1969-2.ch004

Jasimuddin, S. M., Connell, N., & Klein, J. H. (2011). Understanding organizational memory. In D. Schwartz & D. Te'eni (Eds.), *Encyclopedia of knowledge management* (2nd ed.; pp. 1536–1544). Hershey, PA: IGI Global. doi:10.4018/978-1-59904-931-1.ch146

Jennex, M. E. (2010). Do organizational memory and information technology interact to affect organizational information needs and provision? In *Ubiquitous developments in knowledge management: Integrations and trends* (pp. 1–20). Hershey, PA: IGI Global. doi:10.4018/978-1-60566-954-0

Jennex, M. E. (2010). Knowledge sharing model of 24-hour knowledge factory. In *Ubiquitous developments in knowledge management: Integrations and trends* (pp. 141–154). Hershey, PA: IGI Global. doi:10.4018/978-1-60566-954-0

Jennex, M. E. (2010). Operationalizing knowledge sharing for informers. In *Ubiquitous developments in knowledge management: Integrations and trends* (pp. 319–340). Hershey, PA: IGI Global. doi:10.4018/978-1-60566-954-0

Jennex, M. E. (2010). Qualitative pre-processing for semantic search of unstructured knowledge. In *Ubiquitous developments in knowledge management: Integrations and trends* (pp. 252–263). Hershey, PA: IGI Global. doi:10.4018/978-1-60566-954-0

Jennex, M. E. (2010). A specialized evaluation and comparison of sample data mining software. In *Ubiquitous developments in knowledge management: Integrations and trends* (pp. 300–318). Hershey, PA: IGI Global. doi:10.4018/978-1-60566-954-0

Jennex, M. E. (2010). Using soft systems methodology to reveal socio-technical barriers to knowledge sharing and management: A case study from the UK national health service. In *Ubiquitous developments in knowledge management: Integrations and trends* (pp. 215–235). Hershey, PA: IGI Global. doi:10.4018/978-1-60566-954-0

Jennex, M. E. (2011). Knowledge management success models. In D. Schwartz & D. Te'eni (Eds.), *Encyclopedia of knowledge management* (2nd ed.; pp. 763–771). Hershey, PA: IGI Global. doi:10.4018/978-1-59904-931-1.ch074

Jennex, M. E., & Olfman, L. (2011). A model of knowledge management success. In M. Jennex & S. Smolnik (Eds.), *Strategies for knowledge management success: Exploring organizational efficacy* (pp. 14–31). Hershey, PA: IGI Global. doi:10.4018/978-1-60566-709-6.ch002

Jennex, M. E., Smolnik, S., & Croasdell, D. (2011). Towards a consensus knowledge management success definition. In M. Jennex & S. Smolnik (Eds.), *Strategies for knowledge management success: Exploring organizational efficacy* (pp. 1–13). Hershey, PA: IGI Global. doi:10.4018/978-1-60566-709-6.ch001

Jewels, T. (2013). Teaching enterprise information systems in the United Arab Emirates. In F. Albadri (Ed.), *Information systems applications in the Arab education sector* (pp. 322–337). Hershey, PA: IGI Global. doi:10.4018/978-1-4666-1984-5.ch022

Jolly, R., & Wakeland, W. (2011). Using agent based simulation and game theory analysis to study knowledge flow in organizations: The KMscape. In M. Jennex (Ed.), *Global aspects and cultural perspectives on knowledge management: Emerging dimensions* (pp. 19–29). Hershey, PA: IGI Global. doi:10.4018/978-1-60960-555-1.ch002

Joshi, S. (2014). Web 2.0 and its implications on globally competitive business model. In M. Pańkowska (Ed.), *Frameworks of IT prosumption for business development* (pp. 86–101). Hershey, PA: IGI Global. doi:10.4018/978-1-4666-4313-0.ch007

Judge, R. (2011). A simulation system for evaluating knowledge management system (KMS) implementation strategies in small to mid-size enterprises (SME). In M. Jennex (Ed.), *Global aspects and cultural perspectives on knowledge management: Emerging dimensions* (pp. 92–112). Hershey, PA: IGI Global. doi:10.4018/978-1-60960-555-1.ch007

Kalid, K. S. (2011). Transfer knowledge using stories: A Malaysian university case study. In M. Al-Shammari (Ed.), *Knowledge management in emerging economies: Social, organizational and cultural implementation* (pp. 186–198). Hershey, PA: IGI Global. doi:10.4018/978-1-61692-886-5.ch011

Kamau, C. (2010). Strategising impression management in corporations: Cultural knowledge as capital. In D. Harorimana (Ed.), *Cultural implications of knowledge sharing, management and transfer: Identifying competitive advantage* (pp. 60–83). Hershey, PA: IGI Global. doi:10.4018/978-1-60566-790-4.ch004

Kamthan, P., & Fancott, T. (2011). A knowledge management model for patterns. In D. Schwartz & D. Te'eni (Eds.), *Encyclopedia of knowledge management* (2nd ed.; pp. 694–703). Hershey, PA: IGI Global. doi:10.4018/978-1-59904-931-1.ch067

Kamthan, P., & Pai, H. (2011). Knowledge representation in pattern management. In D. Schwartz & D. Te'eni (Eds.), *Encyclopedia of knowledge management* (2nd ed.; pp. 893–904). Hershey, PA: IGI Global. doi:10.4018/978-1-59904-931-1.ch085

Kane, G. C., Schwaig, K. S., & Storey, V. C. (2011). Information privacy: Understanding how firms behave online. In K. Siau (Ed.), *Theoretical and practical advances in information systems development: Emerging trends and approaches* (pp. 81–100). Hershey, PA: IGI Global. doi:10.4018/978-1-60960-521-6.ch005

Kankanhalli, A., Tan, B. C., & Wei, K. (2011). Knowledge producers and consumers. In D. Schwartz & D. Te'eni (Eds.), *Encyclopedia of knowledge management* (2nd ed.; pp. 867–877). Hershey, PA: IGI Global. doi:10.4018/978-1-59904-931-1.ch083

Karagiannis, D., Woitsch, R., & Hrgovcic, V. (2010). Industrialisation of the knowledge work: The knowledge conveyer belt approach. In E. O'Brien, S. Clifford, & M. Southern (Eds.), *Knowledge management for process, organizational and marketing innovation: Tools and methods* (pp. 79–94). Hershey, PA: IGI Global. doi:10.4018/978-1-61520-829-6.ch005

Karlsson, F., & Ågerfalk, P. J. (2011). Towards structured flexibility in information systems development: Devising a method for method configuration. In K. Siau (Ed.), *Theoretical and practical advances in information systems development: Emerging trends and approaches* (pp. 214–238). Hershey, PA: IGI Global. doi:10.4018/978-1-60960-521-6.ch010

Karna, A., Singh, R., & Verma, S. (2010). Knowledge management for an effective sales and marketing function. In M. Russ (Ed.), *Knowledge management strategies for business development* (pp. 324–337). Hershey, PA: IGI Global. doi:10.4018/978-1-60566-348-7.ch015

Kassim, A. M., & Cheah, Y. (2013). SEMblog: An ontology-based semantic blogging tool for knowledge identification, organization, and reuse. In M. Nazir Ahmad, R. Colomb, & M. Abdullah (Eds.), *Ontology-based applications for enterprise systems and knowledge management* (pp. 210–223). Hershey, PA: IGI Global. doi:10.4018/978-1-4666-1993-7.ch012

Kayakutlu, G. (2010). Knowledge worker profile: A framework to clarify expectations. In K. Metaxiotis, F. Carrillo, & T. Yigitcanlar (Eds.), *Knowledge-based development for cities and societies: Integrated multi-level approaches* (pp. 162–178). Hershey, PA: IGI Global. doi:10.4018/978-1-61520-721-3.ch010

Kettunen, J., & Chaudhuri, M. R. (2011). Knowledge management to promote organizational change in India. In M. Al-Shammari (Ed.), *Knowledge management in emerging economies: Social, organizational and cultural implementation* (pp. 308–324). Hershey, PA: IGI Global. doi:10.4018/978-1-61692-886-5.ch019

Khalil, O. E., & Seleim, A. (2012). Culture and knowledge transfer capacity: A cross-national study. In M. Jennex (Ed.), *Conceptual models and outcomes of advancing knowledge management: New technologies* (pp. 305–332). Hershey, PA: IGI Global. doi:10.4018/978-1-4666-0035-5.ch016

Khasawneh, R., & Alazzam, A. (2014). Towards customer knowledge management (CKM): Where knowledge and customer meet. In M. Chilton & J. Bloodgood (Eds.), *Knowledge management and competitive advantage: Issues and potential solutions* (pp. 109–121). Hershey, PA: IGI Global. doi:10.4018/978-1-4666-4679-7.ch007

Kim, J. (2014). Big data sharing among academics. In W. Hu & N. Kaabouch (Eds.), *Big data management, technologies, and applications* (pp. 177–194). Hershey, PA: IGI Global. doi:10.4018/978-1-4666-4699-5.ch008

Kim, S., Felan, J., & Kang, M. H. (2011). An ontological approach to enterprise knowledge modeling in a shipping company. *International Journal of Knowledge Management*, 7(4), 70–84. doi:10.4018/jkm.2011100105

Kim, S., Felan, J., & Kang, M. H. (2013). An ontological approach to enterprise knowledge modeling in a shipping company. In M. Jennex (Ed.), *Dynamic models for knowledge-driven organizations* (pp. 351–363). Hershey, PA: IGI Global. doi:10.4018/978-1-4666-2485-6.ch021

King, W. R. (2011). Knowledge transfer. In D. Schwartz & D. Te'eni (Eds.), *Encyclopedia of knowledge management* (2nd ed.; pp. 967–976). Hershey, PA: IGI Global. doi:10.4018/978-1-59904-931-1.ch092

Kivijärvi, H., Piirainen, K., & Tuominen, M. (2010). Sustaining organizational innovativeness: Advancing knowledge sharing during the scenario process. *International Journal of Knowledge Management, 6*(2), 22–39. doi:10.4018/jkm.2010040102

Kivijärvi, H., Piirainen, K., & Tuominen, M. (2012). Sustaining organizational innovativeness: Advancing knowledge sharing during the scenario process. In M. Jennex (Ed.), *Conceptual models and outcomes of advancing knowledge management: New technologies* (pp. 99–117). Hershey, PA: IGI Global. doi:10.4018/978-1-4666-0035-5.ch006

Knyazhansky, M., & Plotkin, T. (2012). Knowledge bases over algebraic models: Some notes about informational equivalence. *International Journal of Knowledge Management, 8*(1), 22–39. doi:10.4018/jkm.2012010102

Kong, E. (2014). The role of social intelligence in acquiring external knowledge for human capital development, organisational learning, and innovation. In M. Chilton & J. Bloodgood (Eds.), *Knowledge management and competitive advantage: Issues and potential solutions* (pp. 53–70). Hershey, PA: IGI Global. doi:10.4018/978-1-4666-4679-7.ch004

Kor, A., & Orange, G. (2011). A survey of epistemology and its implications on an organisational information and knowledge management model. In A. Eardley & L. Uden (Eds.), *Innovative knowledge management: Concepts for organizational creativity and collaborative design* (pp. 95–124). Hershey, PA: IGI Global. doi:10.4018/978-1-60566-701-0.ch006

Kostrzewa, A., Laaksoharju, M., & Kavathatzopoulos, I. (2011). Management of moral knowledge and ethical processes in organizations. In G. Morais da Costa (Ed.), *Ethical issues and social dilemmas in knowledge management: Organizational innovation* (pp. 199–214). Hershey, PA: IGI Global. doi:10.4018/978-1-61520-873-9.ch012

Kraaijenbrink, J., & Wijnhoven, F. (2011). External knowledge integration. In D. Schwartz & D. Te'eni (Eds.), *Encyclopedia of knowledge management* (2nd ed.; pp. 308–319). Hershey, PA: IGI Global. doi:10.4018/978-1-59904-931-1.ch030

Kraft, T. A., & Steenkamp, A. L. (2012). A holistic approach for understanding project management. In F. Stowell (Ed.), *Systems approach applications for developments in information technology* (pp. 25–39). Hershey, PA: IGI Global. doi:10.4018/978-1-4666-1562-5.ch003

Kulkarni, U., & Freeze, R. (2011). Measuring knowledge management capabilities. In D. Schwartz & D. Te'eni (Eds.), *Encyclopedia of knowledge management* (2nd ed.; pp. 1090–1100). Hershey, PA: IGI Global. doi:10.4018/978-1-59904-931-1.ch104

Kumar, A. S., Alrabea, A., & Sekhar, P. C. (2013). Temporal association rule mining in large databases. In *Data mining: Concepts, methodologies, tools, and applications* (pp. 586–602). Hershey, PA: IGI Global. doi:10.4018/978-1-4666-2455-9.ch029

Laihonen, H., & Koivuaho, M. (2011). Knowledge flow audit: indentifying, measuring and managing knowledge asset dynamics. In B. Vallejo-Alonso, A. Rodriguez-Castellanos, & G. Arregui-Ayastuy (Eds.), *Identifying, measuring, and valuing knowledge-based intangible assets: New perspectives* (pp. 22–42). Hershey, PA: IGI Global. doi:10.4018/978-1-60960-054-9.ch002

Land, F., Amjad, U., & Nolas, S. (2011). Knowledge management processes. In D. Schwartz & D. Te'eni (Eds.), *Encyclopedia of knowledge management* (2nd ed.; pp. 719–727). Hershey, PA: IGI Global. doi:10.4018/978-1-59904-931-1.ch070

Lavanderos, L. P., & Fiol, E. S. (2011). Production cognitive capital as a measurement of intellectual capital. In B. Vallejo-Alonso, A. Rodriguez-Castellanos, & G. Arregui-Ayastuy (Eds.), *Identifying, measuring, and valuing knowledge-based intangible assets: New perspectives* (pp. 112–132). Hershey, PA: IGI Global. doi:10.4018/978-1-60960-054-9.ch006

Lavoué, É., George, S., & Prévôt, P. (2011). A knowledge management tool for the interconnection of communities of practice. *International Journal of Knowledge Management, 7*(1), 55–76. doi:10.4018/jkm.2011010104

Lee, H., Chan, K., & Tsui, E. (2013). Knowledge mining Wikipedia: An ontological approach. In G. Yang (Ed.), *Multidisciplinary studies in knowledge and systems science* (pp. 52–62). Hershey, PA: IGI Global. doi:10.4018/978-1-4666-3998-0.ch005

Leung, N. K. (2011). A re-distributed knowledge management framework in help desk. In D. Schwartz & D. Te'eni (Eds.), *Encyclopedia of knowledge management* (2nd ed.; pp. 1374–1381). Hershey, PA: IGI Global. doi:10.4018/978-1-59904-931-1.ch131

Li, Y., Guo, H., & Wang, S. (2010). A multiple-bits watermark for relational data. In K. Siau & J. Erickson (Eds.), *Principle advancements in database management technologies: New applications and frameworks* (pp. 1–22). Hershey, PA: IGI Global. doi:10.4018/978-1-60566-904-5.ch001

Lin, C. Y. (2013). Intellectual capital explains a country's resilience to financial crisis: A resource-based view. In P. Ordóñez de Pablos, R. Tennyson, & J. Zhao (Eds.), *Intellectual capital strategy management for knowledge-based organizations* (pp. 52–75). Hershey, PA: IGI Global. doi:10.4018/978-1-4666-3655-2.ch005

Lin, Y., & Dalkir, K. (2012). Factors affecting KM implementation in the Chinese community. In M. Jennex (Ed.), *Conceptual models and outcomes of advancing knowledge management: New technologies* (pp. 1–23). Hershey, PA: IGI Global. doi:10.4018/978-1-4666-0035-5.ch001

Lindsey, K. L. (2011). Barriers to knowledge sharing. In D. Schwartz & D. Te'eni (Eds.), *Encyclopedia of knowledge management* (2nd ed.; pp. 49–61). Hershey, PA: IGI Global. doi:10.4018/978-1-59904-931-1.ch006

Liu, K., Tan, H. B., & Chen, X. (2013). Aiding maintenance of database applications through extracting attribute dependency graph. *Journal of Database Management*, *24*(1), 20–35. doi:10.4018/jdm.2013010102

Liu, K., Tan, H. B., & Chen, X. (2013). Automated insertion of exception handling for key and referential constraints. *Journal of Database Management*, *24*(1), 1–19. doi:10.4018/jdm.2013010101

Locuratolo, E., & Palomäki, J. (2013). Ontology for database preservation. In M. Nazir Ahmad, R. Colomb, & M. Abdullah (Eds.), *Ontology-based applications for enterprise systems and knowledge management* (pp. 141–157). Hershey, PA: IGI Global. doi:10.4018/978-1-4666-1993-7.ch008

López-Nicolás, C., & Meroño-Cerdán, Á. L. (2010). A model for knowledge management and intellectual capital audits. In M. Russ (Ed.), *Knowledge management strategies for business development* (pp. 115–131). Hershey, PA: IGI Global. doi:10.4018/978-1-60566-348-7.ch006

Luck, D. (2010). The implications of the development and implementation of CRM for knowledge management. In M. Russ (Ed.), *Knowledge management strategies for business development* (pp. 338–352). Hershey, PA: IGI Global. doi:10.4018/978-1-60566-348-7.ch016

Lukovic, I., Ivancevic, V., Celikovic, M., & Aleksic, S. (2014). DSLs in action with model based approaches to information system development. In *Software design and development: Concepts, methodologies, tools, and applications* (pp. 596–626). Hershey, PA: IGI Global. doi:10.4018/978-1-4666-4301-7.ch029

Luna-Reyes, L. F., & Gil-Garcia, J. R. (2012). Government and inter-organizational collaboration as strategies for administrative reform in Mexico. In T. Papadopoulos & P. Kanellis (Eds.), *Public sector reform using information technologies: Transforming policy into practice* (pp. 79–101). Hershey, PA: IGI Global. doi:10.4018/978-1-60960-839-2.ch005

Lungu, C. I., Caraiani, C., & Dascalu, C. (2013). Sustainable intellectual capital: The inference of corporate social responsibility within intellectual capital. In P. Ordóñez de Pablos, R. Tennyson, & J. Zhao (Eds.), *Intellectual capital strategy management for knowledge-based organizations* (pp. 156–173). Hershey, PA: IGI Global. doi:10.4018/978-1-4666-3655-2.ch009

Ma, Z. M. (2011). Engineering design knowledge management. In D. Schwartz & D. Te'eni (Eds.), *Encyclopedia of knowledge management* (2nd ed.; pp. 263–269). Hershey, PA: IGI Global. doi:10.4018/978-1-59904-931-1.ch025

Maier, R., & Hadrich, T. (2011). Knowledge management systems. In D. Schwartz & D. Te'eni (Eds.), *Encyclopedia of knowledge management* (2nd ed.; pp. 779–790). Hershey, PA: IGI Global. doi:10.4018/978-1-59904-931-1.ch076

Maria, E. D., & Micelli, S. (2010). SMEs and competitive advantage: A mix of innovation, marketing and ICT—The case of "made in Italy". In M. Russ (Ed.), *Knowledge management strategies for business development* (pp. 310–323). Hershey, PA: IGI Global. doi:10.4018/978-1-60566-348-7.ch014

Mariano, S., & Simionato, N. (2010). Where are we looking? A practical approach to managing knowledge captured from eye-tracking experiments: The experience of gulf air. In E. O'Brien, S. Clifford, & M. Southern (Eds.), *Knowledge management for process, organizational and marketing innovation: Tools and methods* (pp. 216–227). Hershey, PA: IGI Global. doi:10.4018/978-1-61520-829-6.ch013

Marques, M. B. (2014). The value of information and information services in knowledge society. In G. Jamil, A. Malheiro, & F. Ribeiro (Eds.), *Rethinking the conceptual base for new practical applications in information value and quality* (pp. 134–161). Hershey, PA: IGI Global. doi:10.4018/978-1-4666-4562-2.ch007

Masrom, M., Mahmood, N. H., & Al-Araimi, A. A. (2014). Exploring knowledge types and knowledge protection in organizations. In M. Chilton & J. Bloodgood (Eds.), *Knowledge management and competitive advantage: Issues and potential solutions* (pp. 271–280). Hershey, PA: IGI Global. doi:10.4018/978-1-4666-4679-7.ch016

Masterson, F. (2013). Knowledge management in practice: Using wikis to facilitate project-based learning. In S. Buckley & M. Jakovljevic (Eds.), *Knowledge management innovations for interdisciplinary education: Organizational applications* (pp. 385–401). Hershey, PA: IGI Global. doi:10.4018/978-1-4666-1969-2.ch019

Mattmann, C. A., Hart, A., Cinquini, L., Lazio, J., Khudikyan, S., & Jones, D. … Robnett, J. (2014). Scalable data mining, archiving, and big data management for the next generation astronomical telescopes. In W. Hu, & N. Kaabouch (Eds.), Big data management, technologies, and applications (pp. 196-221). Hershey, PA: IGI Global. doi:10.4018/978-1-4666-4699-5.ch009

Maule, R. W. (2011). Military knowledge management. In D. Schwartz & D. Te'eni (Eds.), *Encyclopedia of knowledge management* (2nd ed.; pp. 1125–1135). Hershey, PA: IGI Global. doi:10.4018/978-1-59904-931-1.ch108

Mavridis, I. (2011). Deploying privacy improved RBAC in web information systems. *International Journal of Information Technologies and Systems Approach*, *4*(2), 70–87. doi:10.4018/jitsa.2011070105

Mavridis, I. (2012). Deploying privacy improved RBAC in web information systems. In F. Stowell (Ed.), *Systems approach applications for developments in information technology* (pp. 298–315). Hershey, PA: IGI Global. doi:10.4018/978-1-4666-1562-5.ch020

McLaughlin, S. (2011). Assessing the impact of knowledge transfer mechanisms on supply chain performance. In M. Jennex (Ed.), *Global aspects and cultural perspectives on knowledge management: Emerging dimensions* (pp. 157–171). Hershey, PA: IGI Global. doi:10.4018/978-1-60960-555-1.ch011

Medina, J. M., & Spinola, M. D. (2011). Understanding the behavior of knowledge management pathways: The case of small manufacturers of footwear in Peru and Brazil. In M. Al-Shammari (Ed.), *Knowledge management in emerging economies: Social, organizational and cultural implementation* (pp. 261–271). Hershey, PA: IGI Global. doi:10.4018/978-1-61692-886-5.ch016

Meloche, J. A., Hasan, H., Willis, D., Pfaff, C. C., & Qi, Y. (2011). Cocreating corporate knowledge with a wiki. In M. Jennex (Ed.), *Global aspects and cultural perspectives on knowledge management: Emerging dimensions* (pp. 126–143). Hershey, PA: IGI Global. doi:10.4018/978-1-60960-555-1.ch009

Melzer, S. (2013). On the relationship between ontology-based and holistic representations in a knowledge management system. In M. Nazir Ahmad, R. Colomb, & M. Abdullah (Eds.), *Ontology-based applications for enterprise systems and knowledge management* (pp. 292–323). Hershey, PA: IGI Global. doi:10.4018/978-1-4666-1993-7.ch017

Mendes, E., & Baker, S. (2013). Using knowledge management and aggregation techniques to improve web effort estimation. In S. Saeed & I. Alsmadi (Eds.), *Knowledge-based processes in software development* (pp. 64–85). Hershey, PA: IGI Global. doi:10.4018/978-1-4666-4229-4.ch005

Metaxiotis, K. (2011). Healthcare knowledge management. In D. Schwartz & D. Te'eni (Eds.), *Encyclopedia of knowledge management* (2nd ed.; pp. 366–375). Hershey, PA: IGI Global. doi:10.4018/978-1-59904-931-1.ch035

Mikolajuk, Z. (2013). Community-based development of knowledge products. In M. Jennex (Ed.), *Dynamic models for knowledge-driven organizations* (pp. 268–281). Hershey, PA: IGI Global. doi:10.4018/978-1-4666-2485-6.ch016

Mischo, W. H., Schlembach, M. C., Bishoff, J., & German, E. M. (2012). User search activities within an academic library gateway: Implications for web-scale discovery systems. In M. Popp & D. Dallis (Eds.), *Planning and implementing resource discovery tools in academic libraries* (pp. 153–173). Hershey, PA: IGI Global. doi:10.4018/978-1-4666-1821-3.ch010

Mishra, B., & Shukla, K. K. (2014). Data mining techniques for software quality prediction. In *Software design and development: Concepts, methodologies, tools, and applications* (pp. 401–428). Hershey, PA: IGI Global. doi:10.4018/978-1-4666-4301-7.ch021

Moffett, S., Walker, T., & McAdam, R. (2014). Best value and performance management inspired change within UK councils: A knowledge management perspective. In Y. Al-Bastaki & A. Shajera (Eds.), *Building a competitive public sector with knowledge management strategy* (pp. 199–226). Hershey, PA: IGI Global. doi:10.4018/978-1-4666-4434-2.ch009

Mueller, C. E., & Bradley, K. D. (2011). Utilizing the Rasch model to develop and evaluate items for the tacit knowledge inventory for superintendents (TKIS). In M. Jennex (Ed.), *Global aspects and cultural perspectives on knowledge management: Emerging dimensions* (pp. 264–284). Hershey, PA: IGI Global. doi:10.4018/978-1-60960-555-1.ch017

Muhammed, S., Doll, W. J., & Deng, X. (2011). Impact of knowledge management practices on task knowledge: An individual level study. *International Journal of Knowledge Management, 7*(4), 1–21. doi:10.4018/jkm.2011100101

Muhammed, S., Doll, W. J., & Deng, X. (2011). Measuring knowledge management outcomes at the individual level: Towards a tool for research on organizational culture. In M. Jennex (Ed.), *Global aspects and cultural perspectives on knowledge management: Emerging dimensions* (pp. 1–18). Hershey, PA: IGI Global. doi:10.4018/978-1-60960-555-1.ch001

Muhammed, S., Doll, W. J., & Deng, X. (2013). Impact of knowledge management practices on task knowledge: An individual level study. In M. Jennex (Ed.), *Dynamic models for knowledge-driven organizations* (pp. 282–301). Hershey, PA: IGI Global. doi:10.4018/978-1-4666-2485-6.ch017

Murata, K. (2011). Knowledge creation and sharing in Japanese organisations: A socio-cultural perspective on ba. In G. Morais da Costa (Ed.), *Ethical issues and social dilemmas in knowledge management: Organizational innovation* (pp. 1–16). Hershey, PA: IGI Global. doi:10.4018/978-1-61520-873-9.ch001

Murphy, P. (2013). Systems of communication: Information, explanation, and imagination. In G. Yang (Ed.), *Multidisciplinary studies in knowledge and systems science* (pp. 63–78). Hershey, PA: IGI Global. doi:10.4018/978-1-4666-3998-0.ch006

Nach, H. (2013). Structuring knowledge for enterprise resource planning implementation through an ontology. In M. Nazir Ahmad, R. Colomb, & M. Abdullah (Eds.), *Ontology-based applications for enterprise systems and knowledge management* (pp. 25–42). Hershey, PA: IGI Global. doi:10.4018/978-1-4666-1993-7.ch002

Nah, F. F., Hong, W., Chen, L., & Lee, H. (2010). Information search patterns in e-commerce product comparison services. *Journal of Database Management, 21*(2), 26–40. doi:10.4018/jdm.2010040102

Nah, F. F., Hong, W., Chen, L., & Lee, H. (2012). Information search patterns in e-commerce product comparison services. In K. Siau (Ed.), *Cross-disciplinary models and applications of database management: Advancing approaches* (pp. 131–145). Hershey, PA: IGI Global. doi:10.4018/978-1-61350-471-0.ch006

Natarajan, R., & Shekar, B. (2011). Knowledge patterns in databases. In D. Schwartz & D. Te'eni (Eds.), *Encyclopedia of knowledge management* (2nd ed.; pp. 842–852). Hershey, PA: IGI Global. doi:10.4018/978-1-59904-931-1.ch081

Nelson, R. E., & Hsu, H. S. (2011). A social network perspective on knowledge management. In D. Schwartz & D. Te'eni (Eds.), *Encyclopedia of knowledge management* (2nd ed.; pp. 1470–1478). Hershey, PA: IGI Global. doi:10.4018/978-1-59904-931-1.ch140

Neto, R. C., & Souza, R. R. (2010). Knowledge management as an organizational process: From a theoretical framework to implementation guidelines. In E. O'Brien, S. Clifford, & M. Southern (Eds.), *Knowledge management for process, organizational and marketing innovation: Tools and methods* (pp. 16–35). Hershey, PA: IGI Global. doi:10.4018/978-1-61520-829-6.ch002

Newell, S. (2011). Understanding innovation processes. In D. Schwartz & D. Te'eni (Eds.), *Encyclopedia of knowledge management* (2nd ed.; pp. 1525–1535). Hershey, PA: IGI Global. doi:10.4018/978-1-59904-931-1.ch145

Nikabadi, M. S., & Zamanloo, S. (2012). A multidimensional structure for describing the influence of supply chain strategies, business strategies, and knowledge management strategies on knowledge sharing in supply chain. *International Journal of Knowledge Management*, *8*(4), 50–70. doi:10.4018/jkm.2012100103

Nissen, M. (2014). Cyberspace and cloud knowledge. In *Harnessing dynamic knowledge principles in the technology-driven world* (pp. 193–204). Hershey, PA: IGI Global. doi:10.4018/978-1-4666-4727-5.ch012

Nissen, M. (2014). Social media knowledge. In *Harnessing dynamic knowledge principles in the technology-driven world* (pp. 219–227). Hershey, PA: IGI Global. doi:10.4018/978-1-4666-4727-5.ch014

Nissen, M. E. (2014). Harnessing knowledge power for competitive advantage. In M. Chilton & J. Bloodgood (Eds.), *Knowledge management and competitive advantage: Issues and potential solutions* (pp. 20–34). Hershey, PA: IGI Global. doi:10.4018/978-1-4666-4679-7.ch002

Nissen, M. E., & Levitt, R. E. (2011). Knowledge management research through computational experimentation. In D. Schwartz & D. Te'eni (Eds.), *Encyclopedia of knowledge management* (2nd ed.; pp. 728–737). Hershey, PA: IGI Global. doi:10.4018/978-1-59904-931-1.ch071

Nisula, A. (2014). Developing organizational renewal capability in the municipal (city) organization. In Y. Al-Bastaki & A. Shajera (Eds.), *Building a competitive public sector with knowledge management strategy* (pp. 151–172). Hershey, PA: IGI Global. doi:10.4018/978-1-4666-4434-2.ch007

Niu, B., Martin, P., & Powley, W. (2011). Towards autonomic workload management in DBMSs. In K. Siau (Ed.), *Theoretical and practical advances in information systems development: Emerging trends and approaches* (pp. 154–173). Hershey, PA: IGI Global. doi:10.4018/978-1-60960-521-6.ch008

Nobre, F. S., & Walker, D. S. (2011). A dynamic ability-based view of the organization. *International Journal of Knowledge Management, 7*(2), 86–101. doi:10.4018/jkm.2011040105

O'Brien, J. (2014). Lessons from the private sector: A framework to be adopted in the public sector. In Y. Al-Bastaki & A. Shajera (Eds.), *Building a competitive public sector with knowledge management strategy* (pp. 173–198). Hershey, PA: IGI Global. doi:10.4018/978-1-4666-4434-2.ch008

Omari, A. (2013). Supporting companies management and improving their productivity through mining customers transactions. In *Data mining: Concepts, methodologies, tools, and applications* (pp. 1519–1533). Hershey, PA: IGI Global. doi:10.4018/978-1-4666-2455-9.ch079

Onwubiko, C. (2014). Modelling situation awareness information and system requirements for the mission using goal-oriented task analysis approach. In *Software design and development: Concepts, methodologies, tools, and applications* (pp. 460–478). Hershey, PA: IGI Global. doi:10.4018/978-1-4666-4301-7.ch023

Orth, A., Smolnik, S., & Jennex, M. E. (2011). The relevance of integration for knowledge management success: Towards conceptual and empirical evidence. In M. Jennex & S. Smolnik (Eds.), *Strategies for knowledge management success: Exploring organizational efficacy* (pp. 238–261). Hershey, PA: IGI Global. doi:10.4018/978-1-60566-709-6.ch013

Othman, A. K., & Abdullah, H. S. (2011). The influence of emotional intelligence on tacit knowledge sharing in service organizations. In M. Al-Shammari (Ed.), *Knowledge management in emerging economies: Social, organizational and cultural implementation* (pp. 171–185). Hershey, PA: IGI Global. doi:10.4018/978-1-61692-886-5.ch010

Pagallo, U. (2011). The trouble with digital copies: A short KM phenomenology. In G. Morais da Costa (Ed.), *Ethical issues and social dilemmas in knowledge management: Organizational innovation* (pp. 97–112). Hershey, PA: IGI Global. doi:10.4018/978-1-61520-873-9.ch007

Palte, R., Hertlein, M., Smolnik, S., & Riempp, G. (2013). The effects of a KM strategy on KM performance in professional services firms. In M. Jennex (Ed.), *Dynamic models for knowledge-driven organizations* (pp. 16–35). Hershey, PA: IGI Global. doi:10.4018/978-1-4666-2485-6.ch002

Pańkowska, M. (2014). Information technology prosumption acceptance by business information system consultants. In M. Pańkowska (Ed.), *Frameworks of IT prosumption for business development* (pp. 119–141). Hershey, PA: IGI Global. doi:10.4018/978-1-4666-4313-0.ch009

Pankowski, T. (2011). Pattern-based schema mapping and query answering in peer-to-peer XML data integration system. In L. Yan & Z. Ma (Eds.), *Advanced database query systems: Techniques, applications and technologies* (pp. 221–246). Hershey, PA: IGI Global. doi:10.4018/978-1-60960-475-2.ch009

Papoutsakis, H. (2010). New product development based on knowledge creation and technology education. In E. O'Brien, S. Clifford, & M. Southern (Eds.), *Knowledge management for process, organizational and marketing innovation: Tools and methods* (pp. 148–163). Hershey, PA: IGI Global. doi:10.4018/978-1-61520-829-6.ch009

Paquette, S. (2011). Applying knowledge management in the environmental and climate change sciences. In D. Schwartz & D. Te'eni (Eds.), *Encyclopedia of knowledge management* (2nd ed.; pp. 20–26). Hershey, PA: IGI Global. doi:10.4018/978-1-59904-931-1.ch003

Paquette, S. (2011). Customer knowledge management. In D. Schwartz & D. Te'eni (Eds.), *Encyclopedia of knowledge management* (2nd ed.; pp. 175–184). Hershey, PA: IGI Global. doi:10.4018/978-1-59904-931-1.ch017

Parker, K. R., & Nitse, P. S. (2011). Competitive intelligence gathering. In D. Schwartz & D. Te'eni (Eds.), *Encyclopedia of knowledge management* (2nd ed.; pp. 103–111). Hershey, PA: IGI Global. doi:10.4018/978-1-59904-931-1.ch011

Páscoa, C., & Tribolet, J. (2014). Maintaining organizational viability and performance: The organizational configuration map. In G. Jamil, A. Malheiro, & F. Ribeiro (Eds.), *Rethinking the conceptual base for new practical applications in information value and quality* (pp. 266–283). Hershey, PA: IGI Global. doi:10.4018/978-1-4666-4562-2.ch012

Paukert, M., Niederée, C., & Hemmje, M. (2011). Knowledge in innovation processes. In D. Schwartz & D. Te'eni (Eds.), *Encyclopedia of knowledge management* (2nd ed.; pp. 570–580). Hershey, PA: IGI Global. doi:10.4018/978-1-59904-931-1.ch055

Pawlak, P. (2011). Global "knowledge management" in humanist perspective. In G. Morais da Costa (Ed.), Ethical issues and social dilemmas in knowledge management: Organizational innovation (pp. 45-62). Hershey, PA: IGI Global. doi:10.4018/978-1-61520-873-9.ch004

Perry, M. (2013). Strategic knowledge management: A university application. In S. Buckley & M. Jakovljevic (Eds.), *Knowledge management innovations for interdisciplinary education: Organizational applications* (pp. 132–144). Hershey, PA: IGI Global. doi:10.4018/978-1-4666-1969-2.ch007

Pessoa, C. R., Silva, U. P., & Cruz, C. H. (2014). Information management in industrial areas: A knowledge management view. In G. Jamil, A. Malheiro, & F. Ribeiro (Eds.), *Rethinking the conceptual base for new practical applications in information value and quality* (pp. 378–395). Hershey, PA: IGI Global. doi:10.4018/978-1-4666-4562-2.ch017

Peter, H., & Greenidge, C. (2011). An ontology-based extraction framework for a semantic web application. *International Journal of Knowledge-Based Organizations, 1*(3), 56–71. doi:10.4018/ijkbo.2011070104

Peter, H., & Greenidge, C. (2013). An ontology-based extraction framework for a semantic web application. In J. Wang (Ed.), *Intelligence methods and systems advancements for knowledge-based business* (pp. 231–246). Hershey, PA: IGI Global. doi:10.4018/978-1-4666-1873-2.ch013

Pham, Q. T., & Hara, Y. (2011). KM approach for improving the labor productivity of Vietnamese enterprise. *International Journal of Knowledge Management, 7*(3), 27–42. doi:10.4018/jkm.2011070103

Pham, Q. T., & Hara, Y. (2013). KM approach for improving the labor productivity of Vietnamese enterprise. In M. Jennex (Ed.), *Dynamic models for knowledge-driven organizations* (pp. 206–219). Hershey, PA: IGI Global. doi:10.4018/978-1-4666-2485-6.ch012

Philpott, E., & Beaumont-Kerridge, J. (2010). Overcoming reticence to aid knowledge creation between universities and business: A case reviewed. In D. Harorimana (Ed.), *Cultural implications of knowledge sharing, management and transfer: Identifying competitive advantage* (pp. 355–368). Hershey, PA: IGI Global. doi:10.4018/978-1-60566-790-4.ch016

Pike, S., & Roos, G. (2011). Measuring and valuing knowledge-based intangible assets: Real business uses. In B. Vallejo-Alonso, A. Rodriguez-Castellanos, & G. Arregui-Ayastuy (Eds.), *Identifying, measuring, and valuing knowledge-based intangible assets: New perspectives* (pp. 268–293). Hershey, PA: IGI Global. doi:10.4018/978-1-60960-054-9.ch013

Pineda, J. L., Zapata, L. E., & Ramírez, J. (2010). Strengthening knowledge transfer between the university and enterprise: A conceptual model for collaboration. In D. Harorimana (Ed.), *Cultural implications of knowledge sharing, management and transfer: Identifying competitive advantage* (pp. 134–151). Hershey, PA: IGI Global. doi:10.4018/978-1-60566-790-4.ch007

Platonov, V., & Bergman, J. (2013). Cross-border cooperative network in the perspective of innovation dynamics. In J. Wang (Ed.), *Intelligence methods and systems advancements for knowledge-based business* (pp. 150–169). Hershey, PA: IGI Global. doi:10.4018/978-1-4666-1873-2.ch009

Poels, G. (2013). Understanding business domain models: The effect of recognizing resource-event-agent conceptual modeling structures. In K. Siau (Ed.), *Innovations in database design, web applications, and information systems management* (pp. 72–106). Hershey, PA: IGI Global. doi:10.4018/978-1-4666-2044-5.ch004

Poels, G., Decreus, K., Roelens, B., & Snoeck, M. (2013). Investigating goal-oriented requirements engineering for business processes. *Journal of Database Management, 24*(2), 35–71. doi:10.4018/jdm.2013040103

Pomares-Quimbaya, A., & Torres-Moreno, M. E. (2013). Knowledge management processes supported by ontology technologies. In M. Nazir Ahmad, R. Colomb, & M. Abdullah (Eds.), *Ontology-based applications for enterprise systems and knowledge management* (pp. 125–140). Hershey, PA: IGI Global. doi:10.4018/978-1-4666-1993-7.ch007

Ponis, S. T., Vagenas, G., & Koronis, E. (2010). Exploring the knowledge management landscape: A critical review of existing knowledge management frameworks. In D. Harorimana (Ed.), *Cultural implications of knowledge sharing, management and transfer: Identifying competitive advantage* (pp. 1–25). Hershey, PA: IGI Global. doi:10.4018/978-1-60566-790-4.ch001

Powers, S. M., & Salmon, C. (2010). Management of learning space. In D. Wu (Ed.), *Temporal structures in individual time management: Practices to enhance calendar tool design* (pp. 210–219). Hershey, PA: IGI Global. doi:10.4018/978-1-60566-776-8.ch015

Pretorius, A. B., & Coetzee, F. P. (2011). Model of a knowledge management support system for choosing intellectual capital assessment methods. In B. Vallejo-Alonso, A. Rodriguez-Castellanos, & G. Arregui-Ayastuy (Eds.), *Identifying, measuring, and valuing knowledge-based intangible assets: New perspectives* (pp. 336–359). Hershey, PA: IGI Global. doi:10.4018/978-1-60960-054-9.ch016

Pullinger, D. (2011). Mobilizing knowledge in the UK public sector: Current issues and discourse. In G. Morais da Costa (Ed.), *Ethical issues and social dilemmas in knowledge management: Organizational innovation* (pp. 232–249). Hershey, PA: IGI Global. doi:10.4018/978-1-61520-873-9.ch014

Rabaey, M. (2013). Complex adaptive systems thinking approach for intelligence base in support of intellectual capital management. In P. Ordóñez de Pablos, R. Tennyson, & J. Zhao (Eds.), *Intellectual capital strategy management for knowledge-based organizations* (pp. 122–141). Hershey, PA: IGI Global. doi:10.4018/978-1-4666-3655-2.ch007

Rabaey, M., & Mercken, R. (2013). Framework of knowledge and intelligence base: From intelligence to service. In *Data mining: Concepts, methodologies, tools, and applications* (pp. 474–502). Hershey, PA: IGI Global. doi:10.4018/978-1-4666-2455-9.ch023

Radziwill, N. M., & DuPlain, R. F. (2010). Quality and continuous improvement in knowledge management. In M. Russ (Ed.), *Knowledge management strategies for business development* (pp. 353–363). Hershey, PA: IGI Global. doi:10.4018/978-1-60566-348-7.ch017

Rahman, B. A., Saad, N. M., & Harun, M. S. (2010). Knowledge management orientation and business performance: The Malaysian manufacturing and service industries perspective. In K. Metaxiotis, F. Carrillo, & T. Yigitcanlar (Eds.), *Knowledge-based development for cities and societies: Integrated multi-level approaches* (pp. 315–328). Hershey, PA: IGI Global. doi:10.4018/978-1-61520-721-3.ch019

Randles, T. J., Blades, C. D., & Fadlalla, A. (2012). The knowledge spectrum. *International Journal of Knowledge Management, 8*(2), 65–78. doi:10.4018/jkm.2012040104

Real, J. C., Leal, A., & Roldan, J. L. (2011). Measuring organizational learning as a multidimensional construct. In D. Schwartz & D. Te'eni (Eds.), *Encyclopedia of knowledge management* (2nd ed.; pp. 1101–1109). Hershey, PA: IGI Global. doi:10.4018/978-1-59904-931-1.ch105

Rech, J., & Bogner, C. (2010). Qualitative analysis of semantically enabled knowledge management systems in agile software engineering. *International Journal of Knowledge Management*, *6*(2), 66–85. doi:10.4018/jkm.2010040104

Rech, J., & Bogner, C. (2012). Qualitative analysis of semantically enabled knowledge management systems in agile software engineering. In M. Jennex (Ed.), *Conceptual models and outcomes of advancing knowledge management: New technologies* (pp. 144–164). Hershey, PA: IGI Global. doi:10.4018/978-1-4666-0035-5.ch008

Reis, R. S., & Curzi, Y. (2011). Knowledge integration in the creative process of globally distributed teams. In M. Al-Shammari (Ed.), *Knowledge management in emerging economies: Social, organizational and cultural implementation* (pp. 47–65). Hershey, PA: IGI Global. doi:10.4018/978-1-61692-886-5.ch003

Remli, M. A., & Deris, S. (2013). An approach for biological data integration and knowledge retrieval based on ontology, semantic web services composition, and AI planning. In M. Nazir Ahmad, R. Colomb, & M. Abdullah (Eds.), *Ontology-based applications for enterprise systems and knowledge management* (pp. 324–342). Hershey, PA: IGI Global. doi:10.4018/978-1-4666-1993-7.ch018

Reychav, I., Stein, E. W., Weisberg, J., & Glezer, C. (2012). The role of knowledge sharing in raising the task innovativeness of systems analysts. *International Journal of Knowledge Management*, *8*(2), 1–22. doi:10.4018/jkm.2012040101

Reychav, I., & Weisberg, J. (2011). Human capital in knowledge creation, management, and utilization. In D. Schwartz & D. Te'eni (Eds.), *Encyclopedia of knowledge management* (2nd ed.; pp. 389–401). Hershey, PA: IGI Global. doi:10.4018/978-1-59904-931-1.ch037

Rhoads, E., O'Sullivan, K. J., & Stankosky, M. (2011). An evaluation of factors that influence the success of knowledge management practices in US federal agencies. In M. Jennex & S. Smolnik (Eds.), *Strategies for knowledge management success: Exploring organizational efficacy* (pp. 74–90). Hershey, PA: IGI Global. doi:10.4018/978-1-60566-709-6.ch005

Ribière, V. M. (2011). The effect of organizational trust on the success of codification and personalization KM approaches. In M. Jennex & S. Smolnik (Eds.), *Strategies for knowledge management success: Exploring organizational efficacy* (pp. 192–212). Hershey, PA: IGI Global. doi:10.4018/978-1-60566-709-6.ch011

Ribière, V. M., & Román, J. A. (2011). Knowledge flow. In D. Schwartz & D. Te'eni (Eds.), *Encyclopedia of knowledge management* (2nd ed.; pp. 549–559). Hershey, PA: IGI Global. doi:10.4018/978-1-59904-931-1.ch053

Ricceri, F., Guthrie, J., & Coyte, R. (2010). The management of knowledge resources within private organisations: Some European "better practice" illustrations. In E. O'Brien, S. Clifford, & M. Southern (Eds.), *Knowledge management for process, organizational and marketing innovation: Tools and methods* (pp. 36–61). Hershey, PA: IGI Global. doi:10.4018/978-1-61520-829-6.ch003

Riss, U. V. (2011). Pattern-based task management as means of organizational knowledge maturing. *International Journal of Knowledge-Based Organizations, 1*(1), 20–41. doi:10.4018/ijkbo.2011010102

Riss, U. V. (2013). Pattern-based task management as means of organizational knowledge maturing. In J. Wang (Ed.), *Intelligence methods and systems advancements for knowledge-based business* (pp. 1–23). Hershey, PA: IGI Global. doi:10.4018/978-1-4666-1873-2.ch001

Roos, G. (2013). The role of intellectual capital in business model innovation: An empirical study. In P. Ordóñez de Pablos, R. Tennyson, & J. Zhao (Eds.), *Intellectual capital strategy management for knowledge-based organizations* (pp. 76–121). Hershey, PA: IGI Global. doi:10.4018/978-1-4666-3655-2.ch006

Rothberg, H. N., & Klingenberg, B. (2010). Learning before doing: A theoretical perspective and practical lessons from a failed cross-border knowledge transfer initiative. In D. Harorimana (Ed.), *Cultural implications of knowledge sharing, management and transfer: Identifying competitive advantage* (pp. 277–294). Hershey, PA: IGI Global. doi:10.4018/978-1-60566-790-4.ch013

Ruano-Mayoral, M., Colomo-Palacios, R., García-Crespo, Á., & Gómez-Berbís, J. M. (2012). Software project managers under the team software process: A study of competences based on literature. In J. Wang (Ed.), *Project management techniques and innovations in information technology* (pp. 115–126). Hershey, PA: IGI Global. doi:10.4018/978-1-4666-0930-3.ch007

Russell, S. (2010). Knowledge management and project management in 3D: A virtual world extension. In E. O'Brien, S. Clifford, & M. Southern (Eds.), *Knowledge management for process, organizational and marketing innovation: Tools and methods* (pp. 62–78). Hershey, PA: IGI Global. doi:10.4018/978-1-61520-829-6.ch004

Ryan, G., & Shinnick, E. (2011). Knowledge and intellectual property rights: An economics perspective. In D. Schwartz & D. Te'eni (Eds.), *Encyclopedia of knowledge management* (2nd ed.; pp. 489–496). Hershey, PA: IGI Global. doi:10.4018/978-1-59904-931-1.ch047

Sabetzadeh, F., & Tsui, E. (2013). Delivering knowledge services in the cloud. In G. Yang (Ed.), *Multidisciplinary studies in knowledge and systems science* (pp. 247–254). Hershey, PA: IGI Global. doi:10.4018/978-1-4666-3998-0.ch017

Sáenz, J., & Aramburu, N. (2011). Organizational conditions as catalysts for successful people-focused knowledge sharing initiatives: An empirical study. *International Journal of Knowledge-Based Organizations*, *1*(2), 39–56. doi:10.4018/ijkbo.2011040103

Sáenz, J., & Aramburu, N. (2013). Organizational conditions as catalysts for successful people-focused knowledge sharing initiatives: An empirical study. In J. Wang (Ed.), *Intelligence methods and systems advancements for knowledge-based business* (pp. 263–280). Hershey, PA: IGI Global. doi:10.4018/978-1-4666-1873-2.ch015

Sakr, S., & Al-Naymat, G. (2011). Relational techniques for storing and querying RDF data: An overview. In L. Yan & Z. Ma (Eds.), *Advanced database query systems: Techniques, applications and technologies* (pp. 269–285). Hershey, PA: IGI Global. doi:10.4018/978-1-60960-475-2.ch011

Salem, P. J. (2013). The use of mixed methods in organizational communication research. In M. Bocarnea, R. Reynolds, & J. Baker (Eds.), *Online instruments, data collection, and electronic measurements: Organizational advancements* (pp. 24–39). Hershey, PA: IGI Global. doi:10.4018/978-1-4666-2172-5.ch002

Salisbury, M. (2011). A framework for managing the life cycle of knowledge in global organizations. In M. Jennex (Ed.), *Global aspects and cultural perspectives on knowledge management: Emerging dimensions* (pp. 64–80). Hershey, PA: IGI Global. doi:10.4018/978-1-60960-555-1.ch005

Salleh, K. (2014). Drivers, benefits, and challenges of knowledge management in electronic government: Preliminary examination. In Y. Al-Bastaki & A. Shajera (Eds.), *Building a competitive public sector with knowledge management strategy* (pp. 135–150). Hershey, PA: IGI Global. doi:10.4018/978-1-4666-4434-2.ch006

Salleh, K., Ikhsan, S. O., & Ahmad, S. N. (2011). Knowledge management enablers and knowledge sharing process: A case study of public sector accounting organization in Malaysia. In M. Al-Shammari (Ed.), *Knowledge management in emerging economies: Social, organizational and cultural implementation* (pp. 199–211). Hershey, PA: IGI Global. doi:10.4018/978-1-61692-886-5.ch012

Saunders, C. (2011). Knowledge sharing in legal practice. In D. Schwartz & D. Te'eni (Eds.), *Encyclopedia of knowledge management* (2nd ed.; pp. 935–945). Hershey, PA: IGI Global. doi:10.4018/978-1-59904-931-1.ch089

Scarso, E., & Bolisani, E. (2011). Knowledge intermediation. In D. Schwartz & D. Te'eni (Eds.), *Encyclopedia of knowledge management* (2nd ed.; pp. 601–611). Hershey, PA: IGI Global. doi:10.4018/978-1-59904-931-1.ch058

Scarso, E., & Bolisani, E. (2011). Managing professions for knowledge management. *International Journal of Knowledge Management, 7*(3), 61–75. doi:10.4018/jkm.2011070105

Scarso, E., & Bolisani, E. (2013). Managing professions for knowledge management. In M. Jennex (Ed.), *Dynamic models for knowledge-driven organizations* (pp. 238–253). Hershey, PA: IGI Global. doi:10.4018/978-1-4666-2485-6.ch014

Scarso, E., Bolisani, E., & Padova, A. (2011). The complex issue of measuring KM performance: Lessons from the practice. In B. Vallejo-Alonso, A. Rodriguez-Castellanos, & G. Arregui-Ayastuy (Eds.), *Identifying, measuring, and valuing knowledge-based intangible assets: New perspectives* (pp. 208–230). Hershey, PA: IGI Global. doi:10.4018/978-1-60960-054-9.ch010

Schumann, C., & Tittmann, C. (2010). Potentials for externalizing and measuring of tacit knowledge within knowledge nodes in the context of knowledge networks. In D. Harorimana (Ed.), *Cultural implications of knowledge sharing, management and transfer: Identifying competitive advantage* (pp. 84–107). Hershey, PA: IGI Global. doi:10.4018/978-1-60566-790-4.ch005

Schwartz, D. (2011). An Aristotelian view of knowledge for knowledge management. In D. Schwartz & D. Te'eni (Eds.), *Encyclopedia of knowledge management* (2nd ed.; pp. 39–48). Hershey, PA: IGI Global. doi:10.4018/978-1-59904-931-1.ch005

Senaratne, S., & Victoria, M. F. (2014). Building a supportive culture for sustained organisational learning in public sectors. In Y. Al-Bastaki & A. Shajera (Eds.), *Building a competitive public sector with knowledge management strategy* (pp. 118–134). Hershey, PA: IGI Global. doi:10.4018/978-1-4666-4434-2.ch005

Shah, A., Singhera, Z., & Ahsan, S. (2011). Web services for bioinformatics. In M. Al-Mutairi & L. Mohammed (Eds.), *Cases on ICT utilization, practice and solutions: Tools for managing day-to-day issues* (pp. 28–46). Hershey, PA: IGI Global. doi:10.4018/978-1-60960-015-0.ch003

Shajera, A., & Al-Bastaki, Y. (2014). Organisational readiness for knowledge management: Bahrain public sector case study. In Y. Al-Bastaki & A. Shajera (Eds.), *Building a competitive public sector with knowledge management strategy* (pp. 104–117). Hershey, PA: IGI Global. doi:10.4018/978-1-4666-4434-2.ch004

Sharma, A. K., Goswami, A., & Gupta, D. (2011). An extended relational model & SQL for fuzzy multidatabases. In L. Yan & Z. Ma (Eds.), *Advanced database query systems: Techniques, applications and technologies* (pp. 185–219). Hershey, PA: IGI Global. doi:10.4018/978-1-60960-475-2.ch008

Sharma, R., Banati, H., & Bedi, P. (2012). Building socially-aware e-learning systems through knowledge management. *International Journal of Knowledge Management, 8*(3), 1–26. doi:10.4018/jkm.2012070101

Sharma, R. S., Chandrasekar, G., & Vaitheeswaran, B. (2012). A knowledge framework for development: Empirical investigation of 30 societies. In M. Jennex (Ed.), *Conceptual models and outcomes of advancing knowledge management: New technologies* (pp. 244–265). Hershey, PA: IGI Global. doi:10.4018/978-1-4666-0035-5.ch013

Shaw, D. (2011). Mapping group knowledge. In D. Schwartz & D. Te'eni (Eds.), *Encyclopedia of knowledge management* (2nd ed.; pp. 1072–1081). Hershey, PA: IGI Global. doi:10.4018/978-1-59904-931-1.ch102

Sheluhin, O. I., & Atayero, A. A. (2013). Principles of modeling in information communication systems and networks. In A. Atayero & O. Sheluhin (Eds.), *Integrated models for information communication systems and networks: Design and development* (pp. 1–15). Hershey, PA: IGI Global. doi:10.4018/978-1-4666-2208-1.ch001

Sheluhin, O. I., & Garmashev, A. V. (2013). Numerical methods of multifractal analysis in information communication systems and networks. In A. Atayero & O. Sheluhin (Eds.), *Integrated models for information communication systems and networks: Design and development* (pp. 16–46). Hershey, PA: IGI Global. doi:10.4018/978-1-4666-2208-1.ch002

Siau, K., Long, Y., & Ling, M. (2010). Toward a unified model of information systems development success. *Journal of Database Management, 21*(1), 80–101. doi:10.4018/jdm.2010112304

Siau, K., Long, Y., & Ling, M. (2012). Toward a unified model of information systems development success. In K. Siau (Ed.), *Cross-disciplinary models and applications of database management: Advancing approaches* (pp. 80–102). Hershey, PA: IGI Global. doi:10.4018/978-1-61350-471-0.ch004

Simard, A. J., & Jourdeuil, P. (2014). Knowledge manageability: A new paradigm. In Y. Al-Bastaki & A. Shajera (Eds.), *Building a competitive public sector with knowledge management strategy* (pp. 1–52). Hershey, PA: IGI Global. doi:10.4018/978-1-4666-4434-2.ch001

Simonette, M. J., & Spina, E. (2014). Enabling IT innovation through soft systems engineering. In M. Pańkowska (Ed.), *Frameworks of IT prosumption for business development* (pp. 64–72). Hershey, PA: IGI Global. doi:10.4018/978-1-4666-4313-0.ch005

Sivaramakrishnan, S., Delbaere, M., Zhang, D., & Bruning, E. (2012). Critical success factors and outcomes of market knowledge management: A conceptual model and empirical evidence. In M. Jennex (Ed.), *Conceptual models and outcomes of advancing knowledge management: New technologies* (pp. 165–185). Hershey, PA: IGI Global. doi:10.4018/978-1-4666-0035-5.ch009

Small, C. T., & Sage, A. P. (2010). A complex adaptive systems-based enterprise knowledge sharing model. In D. Paradice (Ed.), *Emerging systems approaches in information technologies: Concepts, theories, and applications* (pp. 137–155). Hershey, PA: IGI Global. doi:10.4018/978-1-60566-976-2.ch009

Smedlund, A. (2011). Social network structures for explicit, tacit and potential knowledge. In M. Jennex (Ed.), *Global aspects and cultural perspectives on knowledge management: Emerging dimensions* (pp. 81–90). Hershey, PA: IGI Global. doi:10.4018/978-1-60960-555-1.ch006

Smith, A. D. (2013). Competitive uses of information and knowledge management tools: Case study of supplier-side management. *International Journal of Knowledge-Based Organizations*, *3*(1), 71–87. doi:10.4018/ijkbo.2013010105

Smith, P., & Coakes, E. (2011). Exploiting KM in support of innovation and change. In A. Eardley & L. Uden (Eds.), *Innovative knowledge management: Concepts for organizational creativity and collaborative design* (pp. 242–252). Hershey, PA: IGI Global. doi:10.4018/978-1-60566-701-0.ch015

Smith, T. A., Mills, A. M., & Dion, P. (2010). Linking business strategy and knowledge management capabilities for organizational effectiveness. *International Journal of Knowledge Management*, *6*(3), 22–43. doi:10.4018/jkm.2010070102

Smith, T. A., Mills, A. M., & Dion, P. (2012). Linking business strategy and knowledge management capabilities for organizational effectiveness. In M. Jennex (Ed.), *Conceptual models and outcomes of advancing knowledge management: New technologies* (pp. 186–207). Hershey, PA: IGI Global. doi:10.4018/978-1-4666-0035-5.ch010

Smuts, H., van der Merwe, A., & Loock, M. (2011). Key characteristics relevant for selecting knowledge management software tools. In A. Eardley & L. Uden (Eds.), *Innovative knowledge management: Concepts for organizational creativity and collaborative design* (pp. 18–39). Hershey, PA: IGI Global. doi:10.4018/978-1-60566-701-0.ch002

Soffer, P., & Kaner, M. (2013). Complementing business process verification by validity analysis: A theoretical and empirical evaluation. In K. Siau (Ed.), *Innovations in database design, web applications, and information systems management* (pp. 265–288). Hershey, PA: IGI Global. doi:10.4018/978-1-4666-2044-5.ch010

Soffer, P., Kaner, M., & Wand, Y. (2012). Assigning ontological meaning to workflow nets. In K. Siau (Ed.), *Cross-disciplinary models and applications of database management: Advancing approaches* (pp. 209–244). Hershey, PA: IGI Global. doi:10.4018/978-1-61350-471-0.ch009

Sohrabi, B., Raeesi, I., & Khanlari, A. (2010). Intellectual capital components, measurement and management: A literature survey of concepts and measures. In P. López Sáez, G. Castro, J. Navas López, & M. Delgado Verde (Eds.), *Intellectual capital and technological innovation: Knowledge-based theory and practice* (pp. 1–38). Hershey, PA: IGI Global. doi:10.4018/978-1-61520-875-3.ch001

Sohrabi, B., Raeesi, I., Khanlari, A., & Forouzandeh, S. (2011). A comprehensive model for assessing the organizational readiness of knowledge management. In M. Jennex (Ed.), *Global aspects and cultural perspectives on knowledge management: Emerging dimensions* (pp. 30–48). Hershey, PA: IGI Global. doi:10.4018/978-1-60960-555-1.ch003

Sparrow, J. (2011). Knowledge management in small and medium sized enterprises. In D. Schwartz & D. Te'eni (Eds.), *Encyclopedia of knowledge management* (2nd ed.; pp. 671–681). Hershey, PA: IGI Global. doi:10.4018/978-1-59904-931-1.ch065

Stam, C. D. (2011). Making sense of knowledge productivity. In B. Vallejo-Alonso, A. Rodriguez-Castellanos, & G. Arregui-Ayastuy (Eds.), *Identifying, measuring, and valuing knowledge-based intangible assets: New perspectives* (pp. 133–155). Hershey, PA: IGI Global. doi:10.4018/978-1-60960-054-9.ch007

Stamkopoulos, K., Pitoura, E., Vassiliadis, P., & Zarras, A. (2012). Accelerating web service workflow execution via intelligent allocation of services to servers. In K. Siau (Ed.), *Cross-disciplinary models and applications of database management: Advancing approaches* (pp. 385–416). Hershey, PA: IGI Global. doi:10.4018/978-1-61350-471-0.ch016

Sterling, L. (2011). Applying agents within knowledge management. In D. Schwartz & D. Te'eni (Eds.), *Encyclopedia of knowledge management* (2nd ed.; pp. 12–19). Hershey, PA: IGI Global. doi:10.4018/978-1-59904-931-1.ch002

Su, S., & Chiong, R. (2011). Business intelligence. In D. Schwartz & D. Te'eni (Eds.), *Encyclopedia of knowledge management* (2nd ed.; pp. 72–80). Hershey, PA: IGI Global. doi:10.4018/978-1-59904-931-1.ch008

Su, W. B., Li, X., & Chow, C. W. (2012). Exploring the extent and impediments of knowledge sharing in Chinese business enterprise. In M. Jennex (Ed.), *Conceptual models and outcomes of advancing knowledge management: New technologies* (pp. 266–290). Hershey, PA: IGI Global. doi:10.4018/978-1-4666-0035-5.ch014

Subramanian, D. V., & Geetha, A. (2012). Application of multi-dimensional metric model, database, and WAM for KM system evaluation. *International Journal of Knowledge Management, 8*(4), 1–21. doi:10.4018/jkm.2012100101

Surendran, A., & Samuel, P. (2013). Knowledge-based code clone approach in embedded and real-time systems. In S. Saeed & I. Alsmadi (Eds.), *Knowledge-based processes in software development* (pp. 49–62). Hershey, PA: IGI Global. doi:10.4018/978-1-4666-4229-4.ch004

Takahashi, Y. (2011). The importance of balancing knowledge protection and knowledge interchange. In G. Morais da Costa (Ed.), *Ethical issues and social dilemmas in knowledge management: Organizational innovation* (pp. 180–198). Hershey, PA: IGI Global. doi:10.4018/978-1-61520-873-9.ch011

Talet, A. N., Alhawari, S., & Alryalat, H. (2012). The outcome of knowledge process for customers of Jordanian companies on the achievement of customer knowledge retention. In M. Jennex (Ed.), *Conceptual models and outcomes of advancing knowledge management: New technologies* (pp. 45–61). Hershey, PA: IGI Global. doi:10.4018/978-1-4666-0035-5.ch003

Talet, A. N., Alhawari, S., Mansour, E., & Alryalat, H. (2011). The practice of Jordanian business to attain customer knowledge acquisition. *International Journal of Knowledge Management, 7*(2), 49–67. doi:10.4018/jkm.2011040103

Tanner, K. (2011). The role of emotional capital in organisational KM. In D. Schwartz & D. Te'eni (Eds.), *Encyclopedia of knowledge management* (2nd ed.; pp. 1396–1409). Hershey, PA: IGI Global. doi:10.4018/978-1-59904-931-1.ch133

Tauber, D., & Schwartz, D. G. (2011). Integrating knowledge management with the systems analysis process. In D. Schwartz & D. Te'eni (Eds.), *Encyclopedia of knowledge management* (2nd ed.; pp. 431–441). Hershey, PA: IGI Global. doi:10.4018/978-1-59904-931-1.ch041

Tavana, M., Busch, T. E., & Davis, E. L. (2011). Modeling operational robustness and resiliency with high-level Petri nets. *International Journal of Knowledge-Based Organizations, 1*(2), 17–38. doi:10.4018/ijkbo.2011040102

Taxén, L. (2010). Aligning business and knowledge strategies: A practical approach for aligning business and knowledge strategies. In M. Russ (Ed.), *Knowledge management strategies for business development* (pp. 277–308). Hershey, PA: IGI Global. doi:10.4018/978-1-60566-348-7.ch013

Te'eni, D. (2011). Knowledge for communicating knowledge. In D. Schwartz & D. Te'eni (Eds.), *Encyclopedia of knowledge management* (2nd ed.; pp. 560–569). Hershey, PA: IGI Global. doi:10.4018/978-1-59904-931-1.ch054

Toiviainen, H., & Kerosuo, H. (2013). Development curriculum for knowledge-based organizations: Lessons from a learning network. *International Journal of Knowledge-Based Organizations, 3*(3), 1–18. doi:10.4018/ijkbo.2013070101

Tran, B. (2014). The human element of the knowledge worker: Identifying, managing, and protecting the intellectual capital within knowledge management. In M. Chilton & J. Bloodgood (Eds.), *Knowledge management and competitive advantage: Issues and potential solutions* (pp. 281–303). Hershey, PA: IGI Global. doi:10.4018/978-1-4666-4679-7.ch017

Tsamoura, E., Gounaris, A., & Manolopoulos, Y. (2011). Optimal service ordering in decentralized queries over web services. *International Journal of Knowledge-Based Organizations, 1*(2), 1–16. doi:10.4018/ijkbo.2011040101

Tsamoura, E., Gounaris, A., & Manolopoulos, Y. (2013). Optimal service ordering in decentralized queries over web services. In J. Wang (Ed.), *Intelligence methods and systems advancements for knowledge-based business* (pp. 43–58). Hershey, PA: IGI Global. doi:10.4018/978-1-4666-1873-2.ch003

Tull, J. (2013). Slow knowledge: The case for savouring learning and innovation. In S. Buckley & M. Jakovljevic (Eds.), *Knowledge management innovations for interdisciplinary education: Organizational applications* (pp. 274–297). Hershey, PA: IGI Global. doi:10.4018/978-1-4666-1969-2.ch014

Turner, G., & Minonne, C. (2013). Effective knowledge management through measurement. In S. Buckley & M. Jakovljevic (Eds.), *Knowledge management innovations for interdisciplinary education: Organizational applications* (pp. 145–176). Hershey, PA: IGI Global. doi:10.4018/978-1-4666-1969-2.ch008

Uden, L., & Eardley, A. (2011). Knowledge sharing in the learning process: Experience with problem-based learning. In A. Eardley & L. Uden (Eds.), *Innovative knowledge management: Concepts for organizational creativity and collaborative design* (pp. 215–229). Hershey, PA: IGI Global. doi:10.4018/978-1-60566-701-0.ch013

Upadhyaya, S., Rao, H. R., & Padmanabhan, G. (2011). Secure knowledge management. In D. Schwartz & D. Te'eni (Eds.), *Encyclopedia of knowledge management* (2nd ed.; pp. 1429–1437). Hershey, PA: IGI Global. doi:10.4018/978-1-59904-931-1.ch136

Urbancová, H., & Königová, M. (2013). The influence of the application of business continuity management, knowledge management, and knowledge continuity management on the innovation in organizations. In S. Buckley & M. Jakovljevic (Eds.), *Knowledge management innovations for interdisciplinary education: Organizational applications* (pp. 254–273). Hershey, PA: IGI Global. doi:10.4018/978-1-4666-1969-2.ch013

Van Canh, T., & Zyngier, S. (2014). Using ERG theory as a lens to understand the sharing of academic tacit knowledge: Problems and issues in developing countries – Perspectives from Vietnam. In M. Chilton & J. Bloodgood (Eds.), *Knowledge management and competitive advantage: Issues and potential solutions* (pp. 174–201). Hershey, PA: IGI Global. doi:10.4018/978-1-4666-4679-7.ch010

Vat, K. H. (2011). Knowledge synthesis framework. In D. Schwartz & D. Te'eni (Eds.), *Encyclopedia of knowledge management* (2nd ed.; pp. 955–966). Hershey, PA: IGI Global. doi:10.4018/978-1-59904-931-1.ch091

Vert, S. (2012). Extensions of web browsers useful to knowledge workers. In C. Jouis, I. Biskri, J. Ganascia, & M. Roux (Eds.), *Next generation search engines: Advanced models for information retrieval* (pp. 239–273). Hershey, PA: IGI Global. doi:10.4018/978-1-4666-0330-1.ch011

Wagner, L., & Van Belle, J. (2011). Web mining for strategic competitive intelligence: South African experiences and a practical methodology. In M. Al-Shammari (Ed.), *Knowledge management in emerging economies: Social, organizational and cultural implementation* (pp. 1–19). Hershey, PA: IGI Global. doi:10.4018/978-1-61692-886-5.ch001

Wautelet, Y., Schinckus, C., & Kolp, M. (2010). Towards knowledge evolution in software engineering: An epistemological approach. *International Journal of Information Technologies and Systems Approach, 3*(1), 21–40. doi:10.4018/jitsa.2010100202

Weiß, S., Makolm, J., Ipsmiller, D., & Egger, N. (2011). DYONIPOS: Proactive knowledge supply. In M. Jennex & S. Smolnik (Eds.), *Strategies for knowledge management success: Exploring organizational efficacy* (pp. 277–287). Hershey, PA: IGI Global. doi:10.4018/978-1-60566-709-6.ch015

Welschen, J., Todorova, N., & Mills, A. M. (2012). An investigation of the impact of intrinsic motivation on organizational knowledge sharing. *International Journal of Knowledge Management, 8*(2), 23–42. doi:10.4018/jkm.2012040102

Wickramasinghe, N. (2011). Knowledge creation. In D. Schwartz & D. Te'eni (Eds.), *Encyclopedia of knowledge management* (2nd ed.; pp. 527–538). Hershey, PA: IGI Global. doi:10.4018/978-1-59904-931-1.ch051

Wickramasinghe, N. (2011). Knowledge management: The key to delivering superior healthcare solutions. In A. Eardley & L. Uden (Eds.), *Innovative knowledge management: Concepts for organizational creativity and collaborative design* (pp. 190–203). Hershey, PA: IGI Global. doi:10.4018/978-1-60566-701-0.ch011

Wijnhoven, F. (2011). Operational knowledge management. In D. Schwartz & D. Te'eni (Eds.), *Encyclopedia of knowledge management* (2nd ed.; pp. 1237–1249). Hershey, PA: IGI Global. doi:10.4018/978-1-59904-931-1.ch118

Williams, R. (2011). A Knowledge Process Cycle. In D. Schwartz & D. Te'eni (Eds.), *Encyclopedia of knowledge management* (2nd ed.; pp. 853–866). Hershey, PA: IGI Global. doi:10.4018/978-1-59904-931-1.ch082

Wilson, R. L., Rosen, P. A., & Al-Ahmadi, M. S. (2011). Knowledge structure and data mining techniques. In D. Schwartz & D. Te'eni (Eds.), *Encyclopedia of knowledge management* (2nd ed.; pp. 946–954). Hershey, PA: IGI Global. doi:10.4018/978-1-59904-931-1.ch090

Wimmer, H., Yoon, V., & Rada, R. (2013). Integrating knowledge sources: An ontological approach. *International Journal of Knowledge Management, 9*(1), 60–75. doi:10.4018/jkm.2013010104

Woods, S., Poteet, S. R., Kao, A., & Quach, L. (2011). Knowledge dissemination in portals. In D. Schwartz & D. Te'eni (Eds.), *Encyclopedia of knowledge management* (2nd ed.; pp. 539–548). Hershey, PA: IGI Global. doi:10.4018/978-1-59904-931-1.ch052

Worden, D. (2010). Agile alignment of enterprise execution capabilities with strategy. In M. Russ (Ed.), *Knowledge management strategies for business development* (pp. 45–62). Hershey, PA: IGI Global. doi:10.4018/978-1-60566-348-7.ch003

Wu, D. (2010). Who are effective time managers? Bivariate correlation analysis and hypotheses testing. In D. Wu (Ed.), *Temporal structures in individual time management: Practices to enhance calendar tool design* (pp. 116–138). Hershey, PA: IGI Global. doi:10.4018/978-1-60566-776-8.ch009

Wu, J., Du, H., Li, X., & Li, P. (2010). Creating and delivering a successful knowledge management strategy. In M. Russ (Ed.), *Knowledge management strategies for business development* (pp. 261–276). Hershey, PA: IGI Global. doi:10.4018/978-1-60566-348-7.ch012

Wu, J., Liu, N., & Xuan, Z. (2013). Simulation on knowledge transfer processes from the perspectives of individual's mentality and behavior. In G. Yang (Ed.), *Multidisciplinary studies in knowledge and systems science* (pp. 233–246). Hershey, PA: IGI Global. doi:10.4018/978-1-4666-3998-0.ch016

Wu, J., Wang, S., & Pan, D. (2013). Evaluation of technological influence power of enterprises through the enterprise citation network. In G. Yang (Ed.), *Multidisciplinary studies in knowledge and systems science* (pp. 34–44). Hershey, PA: IGI Global. doi:10.4018/978-1-4666-3998-0.ch003

Wu, S. T. (2011). Innovation in new technology and knowledge management: Comparative case studies of its evolution during a quarter century of change. In A. Eardley & L. Uden (Eds.), *Innovative knowledge management: Concepts for organizational creativity and collaborative design* (pp. 77–93). Hershey, PA: IGI Global. doi:10.4018/978-1-60566-701-0.ch005

Xiao, L., & Pei, Y. (2013). A task context aware physical distribution knowledge service system. In G. Yang (Ed.), *Multidisciplinary studies in knowledge and systems science* (pp. 18–33). Hershey, PA: IGI Global. doi:10.4018/978-1-4666-3998-0.ch002

Xu, D., & Wang, H. (2011). Integration of knowledge management and e-learning. In D. Schwartz & D. Te'eni (Eds.), *Encyclopedia of knowledge management* (2nd ed.; pp. 442–451). Hershey, PA: IGI Global. doi:10.4018/978-1-59904-931-1.ch042

Yaniv, E., & Schwartz, D. G. (2011). Organizational attention. In D. Schwartz & D. Te'eni (Eds.), *Encyclopedia of knowledge management* (2nd ed.; pp. 1270–1279). Hershey, PA: IGI Global. doi:10.4018/978-1-59904-931-1.ch121

Yeung, C. L., Cheung, C. F., Wang, W. M., & Tsui, E. (2013). A study of organizational narrative simulation for decision support. In G. Yang (Ed.), *Multidisciplinary studies in knowledge and systems science* (pp. 179–192). Hershey, PA: IGI Global. doi:10.4018/978-1-4666-3998-0.ch013

Yigitcanlar, T. (2013). Moving towards a knowledge city? Brisbane's experience in knowledge-based urban development. In J. Wang (Ed.), *Intelligence methods and systems advancements for knowledge-based business* (pp. 114–131). Hershey, PA: IGI Global. doi:10.4018/978-1-4666-1873-2.ch007

Yigitcanlar, T., & Martinez-Fernandez, C. (2010). Making space and place for knowledge production: Socio-spatial development of knowledge community precincts. In K. Metaxiotis, F. Carrillo, & T. Yigitcanlar (Eds.), *Knowledge-based development for cities and societies: Integrated multi-level approaches* (pp. 99–117). Hershey, PA: IGI Global. doi:10.4018/978-1-61520-721-3.ch006

Yıldırım, A. A., Özdoğan, C., & Watson, D. (2014). Parallel data reduction techniques for big datasets. In W. Hu & N. Kaabouch (Eds.), *Big data management, technologies, and applications* (pp. 72–93). Hershey, PA: IGI Global. doi:10.4018/978-1-4666-4699-5.ch004

Yoon, K. S. (2012). Measuring the influence of expertise and epistemic engagement to the practice of knowledge management. *International Journal of Knowledge Management, 8*(1), 40–70. doi:10.4018/jkm.2012010103

Yusof, Z. M., & Ismail, M. B. (2011). Factors affecting knowledge sharing practice in Malaysia: A preliminary overview. In M. Al-Shammari (Ed.), *Knowledge management in emerging economies: Social, organizational and cultural implementation* (pp. 157–170). Hershey, PA: IGI Global. doi:10.4018/978-1-61692-886-5.ch009

Zapata-Cantú, L., Ramírez, J., & Pineda, J. L. (2011). HRM adaptation to knowledge management initiatives: Three Mexican cases. In M. Al-Shammari (Ed.), *Knowledge management in emerging economies: Social, organizational and cultural implementation* (pp. 273–293). Hershey, PA: IGI Global. doi:10.4018/978-1-61692-886-5.ch017

Zarri, G. P. (2011). Knowledge representation. In D. Schwartz & D. Te'eni (Eds.), *Encyclopedia of knowledge management* (2nd ed.; pp. 878–892). Hershey, PA: IGI Global. doi:10.4018/978-1-59904-931-1.ch084

Zarri, G. P. (2011). RDF and OWL for knowledge management. In D. Schwartz & D. Te'eni (Eds.), *Encyclopedia of knowledge management* (2nd ed.; pp. 1355–1373). Hershey, PA: IGI Global. doi:10.4018/978-1-59904-931-1.ch130

Zhang, Y., Wang, Y., Colucci, W., & Wang, Z. (2013). The paradigm shift in organizational research. In J. Wang (Ed.), *Intelligence methods and systems advancements for knowledge-based business* (pp. 60–74). Hershey, PA: IGI Global. doi:10.4018/978-1-4666-1873-2.ch004

Zhang, Z. J. (2011). Managing customer knowledge with social software. In D. Schwartz & D. Te'eni (Eds.), *Encyclopedia of knowledge management* (2nd ed.; pp. 1046–1053). Hershey, PA: IGI Global. doi:10.4018/978-1-59904-931-1.ch099

Zyngier, S. (2011). Governance of knowledge management. In D. Schwartz & D. Te'eni (Eds.), *Encyclopedia of knowledge management* (2nd ed.; pp. 354–365). Hershey, PA: IGI Global. doi:10.4018/978-1-59904-931-1.ch034

Zyngier, S. (2011). Knowledge management: Realizing value through governance. *International Journal of Knowledge Management*, *7*(1), 35–54. doi:10.4018/jkm.2011010103

Zyngier, S. (2013). Knowledge management: Realizing value through governance. In M. Jennex (Ed.), *Dynamic models for knowledge-driven organizations* (pp. 36–55). Hershey, PA: IGI Global. doi:10.4018/978-1-4666-2485-6.ch003

Compilation of References

Acharya, S., Poosala, V., & Ramaswamy, S. (1999). Selectivity estimation in spatial databases. *SIGMOD '99 Proceedings of the 1999 ACM SIGMOD international conference on Management of data*, 13–24. 10.1145/304182.304184

Addison, P.S. (2002). *The illustrated wavelet transform Hand-Book.* London: Institute of Physics Publishing.

Agafonkin, V. (2015, September 3). *Leaflet*. Retrieved September 3, 2015, from http://leafletjs.com/

AGDesktop Doc. (2011). *WFS To Feature Class Conversion Tool*. Retrieved September 6, 2015, from http://help.arcgis.com/EN/arcgisdesktop/10.0/help/index.html#//00120000000q000000

AGDesktop Doc. (2013). *ArcGIS Help 10.2 - Geodatabase XML*. Retrieved September 6, 2015, from http://resources.arcgis.com/EN/HELP/MAIN/10.2/index.html#//018r00000005000000

AGDesktop Doc. (2015, September 3). *SQL with ST_Geometry*. Retrieved September 3, 2015, from http://resources.arcgis.com/EN/HELP/MAIN/10.2/index.html#//006z0000001v000000

AGJavaScript Doc. (2015, September 3). *Why Dojo?* Retrieved September 3, 2015, from https://developers.arcgis.com/javascript /jshelp/why_dojo.html

AGLG Gallery. (2015, September 3). *Gallery | ArcGIS Solutions*. Retrieved September 3, 2015, from http://solutions.arcgis.com/gallery/#s=0

Akbari, M., & Rajabi, M. A. (2013). Evaluation of Desktop free/Open source GIS software based on functional and non-functional capabilities. *Technical Gazette*, *20*(5), 755–764.

Alsubaiee, S., & Li, C. (2010). Fuzzy keyword search on spatial data. In *International Conference on Database Systems for Advanced Applications. DASFAA*. Springer. 10.1007/978-3-642-12098-5_51

Alsubaiee, S., Behm, A., & Li, C. (2010) Supporting Location-Based Approximate-Keyword Queries. *Proceeding GIS '10 Proceedings of the 18th SIGSPATIAL International Conference on Advances in Geographic Information Systems*, 61-70.

Apache, D. S. (2015). *Apache Directory Website*. Retrieved September 6, 2015, from https://directory.apache.org/

Arasu, A., Ganti, V., & Kaushik, R. (2006) Efficient exact set-similarity joins. *Proceeding VLDB '06 Proceedings of the 32nd international conference on Very large data bases*, 918–929.

ArcNews. (2011). *Open Source Technology and Esri | ArcNews Online*. Retrieved August 28, 2015, from http://www.esri.com/news/arcnews/spring11articles/open-source-technology-and-esri.html

ArcObjects Doc. (2015, September 2). *Creating class extensions*. Retrieved September 2, 2015, from http://resources.arcgis.com/en/help/arcobjects-net/conceptualhelp/index.html#/Creating_class_extensions/000100000201000000/

Arias Muñoz, C., Brovelli, M. A., Corti, S., Micotti, M., Weber, E., & Soncini Sessa, R. (2015). A FOSS approach to Integrated Water Resource Management: The case study of Red-Thai Binh rivers system in Vietnam. *Geomatics Workbooks*, *12*, 471–476.

Armbrust, M., Fox, A., Griffith, R., Joseph, A. D., Katz, R., Konwinski, A., ... Stoica, I. (2010). A view of cloud computing. *Communications of the ACM*, *53*(4), 50–58. doi:10.1145/1721654.1721672

Ashford, R. (2010). QR codes and academic libraries Reaching mobile users. *College & Research Libraries News*, *71*(10), 526–530. doi:10.5860/crln.71.10.8454

Barrera, J. F., Mira, A., & Torroba, R. (2013). Optical encryption and QR codes: Secure and noise-free information retrieval. *Optics Express*, *21*(5), 5373–5378. doi:10.1364/OE.21.005373 PMID:23482108

Beckmann, N., Kriegel, H. P., Schneider, R., & Seeger, B. (1990). The R*-tree: An efficient and robust access method for points and rectangles. *SIGMOD '90 Proceedings of the 1990 ACM SIGMOD international conference on Management of data*, 322-331.

Bédard, Y. B. (1997). *Spatial OLAP*. Academic Press.

Bédard, Y., Merrett, T., & Han, J. (2001). Fundamentals of Spatial Data Warehousing for Geographic Knowledge Discovery. Geographic Data Mining and Knowledge Discovery, 2, 53-73.

Bédard, Y., Gosselin, P., Rivest, S., Proulx, M. J., Nadeau, M., Lebel, G., & Gagnon, M. F. (2003). Integrating GIS components with knowledge discovery technology for environmental health decision support. *International Journal of Medical Informatics*, *70*(1), 79–94. doi:10.1016/S1386-5056(02)00126-0 PMID:12706184

Bédard, Y., & Han, J. (2009). *Fundamentals of Spatial Data Warehousing for Geographic Knowledge Discovery*. Taylor & Francis Group. doi:10.1201/9781420073980.ch3

Bennett, J. (2010). *OpenStreetMap*. Birmingham, UK: Packt Publishing.

Berkhin, P. (2011). Bookmark-coloring algorithm for personalized page rank computing. *Internet Mathematics*, *3*(1), 41–62. doi:10.1080/15427951.2006.10129116

Beven, K., & Kirkby, M. J. (1979). A physically based, variable contributing area model of basin hydrology. *Hydrological Sciences Bulletin*, *24*(1), 43–69. doi:10.1080/02626667909491834

Bhattacharya, B., Lobbrecht, A. H., & Solomatine, D. P. (2003). Neural networks and reinforcement learning in control of water systems. *J Water Res Pl-Asce*, *129*(6), 458–465. doi:10.1061/(ASCE)0733-9496(2003)129:6(458)

Bian, L., & Walsh, S. J. (1993). Scale dependencies of vegetation and topography in a mountainous environment of Montana. *The Professional Geographer*, *45*(1), 1–11. doi:10.1111/j.0033-0124.1993.00001.x

Bimonte, S., Pinet, F., Miralles, A., & Papajorgji, P. (2014). Spatial data warehouses and SOLAP. *GeoInformatica*, *18*(2), 269–272. doi:10.100710707-013-0196-9

Biswas, A., Cresswell, H., Viscarra Rossel, R., & Si, C. B. (2012, April). *Application of curvelet transform in denoising digital elevation model*. In the 5th Global Workshop on Digital Soil Mapping, Sydney, Australia.

Bjorke, J. T., & Nilsen, S. (2003). Wavelets applied to simplification of digital terrain models. *International Journal of Geographical Information Science*, *17*(7), 601–621. doi:10.1080/1365881031000135500

Brasington, J., & Richards, K. (1998). Interactions between model predictions, parameters, and DTM Scales for TOPMODEL. *Computers & Geosciences*, *24*(4), 299–314. doi:10.1016/S0098-3004(97)00081-2

Broder, A. Z., Charikar, M., Frieze, A. M., & Mitzenmacher, M. (1998). *Min wise independent permutations*. STOC. Retrieved from http://citeseerx.ist.psu.edu/viewdoc/download?doi=10.1.1.121.8215&rep=rep1&type=pdf

Brovelli, M. A., Valentini, L., Zamboni, G. 2012b Multi-frame and multi-dimensional historical digital cities: The Como example. *International Journal of Digital Earth*. DOI:10.1080/17538947.2012.739662

Brovelli, M. A., & Zamboni, G. (2012). Virtual globes for 4D environmental analysis. *Applied Geomatics*, *4*(3), 163–172. doi:10.100712518-012-0091-3

Bürger, C. M., Kollet, S., Schumacher, J., & Bösel, D. (2012). Introduction of a web service for cloud computing with the integrated hydrologic simulation platform ParFlow. *Computers & Geosciences*, *48*, 334–336. doi:10.1016/j.cageo.2012.01.007

Burrough, P. A. (1986). *Principles of geographic information systems for earth resources assessment*. Oxford, UK: Clarendon Press.

Ca^mara, G., & Onsrud, H. (2004). Open-source geographic information systems software: myths and realities. In *Open Access and the Public Domain in Digital Data and Information for Science: Proceedings of an International Symposium*. National Research Council, U.S. National Committee for CODATA.

California Culvert Practice. (1955). Department of Public Works, Division of Highways, Sacramento, CA.

Câmara, G. (2003). *Why Open Source GIS Software?* Retrieved from http://www.directionsmag.com/entry/why-open-source-gis-software/123925

Cao, X., Cong, G., & Jensen, C. S. (2010). Retrieving top-k prestige-based relevant spatial web objects. *Proceedings of the VLDB Endowment International Conference on Very Large Data Bases*, *3*(1).

Carr, N. G. (2008). *The big switch: Rewiring the world, from Edison to Google*. WW Norton & Company.

Cary, A., Wolfson, O., & Rishe, N. (2010). Efficient and scalable method for processing top-k spatial boolean queries. *22nd International Conference on Scientific and Statistical Database Management, SSDBM 2010*, 87-95. 10.1007/978-3-642-13818-8_8

CESS. (2004). *Bharathapuzha and Its Problems with Special Preference to Sand Mining from the River Stretch between Chamravattom and Thirunavaya*. Trivandrum: Centre for Earth Science Studies.

Chang, S. G., Yu, B., & Vetterli, M. (2000). Adaptive wavelet thresholding for image denoising and ccompression. *IEEE Transactions on Image Processing*, *9*(9), 1532–1546. doi:10.1109/83.862633 PMID:18262991

Chang, Y.-J., Tsai, S.-K., Chang, Y.-S., & Wang, T.-Y. (2007). A novel wayfinding system based on geo-coded QR codes for individuals with cognitive impairments. *Proceedings of the 9th international ACM SIGACCESS conference on Computers and accessibility*. 10.1145/1296843.1296887

Chaturvedi, M. C. (2010). *India's Waters: Advances in Development and Management*. CRC Press.

Chaudhuri, S., Ganti, V., & Kaushik, R. (2006). A primitive operator for similarity joins in data cleaning. *22nd International Conference on Data Engineering (ICDE'06)*, 5–16. 10.1109/ICDE.2006.9

Chen, D., Shams, S., Carmona-Moreno, C., & Leone, A. (2010, October). Assessment of Open Source GIS Software for Water Resources Management in Developing Countries. *Journal of Hydro-environment Research*, *4*(3), 253–264. doi:10.1016/j.jher.2010.04.017

Cheng, J., & Greiner, R. (2011). *Learning Bayesian Belief Network Classifiers*. University of Alberta.

Chen, Y., Zhu, D., & Zhao, J. (2004, September). Small basin flash flood simulation with TOPMODEL. In *International Conference of GIS and Remote Sensing in Hydrology, Water Resources and Environment (ICGRSHWE), Three Gorges Dam* (pp. 41-49). Academic Press.

Choi, J., Engel, B., & Farnsworth, R. (2005). Web-based GIS and spatial decision support system for watershed management. *Journal of Hydroinformatics*, *7*, 165–174.

Chouhan, T. S. (2005). *Degree, Extent and Treatment of Desertification Hazards in India. Sociedade & Natureza, 1*(1), 901–919.

Cohn, J. P. (2008). Citizen science: Can volunteers do real research? *Bioscience*, *58*(3), 192–197. doi:10.1641/B580303

Comité, F., & Delepoulle, S. (2005). PIQLE: a platform for implementation of Q-learning experiments. *NIPS workshop: reinforcement learning benchmarks and bake-offs II.*

Conde-Lagoa, D., Costa-Montenegro, E., Gonzalez-Castao, F., & Gil-Castieira, F. (2010). *Secure eTickets based on QR-Codes with user-encrypted content.* Paper presented at the Consumer Electronics (ICCE), 2010 Digest of Technical Papers International Conference on. 10.1109/ICCE.2010.5418880

Confino, J. P., & Laplante, P. A. (2010). An open Source Software Evaluation Model. *International Journal of Strategic Information Technology and Applications*, *1*(1), 60–77. doi:10.4018/jsita.2010101505

Cong, G., Jensen, C. D., & Wu, D. (2009). Efficient retrieval of the top-k most relevant spatial web objects. *Proceedings of the VLDB Endowment International Conference on Very Large Data Bases*, *2*(1), 337–348. doi:10.14778/1687627.1687666

Coolidge, T. (2015). *APDM.net | ArcGIS Pipeline Data Model (APDM) Clearinghouse*. Retrieved from http://apdm.net/

Cracknell, A. P., & Hayes, L. W. B. (1991). *Introduction to remote sensing*. London: Taylor and Francis.

Crawford, N. H., & Linsley, R. K. (1966). *Digital simulation in hydrology: Stanford Watershed Model IV.* Department of Civil Engineering Technical Report 39. Palo Alto, CA: Stanford University.

Crothers, H. Esri. (2011). *Benefits of the Local Government Information Model for Water Utilities | ArcGIS Blog*. Retrieved from http://blogs.esri.com/esri/arcgis/2011/10/11/benfits-of-the-local-government-information-model-for-water-utilities/

Daly, R. S. Q. (2011). *Learning Bayesian networks: approaches and.* Glasgow, UK: Cambridge University Press.

De Cola, L. (1989). Fractal analysis of a classified landsat scene. *Photogrammetric Engineering and Remote Sensing*, *55*(5), 601–610.

Delipetrev, B., Jonoski, A., & Solomatine, D. P. (2014). A novel nested dynamic programming algorithm for optimal reservoir operation. Hydroinformatic conference.

Delipetrev, B., Jonoski, A., & Solomatine, D. P. (2014). Development of a web application for water resources based on open source software. *Comput Geosci-Uk*, *62*, 35–42. doi:10.1016/j.cageo.2013.09.012

Delipetrev, B., Mihajlov, D., Delipetrev, M., & Delipetrov, T. (2010). Model of the Hydro-Information System of the Republic of Macedonia. *CIT. Journal of Computing and Information Technology*, *18*(2), 201–204. doi:10.2498/cit.1001804

Doc, E. A. (2015, August 28). *Local MBTiles | ArcGIS for Developers*. Retrieved August 28, 2015, from https://developers.arcgis.com/android/sample-code/local-mbtiles/

Doc, S. S. (2015, September 3). *Spatial Statistics toolbox*. Retrieved September 3, 2015, from http://webhelp.esri.com/arcgisdesktop/9.3/index.cfm?TopicName=An_overview_of_the_Spatial_Statistics_toolbox

Does PostgreSQL perform on Windows as well as on Linux - Server Fault. (n.d.). Retrieved August 27, 2015, from http://serverfault.com/questions/674609/does-PostgreSQL-perform-on-windows-as-well-as-on-linux

Donnelly, F. P. (2010). Evaluating open source GIS for libraries. *Library Hi Tech*, *28*(1), 131–151. doi:10.1108/07378831011026742

Donoho, D. L. (1995). De-Noising by soft-thresholding. *IEEE Transactions on Information Theory*, *41*(3), 613–627. doi:10.1109/18.382009

Donoho, D. L., & Johnstone, I. M. (1994). Ideal spatial adaptation via wavelet shrinkage. *Biometrika*, *81*(3), 425–455. doi:10.1093/biomet/81.3.425

Elvidge, C. D., & Tuttle, B. T. 2008 How virtual globes are revolutionizing earth observation data access and integration. The International Archives of the Photogrammetry, Remote Sensing and Spatial Information Sciences, 37, 137-140.

Esri GitHub. (2015, August 27). *Esri on Github*. Retrieved August 27, 2015, from https://github.com/Esri

ESRI. (2015). Retrieved 2015, from http://support.esri.com

Ester, M., Kriegel, H. P., & Sander, J. (1997). Spatial Data Mining: A Database Approach. *Symposium on Large Spatial Databases*, 47-66. 10.1007/3-540-63238-7_24

Fairservis, W. A. (1967). The origin, character, and decline of an early civilization. *American Museum Novitates*, 2302.

Falorni, G., Teles, V., Vivoni, E. R., Bras, R. L., & Amaratunga, K. S. (2005). Analysis and characterization of the vertical accuracy of digital elevation models from the Shuttle Radar Topography Mission. *Journal of Geophysical Research*, *110*(F2), F02005. doi:10.1029/2003JF000113

Farajalla, N. S., & Vieux, B. E. (1995). Capturing the essential spatial variability in distributed hydrological modeling: Infiltration parameters. *Hydrological Processes*, *8*(1), 55–68. doi:10.1002/hyp.3360090106

Felipe, I. D., Hristidis, V., & Rishe, N. (2008). *Keyword search on spatial databases*. ICDE.

Fernando. (2000). *Geomatic Methods for the Analysis of Data in the Earth Sciences New*. Academic Press.

Finkel, R. A., & Bentley, J. L. (1974). Quad trees: A data structure for retrieval on composite keys. *Acta Informatica*, *4*(1), 1–9. doi:10.1007/BF00288933

Foufoula-Georgiou, E., & Kumar, P. (1995). *Wavelet in geophysics*. New York: Academic Press.

Fraley, C. (2015). *mclust: Normal Mixture Modelling for Model-Based Clustering, Classification, and Density Estimation* (Version 5.0.2). Retrieved from https://cran.r-project.org/web/packages/mclust/index.html

Free Software Foundation. (n.d.). Retrieved from http://www.fsf.org/licensing/essays/free-sw.html

Frommelt. (1998). Algorithms for Characterization and Trend Detection. *KDD: Proceedings / International Conference on Knowledge Discovery & Data Mining. International Conference on Knowledge Discovery & Data Mining*, 44–50.

GADM database. (2015). *Global Administrative Areas*. Retrieved August 30, 2015, from http://www.gadm.org

Galli, P. (n.d.). *Open Source Is the Big Disruptor*. Retrieved from http://www.eweek.com/article2/0,1895,2186932,00.asp

Gernet, J. (1996). *A History of Chinese Civilization*. Cambridge University Press.

Girotra, M., Nagpal, K., Minocha, S., & Sharma, N. (2013). Comparative Survey on Association Rule Mining Algorithms. India. *International Journal of Computers and Applications*, *84*(10), 18–22. doi:10.5120/14612-2862

GIS StackExchange. (2015, September 3). *arcpy tag wiki*. Retrieved September 3, 2015, from http://gis.stackexchange.com/tags/arcpy/info

Goodchild, M. F. (2007a). Citizens as Voluntary Sensors: Spatial Data Infrastructure in the World of Web 2.0. *International Journal of Spatial Data Infrastructures Research*, *2*, 24–32.

Goodchild, M. F. (2007b). Citizens as Sensors: The world of volunteered geography. *GeoJournal*, *69*(4), 211–221. doi:10.100710708-007-9111-y

Gravano, L., Ipeirotis, P. G., Jagadish, H. V., Koudas, N., Muthukrishnan, S., & Srivastava, D. (2001). Approximate string joins in a database. *Proceedings of the 27th VLDB Conference*, 491–500.

Greiner, J. C. R. (2005). *Comparing Bayesian Network Classifiers*. Academic Press.

Guo, H., & Hsu, W. (2002). *A Survey of Algorithms for Real-Time Bayesian Network Inference*. Kansas State, AAAI Technical Report WS-02-15.

Guo, D. (2012). *Spatial data mining and geographic knowledge discovery—An introduction*. ESRI.

Guttman, A. (1984). R-trees: A dynamic index structure for spatial searching. *SIGMOD '84 Proceedings of the 1984 ACM SIGMOD international conference on Management of data*, 47–57.

Haan, C. T., Barfield, B. J., & Hayes, J. C. (1994). *Design hydrology and sedimentology for small catchments*. Academic Press.

Hadjieleftheriou, M., Chandel, A., Koudas, N., & Srivastava, D. (2008a). *Set similarity selection queries at interactive speeds*. ICDE.

Hadjieleftheriou, M., Chandel, A., Koudas, N., & Srivastava, D. (2008b). Fast indexes and algorithms for set similarity selection queries. *Proceeding ICDE '08 Proceedings of the 2008 IEEE 24th International Conference on Data Engineering*, 267-276. 10.1109/ICDE.2008.4497435

Hagander, M. (2015, August 27). *Is PostgreSQL suited to one OS? Is it better on Linux than Windows? - Server Fault*. Retrieved August 27, 2015, from http://serverfault.com/questions/222430/is-PostgreSQL-suited-to-one-os-is-it-better-on-linux-than-windows

Haklay, M., Singleton, A., & Parker, C. (2008). Web Mapping 2.0: The Neogeography of the GeoWeb. *Geography Compass*, *2*(6), 2011–2039. doi:10.1111/j.1749-8198.2008.00167.x

Hariharan, R., Hore, B., Li, C., & Mehrotra, S. (2007). Processing spatial keyword queries in geographic information retrieval systems. *19th International conference on scientific and statistical database management. SSDBM.*

Hartmann, D. L. (1994). *Global Physical Climatology*. San Diego, CA: Academic Press.

Hartnett, P., & Bertolotto, M. (2003). Developing interoperable web-based geo- spatial Digital. GIRSUK 2003. *Session*, *5b*, 185–188.

Harvey, M., & Han, J. (2001). Geographic data mining and knowledge discovery: An overview. In Geographic Data Mining and Knowledge Discovery (pp. 3-32). Academic Press.

Hemalatha, M., & Naga Saranya, N. (2011). A Recent Survey on Knowledge Discovery in Spatial Data. *International Journal of Computer Science Issues, 8*(3), 473-479.

Herring, C. (1994). *An Architecture of Cyberspace: Spatialization of the Internet*. U.S. Army Construction Engineering Research Laboratory. Retrieved from http://citeseerx.ist.psu.edu/viewdoc/download?doi=10.1.1.37.4604&rep=rep1&type=pdf

Horak, J., Orlik, A., & Stromsky, J. (2008). Web services for distributed and interoperable hydro-information systems. *Hydrology and Earth System Sciences*, *12*(2), 635–644. doi:10.5194/hess-12-635-2008

Horton, R. E. (1945). Erosional Development of Streams and their Drainage Basins: Hydro-Physical Approach to Quantitative Morphology. *Geological Society of America Bulletin*, *56*(3), 275–370. doi:10.1130/0016-7606(1945)56[275:EDOSAT]2.0.CO;2

Howe, J. (2006). The rise of crowdsourcing. *Wired Magazine*, *14*(6), 176–183. doi:10.1016/j.isprsjprs.2010.06.005

Ibbitt, R., & Woods, R. (2004). Re-Scaling the topographic index to improve the representation of physical processes in catchment models. *Journal of Hydrology (Amsterdam)*, *293*(1-4), 205–218. doi:10.1016/j.jhydrol.2004.01.016

Ichoku, C., Karnieli, A., & Verchovsky, I. (1996). Application of fractal techniques to the comparative evaluation of two methods of extracting channel networks from digital elevation models. *Water Resources Research, 32*(2), 389–399. doi:10.1029/95WR02762

Ilyas, I.F., Beskales, G., & Soliman, M.A. (2008). A survey of top-k query processing techniques in relational database systems. *ACM Computing Surveys, 40*(4), 1–58.

Infrastructures, D.S.D. (2004). *The SDI Cookbook.* GSDI/Nebert.

Inmon, W. (2002). *Building the Data Warehouse* (3rd ed.). Wiley.

Is PostgreSQL suited to one OS? Is it better on Linux than Windows? - Server Fault. (n.d.). Retrieved August 27, 2015, from http://serverfault.com/questions/222430/is-PostgreSQL-suited-to-one-os-is-it-better-on-linux-than-windows

Jagadish, H. V., Ooi, B. C., Tan, K., Yu, C., & Zhang, R. (2005). iDistance: An adaptive B+ tree based indexing method for nearest neighbor search. *ACM Transactions on Database Systems, 30*(2), 364–397. doi:10.1145/1071610.1071612

Jain, A., & Dubes, R. (1988). Algorithms for Clustering Data. Prentice-Hall.

James, S.M. (1983). *South Fork Eel Watershed erosion investigation.* Northern District, California Department of Water Resources.

Jansen, M. (2001). *Noise reduction by wavelet thresholding. 161 of Lecture Notes in Statistics.* New York: Springer-Verlag. doi:10.1007/978-1-4613-0145-5

Jassar, K. K., & Dhindsa, K. S. (2015). Comparative Study of Spatial Data Mining Technique. *International Journal of Computers and Applications.*

Jeh, G., & Widom, J. (2003). *Scaling personalized web search.* Retrieved from: http://infolab.stanford.edu/~glenj/spws.pdf

Jensen, S., & Domingue, J. O. (1988). Extracting topographic structure from digital elevation model data for geographic information system analysis. *Photogrammetric Engineering and Remote Sensing, 54*(11), 1593–1600.

Jiang, L., & Wong, W. S. (2010). Effects of DEM Source on Hydrologic applications. *Computers, Environment and Urban Systems, 34*(3), 251–261. doi:10.1016/j.compenvurbsys.2009.11.002

Joachims, T. (2001). A statistical learning model of text classification for support vector machines. *Proc. of ACM*, 128–136. 10.1145/383952.383974

Jolma, A., Lehikoinen, A., & Helle, I. (2011). Coupling Bayesian networks and geospatial software for. *19th International Congress on Modelling and Simulation*, Perth, Australia.

Jones, K. H. (1998). A Comparison of Two Approaches to Ranking Algorithms Used to Compute Hill Slopes. *GeoInformatica*, 2(3), 235–256. doi:10.1023/A:1026472421094

Jonoski, A., Alfonso, L., Almoradie, A., Popescu, I., van Andel, S. J., & Vojinovic, Z. (2012). Mobile phone applications in the water domain. *Environmental Engineering and Management Journal*, 11.

Jonoski, A., Almoradie, A., Khan, K., Popescu, I., & Van Andel, S. (2013). Google Android mobile phone applications for water quality information management. *Journal of Hydroinformatics*, *15*(4), 1137–1149. doi:10.2166/hydro.2012.147

Kachelriess, R. Esri. (2015a, August 28). *Esri/geoportal-server*. Retrieved August 28, 2015, from https://github.com/Esri/geoportal-server/

Kachelriess, R. Esri. (2015b, August 28). *Geoportal server as a broker · Esri/geoportal-server Wiki*. Retrieved August 28, 2015, from https://github.com/Esri/geoportal-server/wiki/Geoportal-server-as-a-broker

Kaelbling, L. P. (1996). Special issue on reinforcement learning - Introduction. *Machine Learning*, *22*(1-3), 7–9. doi:10.1007/BF00114721

Kagan, A. M., Linnik Yu, V., & Rao, C. R. (1973). *Characterization problems in mathematical statistics*. New York: Wiley.

Kane, D., & Schneidewind, J. (2011). QR codes as finding aides: Linking electronic and print library resources. *Public Services Quarterly*, *7*(3-4), 111–124. doi:10.1080/15228959.2011.623599

Kan, T.-W., Teng, C.-H., & Chou, W.-S. (2009). Applying QR code in augmented reality applications. *Proceedings of the 8th International Conference on Virtual Reality Continuum and its Applications in Industry*. 10.1145/1670252.1670305

Kanth, K. V. R., Ravada, S., & Abugov, D. (2002). Quadtree and r-tree indexes in oracle spatial: a comparison using gisdata. *SIGMOD '02 Proceedings of the 2002 ACM SIGMOD international conference on Management of data*, 546-557.

Kaufman, L., & Rousseeuw, P. (1990). Finding Groups in Data: An Introduction to Cluster Analysis. New York, NY: John Wiley and Sons.

Khan, S. & Mustafa, J. (2014). Effective Semantic search using thematic similarity. *Journal of King Saud University – Computer and Information Sciences, 26*(2), 161-169.

Khodaei, A., Shahabi, C., & Li, C. (2010). *Hybrid indexing and seamless ranking of spatial and textual features of web documents*. DEXA. doi:10.1007/978-3-642-15364-8_37

Kienzle, S. (2004). The effect of DEM raster resolution on first order, second order and compound terrain derivatives. *Transactions in GIS*, *8*(1), 83–111. doi:10.1111/j.1467-9671.2004.00169.x

Kim, B. J. (2013). A System for detecting the Stray of Objects within User-defined Region using Location-Based Services. *International Journal of Software Engineering and Its Applications*.

Kivistö, S. (2015, August 27). *PostgreSQL on Windows and Linux*. Retrieved August 27, 2015, from http://serverfault.com/questions/674609/does-PostgreSQL-perform-on-windows-as-well-as-on-linux

Koutsoyiannis, D. (2005). Uncertainty, entropy, scaling and hydrological statistics. 1. Marginal distributional properties of hydrological processes and state scaling. *Hydrological Sciences Journal*, *50*(3), 381–404. doi:10.1623/hysj.50.3.381.65031

Kuba, M. (2015, August 27). *Java Performance On Various Platforms*. Retrieved August 27, 2015, from http://dior.ics.muni.cz/~makub/java/speed.html

Kuba, P. (2001). *Data structures for spatial data mining*. Masaryk University Brno.

Kuchment, L. S., & Gelfan, A. N. (2007). Statistical self-similarity of spatial variations of snow cover and its application for modelling snowmelt runoff generation in basins with a sparse snow measurement network. *Predictions in ungauged basins: PUB Kick-off, 309,* 156-163.

Kuhn, W. (2012). Core concepts of spatial information for transdisciplinary research. *International Journal of Geographical Information Science*, *26*(12), 2267–2276. doi:10.1080/13658816.2012.722637

Kundassery & Babu. (2013). RiverineGIS-A Geo-visual Framework for enhancing River Literacy. *Indian Cartographer*, *33*, 453–458.

Kundassery & Babu. (2015). An Oceanographic Exploration of the Solar System. *PLANEX*, 7-11.

Kuo, W. L., Steenhuis, T. S., McCulloch, C. E., Mohler, C. L., Weinstei, D. A., DeGloria, S. T., & Swaney, D. (1999). Effect of grid size on runoff and soil moisture for a variable-source-area hydrology model. *Water Resources Research*, *35*(11), 3419–3428. doi:10.1029/1999WR900183

Labat, D. (2005). Recent advances in wavelet analyses: Part 1–A Review of Concepts. *Journal of Hydrology (Amsterdam)*, *314*(1-4), 275–288. doi:10.1016/j.jhydrol.2005.04.003

Lam, N. S. (1990). Description and measurement of landsat TM images using fractals. *Photogrammetric Engineering and Remote Sensing*, *56*(2), 187–195.

Language, W. O. (2015, September 4). In *Wikipedia, the free encyclopedia*. Retrieved from https://en.wikipedia.org/w/index.php?title=Web_Ontology_Language&oldid=679385402

Lashermes, B., Foufoula-Georgiou, E., & Dietrich, W. E. (2007). Channel network extraction from high resolution topography using wavelets. *Geophysical Research Letters*, *34*(23), n/a. doi:10.1029/2007GL031140

Lee, J. S. (1981). Speckle analysis and smoothing of synthetic aperture radar images. *Computer Graphics and Image Processing*, *17*(1), 24–32. doi:10.1016/S0146-664X(81)80005-6

Lee, J.-G. A. (2015). *Geospatial Big Data: Challenges and Opportunities*. Elsevier Inc.

Levenshtein, V. (1965). Binary codes capable of correcting spurious insertions and deletions of ones. *Soviet Physics, 10*(8).

Li, C., Lu, J., & Lu, Y. (2008). Efficient merging and filtering algorithms for approximate string searches. *2008 IEEE 24th International Conference on Data Engineering*.

Li, C., Wang, B., & Yang, X. (2007). VGRAM: improving performance of approximate queries on string collections using variable length grams. *Proceeding VLDB '07 Proceedings of the 33rd international conference on Very large data bases*, 303–314.

Li, Z., Zhu, Q., & Gold, C. (2005). *Digital terrain modeling: Principles and methodology*. Boca Raton, FL: CRC Press.

Loucks, E. B. (2007). *Water Resources Systems Planning and Management - an introduction to methods, models and applications*. UNESCO Publishing.

Lounge, G. (2014). *What is the Difference Between GIS and Geospatial?* Academic Press.

Madhav Gadgil Commission. (2011). *Report of the Western Ghats Ecology Expert Panel*. The Ministry of Environment and Forests, Government of India.

Maguire, D. J. (2007). GeoWeb 2.0 and Volunteered GI. *Workshop on Volunteered Geographic Information*, 104-106.

Maidment, D. R. (1993). GIS and hydrologic modeling, in environmental modeling with GIS. In M. F. Goodchild, B. O. Parks, & L. T. Steyaert (Eds.), (pp. 147–167). New York: Oxford University Press.

Mallat, S. G. (1998). *A wavelet tour of signal processing* (2nd ed.). San Diego, CA: Academic Press.

Mallat, S., & Hwang, W. L. (1992). Singularity detection and processing with wavelets. *IEEE Transactions on Information Theory*, *38*(2), 617–643. doi:10.1109/18.119727

Mandelbrot, B. B. (1982). The fractal geometry of nature. W. H. Freeman and Company.

Mandelbrot, B. B. (1967). How long is the coast of Britain? Statistical self-Similarity and fractional dimension. *Science. New Series*, *156*(3775), 636–638. PMID:17837158

Mapbox. (2015a, August 28). *Mbtile | Mapbox*. Retrieved August 28, 2015, from https://www.mapbox.com/guides/an-open-platform/#mbtiles

Mapbox. (2015b, August 28). *Vector Tile | Mapbox*. Retrieved August 28, 2015, from https://www.mapbox.com/guides/an-open-platform/#vector-tiles

Marmulla, G. (Ed.). (2001). *Dams, fish and fisheries-Opportunities, challenges and conflict resolution. FAO Fisheries Technial Paper, Issue 419*. Food and Agriculture Organization of the United Nations.

Marquez, A. (2015). *PostGIS essentials*. Birmingham, UK: Packt Publishing.

Martin Ester, H.-p. K. (2001). *Algorithms and applications for spatial data mining*. Geographic data mining and knowledge discovery .

Maruyama, T., Kawachi, T., & Singh, V. P. (2005). Entropy-Based assessment and clustering of potential water resources availability. *Journal of Hydrology (Amsterdam)*, *309*(1–4), 104–113. doi:10.1016/j.jhydrol.2004.11.020

Massis, B. E. (2011). QR codes in the library. *New Library World*, *112*(9/10), 466–469. doi:10.1108/03074801111182058

Mazeika, A., Bohlen, A. M., Koudas, N., & Srivastava, D. (2007). Estimating the selectivity of approximate string queries. *ACM TODS*, *32*(2), 12–52, es. doi:10.1145/1242524.1242529

McKnee. (2004). *The Spatial Web. OGC White Papers*. Retrieved from portal.opengeospatial.org/files/?artifact_id=3859

McLachlan, G., & Basford, K. (1988). Mixture Models: Inference and Applications to Clustering. Marcel Dekker.

Mell, P., & Grance, T. (2009). The NIST definition of cloud computing. *National Institute of Standards and Technology*, *53*, 50.

Melton, M. A. (1957). *An analysis of the relations among elements of climate, surface properties, and geomorphology* (Tech. Rep. No. 3). Office of Naval Research Project NR389-042. New York: Columbia University, Department of Geology.

Mendicino, G., & Sole, A. (1997). The information content theory for the estimation of the topographic index distribution used in TOPMODEL. *Hydrological Processes*, *11*(9), 1099–1114. doi:

Mihajlovic, V. P. (2012). *Dynamic Bayesian Networks:A State of the Art*. University of Twente.

Miller, K. W., Voas, J., & Costello, T. (2010). Free and Open Source Software. *IT Professional*, *12*(6), 14–16. doi:10.1109/MITP.2010.147

Mithal, N. G. (2010). *Spatial Data warehouses.* Academic Press.

Mithal, N., & Garg, S. (2010). *Spatial Data warehouses.* Academic Press.

Model, E. I. (2015). *Local Government Information Model | ArcGIS for Local Government.* Retrieved September 6, 2015, from http://solutions.arcgis.com/local-government/help/local-government-information-model/

Moore, I. D. (1988). A contour-Based terrain analysis program for the environmental sciences (TAPES). *Transactions - American Geophysical Union*, *69*(16), 345.

Moore, I. D., Gessler, P. E., & Nielsen, G. A. (1993). Terrain attributes: estimation methods and scale effects. In A. J. Jakeman, M. B. Beck, & M. McAleer (Eds.), *Topics in Modeling Change in Environmental Systems* (pp. 189–214). London: Wiley.

Moore, I. D., Grayson, R. B., & Ladson, A. R. (1991). Digital terrain modeling: A review of hydrological, geomorphological and biological applications. *Hydrological Processes*, *5*(1), 3–30. doi:10.1002/hyp.3360050103

Morrison, K. W., & Purves, R. S. (2002). Customizable landscape visualizations. Implementation, application and testing of a web-based tool. *Computers, Environment and Urban Systems*, *26*(2-3), 163–183. doi:10.1016/S0198-9715(01)00033-3

Nascimento, D. M., & Raghavan, V. (2008). Participatory Free and Open Source GIS in the Web 2.0 - Exploring trends in GIS in times of Collaborative Creation. *Proceedings of the FOSS4G 2008*. Retrieved from http://wgrass.media.osaka-cu.ac.jp/gisideas10/papers/c89a8e86e7c71486cbea29a3ac62.pdf

National Research Council (NRC). (2006). *Learning to Think Spatially: GIS as a Support System in the K-12 Curriculum*. Washington, DC: The National Academies Press.

Navarro, G. (2001). A guided tour to approximate string matching. *ACM Computing Surveys*, *33*(1), 31–88. doi:10.1145/375360.375365

NGSS Lead States. (2013). *Next Generation Science Standards: For States, By States*. Washington, DC: The National Academies Press.

Niedda, M. (2004). Up scaling hydraulic conductivity by means of entropy of terrain curvature representation. *Water Resources Research*, *40*(4). doi:10.1029/2003WR002721

Nourani, V. (2011). Application of continues wavelet transform to investigate the resolution effect of digital elevation model on topographic index distribution. *International Journal of Soft Computing and Engineering*, *1*(5), 405–410.

Nourani, V., Komasi, M., & Mano, A. (2009). A multivariate ANN-Wavelet approach for rainfall-runoff modeling. *Water Resources Research*, *23*(14), 2877–2894.

Nourani, V., & Mano, A. (2007). Semi-distributed flood runoff model at the subcontinental scale for southwestern Iran. *Hydrological Processes*, *21*(23), 3173–3180. doi:10.1002/hyp.6549

Nourani, V., & Zanardo, S. (2013). Wavelet-Based regularization of the extracted topographic index from high-resolution topography for hydro-geomorphic applications. *Hydrological Processes*, *28*(3), 1345–1357. doi:10.1002/hyp.9665

O'Callaghan, J. F., & Mark, D. M. (1984). The extraction of drainage networks from digital elevation data. *Computer Vision Graph*, *28*(3), 323–344. doi:10.1016/S0734-189X(84)80011-0

O'Loughlin, E. M. (1986). Predictions of surface saturation zones in natural catchments by topographic analysis. *Water Resources Research*, *22*(5), 794–804. doi:10.1029/WR022i005p00794

O'Neill, R. V., Krummel, J. R., Gardner, R. H., Sugihara, G., Jackson, B., DeAngelis, D. L., ... Graham, R. L. (1988). Indices of landscape pattern. *Landscape Ecology*, *1*(3), 153–162. doi:10.1007/BF00162741

O'Reilly, T. (2005). *What is Web 2.0: Design patterns and business models for the next generation of software*. Retrieved from http://oreilly.com/pub/a/web2/archive/what-is-web-20.html

OGC GeoPackage. (2015, August 28). Retrieved August 28, 2015, from http://www.geopackage.org/

OGC. (2015, August 27). *OGC Compliant Products*. Retrieved August 27, 2015, from http://www.opengeospatial.org/resource/products/compliant#ESRI

Open Source Technology and Esri | ArcNews Online. (2011). Retrieved August 28, 2015, from http://www.esri.com/news/arcnews/spring11articles/open-source-technology-and-esri.html

Pagel, B. U., Six, H. U., Toben, H., & Widmayer, P. (1993). Towards an analysis of range query performance in spatial data structures. *PODS '93 Proceedings of the twelfth ACM SIGACT-SIGMOD-SIGART symposium on Principles of database systems*, 214-221. 10.1145/153850.153878

Pan, F., & King, A. W. (2012). Downscaling 1-km topographic index distributions to a finer resolution for the TOPMODEL-Based GCM hydrological modeling. *Journal of Hydrologic Engineering*, *17*(2), 243–251. doi:10.1061/(ASCE)HE.1943-5584.0000438

Panuska, J. C., Moore, I. D., & Kramer, L. A. (1991). Terrain analysis: Integration into the agriculture nonpoint source (AGNPS) pollution model. *Journal of Soil and Water Conservation*, *46*(1), 59–64.

Papadias, D., Mamoulis, N., & Theodoridis, Y. (1999). Processing and optimization of multi way spatial joins using R-trees. *Proceeding PODS '99 Proceedings of the eighteenth ACM SIGMOD-SIGACT-SIGART symposium on Principles of database systems*, 44-55. 10.1145/303976.303981

Passalacqua, P., Tarolli, P., & Foufoula-Georgiou, E. (2010). Testing space-scale methodologies for automatic geomorphic feature extraction from LIDAR in a complex mountainous landscape. *Water Resources Research*, *46*(11). doi:10.1029/2009WR008812

Peckham, S. D., & Goodall, J. L. (2013). Driving plug-and-play models with data from web services: A demonstration of interoperability between CSDMS and CUAHSI-HIS. *Comput Geosci-Uk*, *53*, 154–161. doi:10.1016/j.cageo.2012.04.019

Peng, Z. R., & Tsou, M. H. (2003). *Internet GIS*. Hoboken, NJ: John Wiley &Sons.

Petek, T. (2011). *Geospatial Data Infrastructure for Economic Development.* Academic Press.

Planning Commission. (2007). *Report of the Steering Committee on Water Resources for Eleventh Five Year Plan (2007-2012).* Government of India.

Ponte, J. M., & Croft, W. B. (1998). *A language modeling approach to information retrieval.* SIGIR. doi:10.1145/290941.291008

Project, R. (2015, September 2). *R: What is R?* Retrieved September 2, 2015, from https://www.r-project.org/about.html

QGIS. (2015). *Background of QGIS.* Retrieved on August 30, 2015, from https://en.wikipedia.org/wiki/QGIS

Quinn, P. F., Beven, K. J., & Lamb, R. (1995). The ln (a/tanb) index: How to calculate it and how to use it within the TOPMODEL framework. *Hydrological Processes*, *9*(2), 161–182. doi:10.1002/hyp.3360090204

Quiroga, V. M., Popescu, I., Solomatine, D., & Bociort, L. (2013). Cloud and cluster computing in uncertainty analysis of integrated flood models. *Journal of Hydroinformatics*, *15*(1), 55–70. doi:10.2166/hydro.2012.017

Raju, P. L. N. (2003). Fundamentals of Geographic Information System. *Proc. of the Training Workshop, held at Dehra Dun*, 103-120.

Ramachandran Nair, K. K. (1986). *Kerala State Gazetteer* (Vol. 1). Thiruvananthapuram, Kerala: Kerala Council for Historical Research.

Reed, P. M., Bills, B., Anderson, M., Ketchum, B., & Piasecki, M. (2010). The Mid-Atlantic Watershed Atlas (MAWA): Open access data search & watershed-based community building. *Environmental Modelling & Software*, *25*(6), 808–812. doi:10.1016/j.envsoft.2009.12.004

Rivest, S., Bédard, Y. M., & Proulx, T. B. (2005). SOLAP technology:Merging Hypermedia GIS with Spatial On-Line Analytical Processing: Towards Hypermedia SOLAP. *Geographic Hypermedia: Concepts and Systems.*

Rivest, S., Bédard, Y., Proulx, M. J., & Nadeau, M. (2003). *SOLAP: A new type of user interface to support spatio-temporal multidimensional data exploration and analysis*. Academic Press.

Rocha-Junior, J. B., Gkorgkas, O., Jonassen, S., & Nørvag, K. (2011). Efficient processing of top-k spatial keyword queries. *International Symposium on Spatial and Temporal Databases*, 205–222. 10.1007/978-3-642-22922-0_13

Rocha-Junior, J. B., Vlachou, A., Doulkeridis, C., & Nørvag, K. (2010). Efficient processing of top-k spatial preference queries. *Proceedings of the VLDB Endowment International Conference on Very Large Data Bases*, *4*(2), 93–104. doi:10.14778/1921071.1921076

Rodriguez-Iturbe, I., Rinaldo, A., Rigon, R., Bras, R. L., Ijjasz-Vasquez, E., & Marani, A. (1992). Fractal structures as least energy patterns: The case of river networks. *Geophysical Research Letters*, *19*(9), 889–892. doi:10.1029/92GL00938

Roussopoulos, N., Kelley, S., & Vincent, F. C. (1995a). *Nearest Neighbor Queries*. San Jose, CA: SIGMOD ACM Intl. Conf. on Management of Data. doi:10.1145/223784.223794

Roussopoulos, N., Kelley, S., & Vincent, F. C. (1995b). Nearest neighbour queries. *Proc. SIGMOD*, 71–79.

Ruiz, C. V., & Times, V. C. (2009). A Taxonomy of SOLAP Operators. *SBBD*, 151-165.

Sanchez.LA. (2013). *A Bayesian network-based method for spatial statistics*. Madrid: Computational Intelligence Group School of Computer Science, Technical University of Madrid.

Sang, Y. F., Wang, D., Wu, J. C., Zhu, Q. P., & Wang, L. (2009). Entropy-based wavelet de-Noising method for time series analysis. *Entropy (Basel, Switzerland)*, *11*(4), 1123–1147. doi:10.3390/e11041123

Sarafian, A. R., Nielsen, S. G., Marschall, H. R., McCubbin, F. M., & Monteleone, B. D. (2014). Early accretion of water in the inner solar system from a carbonaceous chondrite-like source. *Science*, *346*(6209), 623–626. doi:10.1126cience.1256717 PMID:25359971

Sarawagi, S., & Kirpal, A. (2004) Efficient set joins on similarity predicate. *Proceeding SIGMOD '04 Proceedings of the 2004 ACM SIGMOD international conference on Management of data*, 743-754. 10.1145/1007568.1007652

Saulnier, G., Obled, C., & Beven, K. (1997). Analytical compensation between DTM grid resolution and effective values of saturated hydraulic conductivity within the TOPMODEL framework. *Hydrological Processes*, *11*(9), 1331–1346. doi:

Schaefer, G., & Kennedy, J. T. (2011). *Billing with QR Codes*. Google Patents.

Scharl, A., & Tochtermann, K. (2007). *The Geospatial Web, Advanced Information and Knowledge Processing Series*. London: Springer.

Schneider, G. V. (2011). *On the Requirements for User-Centric Spatial*. Academic Press.

Schultz, R. B., Kerski, J. J., & Patterson, T. C. (2008). The use of virtual globes as a spatial teaching tool with suggestions for metadata standards. *The Journal of Geography*, *107*(1), 27–34. doi:10.1080/00221340802049844

Shannon, C. E. (1948). A mathematical theory of communications I and II. *The Bell System Technical Journal*, *27*(3), 379–443. doi:10.1002/j.1538-7305.1948.tb01338.x

Shapiro, J. M. (1993). Embedded image coding using zero-trees of wavelet coefficients. *Transactions on Signal Processing*, *11*, 3115–3162.

Shekhar, S., Vatsavai, R. R., Sahay, N., Burk, T. E., & Lime, S. (2001). WMS and GML based interoperable web mapping system. *Proceedings of the ninth ACM international symposium on Advances in geographic information systems*, 106-111. 10.1145/512161.512185

Sherman, G. E. (2008). *Desktop GIS: Mapping the Planet with Open Source Tools*. The Pragmatic Programmers LLC.

Shiklomanov, I. (1993). World fresh water resources. In Water in Crisis: A Guide to the World's Fresh Water Resources. Oxford University Press.

Silvertown, J. (2009). A new dawn for citizen science. *Trends in Ecology & Evolution*, *24*(9), 467–471. doi:10.1016/j.tree.2009.03.017 PMID:19586682

Simoncelli, E. P., & Adelson, H. E. (1996). Noise removal via bayesian wavelet coring. *Proceedings - International Conference on Image Processing*, *1*, 379–382.

Singh, V. P. (2011). Hydrologic synthesis using entropy theory. *Journal of Hydrologic Engineering*, *16*(5), 421–433. doi:10.1061/(ASCE)HE.1943-5584.0000332

Sorensen, R., & Seibert, J. (2007). Effects of DEM resolution on the calculation of topographical indices: TWI and its components. *Journal of Hydrology (Amsterdam)*, *347*(1-2), 79–89. doi:10.1016/j.jhydrol.2007.09.001

SPACIT. (2015). Retrieved on August 30, 2015, from http://www.spatialcitizenship.org

SQLite and ArcGIS. (2015). Retrieved August 28, 2015, from https://desktop.arcgis.com/en/desktop/latest/manage-data/databases/sqlite-and-arcgis.htm#ESRI_SECTION1_E4F82E510ED443088D87F6226E982EBA

Steiniger, S., & Bocher, E. (2009). An overview on current free and open source desktop GIS developments. *International Journal of Geographical Information Science*, *23*(10), 1345–1370. doi:10.1080/13658810802634956

Sui, D. Z. (2008). 2008 The wikification of GIS and its consequences: Or Angelina Jolie's new tattoo and the future of GIS. *Computers, Environment and Urban Systems*, *32*(1), 1–5. doi:10.1016/j.compenvurbsys.2007.12.001

Surya & Azhagusundari. (2014). Spatial Outlier Detection Approaches and Methods: A Survey. *International Journal of Innovative Research and Development, 3*(4), 24-29.

Sutton, R., & Barto, A. (1998). *Reinforcement learning: An introduction*. The MIT press.

Tarboton, D. G., Bras, R. L., & Rodriguez-Iturb, I. (1988). The fractal nature of river networks. *Water Resources Research*, *24*(8), 1317–1322. doi:10.1029/WR024i008p01317

Tera, F. (1980). Reassessment of the Age of the Earth. *Carnegie Institution of Washington Year Book*, *79*, 524–531.

Thomas, L. Esri. (2014). *Esri/tax-parcel-viewer*. Retrieved September 6, 2015, from https://github.com/Esri/tax-parcel-viewer

Tiemann, M. (2009). *How Open Source Software can Save the ICT Industry One Trillion Dollars per Year.* White paper. Retrieved from http://www.opensource.org/files/OSS-2010.pdf

Todini, E. (1996). The ARNO rainfall-runoff model. *Journal of Hydrology (Amsterdam)*, *175*(1-4), 339–382. doi:10.1016/S0022-1694(96)80016-3

Tokunaga, E. (2003). Tiling properties of drainage basins and their physical bases. In I. S. Evans, R. Dikau, E. Tokunaga, H. Ohmori, & M. Hirano (Eds.), *Topics in Concepts and Modelling in Geomorphology: International Perspectives* (pp. 147–166). Tokyo, Japan: Terrapub.

Turner, A. (2015, August 28). *Vector Tiles preview | ArcGIS Blog*. Retrieved August 28, 2015, from http://blogs.esri.com/esri/arcgis/2015/07/20/vector-tiles-preview/

Turner, A. 2007 Introduction to Neogeography, Sebastopol, CA: O'Reilly Media Short Cuts.

UNICEF, FAO, & SaciWATERs. (2013). *Water in India: Situation and Prospects*. Author.

United Nations General Assembly Resolution. (2004). *A/RES/58/217, International Decade for Action, "Water for Life", 2005-2015*. Author.

United Nations General Assembly Resolution. (2011). *A/RES/65/154, International Year of Water Cooperation, 2011*. Author.

Ur, J. A. (2010). Cycles of Civilization in Northern Mesopotamia, 4400–2000 BC. *Journal of Archaeological Research, 18*(4), 387–431. doi:10.100710814-010-9041-y

Vaidyanathan, P. H. (1985). Water Resources of Kerala: A Compendium of Advance Report. Kerala, India: Public Works Department, Govt. of Kerala, Trivandrum.

Valdez, S. I. H. (2000). *Efficient Estimation of Distribution Algorithms by using the Empirical Selection Distribution.* 2Centre for Research in Mathematics A.C.

Vangelo, E. S. (2015, September 3). *libst_raster_pg.dll for PostgreSQL 9.2*. Retrieved September 3, 2015, from https://geonet.esri.com/thread/114964

Vieux, B. E. (1993). DEM Aggregation and smoothing effects on surface runoff modeling. *Journal of Computing in Civil Engineering, 7*(3), 310–338. doi:10.1061/(ASCE)0887-3801(1993)7:3(310)

Walbridge, S. Esri. (2015a). *R-ArcGIS/r-bridge-install*. Retrieved September 6, 2015, from https://github.com/R-ArcGIS/r-bridge-install

Walbridge, S. Esri. (2015b). *R-ArcGIS/r-sample-tools*. Retrieved September 6, 2015, from https://github.com/R-ArcGIS/r-sample-tools

Wang, X. X. Z. (2013). Spatiotemporal Data Modeling and Management: A Survey. The University of Queensland.

Wang, D., & Cai, X. (2006). Scale and resolution effects of topographic index by 2-D continuous wavelet transform. In *Proceedings of Chinese PUB International Symposium* (vol. 322, pp. 108-114). Beijing: IAHS Publication.

Wang, Z., Batelaan, O., & De Smedt, F. (1996). *Distributed model for water and energy transfer* between soil, plants and atmosphere (WetSpa). *Physics and Chemistry of the Earth, 21*(3), 189–193. doi:10.1016/S0079-1946(97)85583-8

White, C. Esri. (2015). *Esri/arcgis-osm-editor*. Retrieved September 2, 2015, from https://github.com/Esri/arcgis-osm-editor

Wiki, G. D. A. L. (2015, August 14). GDAL. In *Wikipedia, the free encyclopedia*. Retrieved from https://en.wikipedia.org/w/index.php?title=GDAL&oldid=676030336

Wilson, J. P., & Gallant, J. C. (2000). *Digital terrain analysis. In terrain analysis: principles and applications*. John Wiley & Sons.

Wolock, D. M., & Price, C. V. (1994). Effects of digital elevation model map scale and data resolution on a topography-based watershed model. *Water Resources Research, 30*(11), 3041–3052.

Wood, E. F., Roundy, J. K., Troy, T. J., van Beek, L. P. H., Bierkens, M. F. P., Blyth, E., ... Whitehead, P. (2011). Hyperresolution global land surface modeling: Meeting a grand challenge for monitoring Earth's terrestrial water. *Water Resources Research*, *47*(5). doi:10.1029/2010WR010090

Worden, G. (2010). *Computer Security Awareness at Home.* Academic Press.

Xiaoshenga, L., & Haifenga, R. (2010). *Extract spatial association rules by the method of spatial analysis*. Academic Press.

Xu, H. Y. J. (2013). *Effective Structure Learning for Estimation of Distribution Algorithms.* Creative Commons Attribution License.

Yang, X., Wang, B., & Li, C. (2008). *Cost-based variable-length gram selection for string collections to support approximate queries efficiently*. SIGMOD. doi:10.1145/1376616.1376655

Yao, B., Tang, M., & Hadjieleftheriou, M. (2013). Spatial approximate string search. *IEEE Transactions on Knowledge and Data Engineering.*

Youngblood, B., & Iacovella, S. (2013). *GeoServer Beginner's Guide*. Birmingham, UK: Packt Publishing.

Zhang, D., Chee, Y. M., Mondal, A., Tung, A. K. H., & Kitsuregawa, H. (2009). Keyword search spatial databases: Towards searching by document. *2009 IEEE 25th International Conference on Data Engineering*, 688–699.

Zhang, D. X., Tan, K.-L., & Tung, A. K. H. (2013). *Scalable top-k spatial keyword search*. ACM.

Zhang, W., & Montgomery, D. R. (1994). Digital elevation model grid size, landscape representation, and hydrologic simulations. *Water Resources Research*, *30*(4), 1019–1028. doi:10.1029/93WR03553

Zhao, Q. (2003). Association Rule Mining: A Survey. Singapore: Technical Report, CAIS, Nanyang Technological University.

Zhou, X., Lin, D., Yang, H., Chen, H., Sun, L., Yang, G., ... Brown, L. (2002). Use of landsat TM satellite surveillance data to measure the impact of the 1998 flood on snail intermediate host dispersal in the lower Yangtze River Basin. *Acta Tropica*, *82*(2), 199–205. doi:10.1016/S0001-706X(02)00011-6 PMID:12020893

Zhou, Y., Xie, X., Wang, C., Gong, Y., & Ma, W. Y. (2005). *Hybrid index structures for location-based web search* (pp. 155–162). CIKM. doi:10.1145/1099554.1099584

Zobel, J., & Moffat, A. (2006). Inverted files for text search engines. *ACM Computing Surveys*, *8*(2), 1–56.

About the Contributors

Naveenchandra N. Srivastava is currently working as an Assistant Professor in Environment area at the Center for Energy, Environment, Urban Governance and Infrastructure Development of Administrative Staff College of India (ASCI), Hyderabad. He is a Ph.D. in Engineering Geology from Indian Institute of Technology – Roorkee, MTech. in Applied Geology from Indian Institute of Technology – Roorkee and MBA from IGNOU. He qualified NET (Joint CSIR-UGC) test. Naveenchandra N. Srivastava has joined ASCI recently in February, 2017. Earlier, he worked for Bhaskaracharya Institute for Space Applications and Geo-informatics (BISAG), Department of Science and Technology, Government of Gujarat as a Project Manager. He also worked at National Innovation Foundation, Ahmedabad as a Research Associate. His team handled many projects concerning DST (Department of Science and Technology, Government of India), MPEDA (Marine Products Export Development Authority), GSPC (Gujarat State Petroleum Corporation, NRSC (National Remote Sensing Centre), Govt. of Gujarat etc. He prepared and got approved many project proposals. He was deputed to the Institute of Seismological Research, Gandhinagar for the purpose of establishing the Institute. His team developed many utility software. He has received best paper award from ISRMTT (Indian Society of Rock Mechanics and Tunneling Technology). He was involved in development of course structure / syllabus MTech. and PG Diploma Course in Geo-informatics. He was also involved in preparation of vision documents, ISO manuals, annual report, business development plans etc. He played significant role in conducting short courses/training programs etc. at BISAG. He has edited this book titled 'Emerging Trends in Open Source Geographic Information Systems' for a USA based publisher IGI Global, an international publisher of progressive academic research. He reviewed many papers for different journals. He is well experienced in the areas of Geo-informatics, Natural Resource Management, Project Management, Engineering Geology, Geotechnical Investigation, Seismology, Open Source Software development, Java and VC++ based software development,

Project Management, ISO and ISMS Audit etc. He has guided many UG and PG students for their project/dissertation work. He has many national and international publications to his credit.

* * *

C. A. Babu is a Professor at Cochin University of Science and Technology for the last 25 years. He has published more than 60 peer-reviewed publications, two books and one book chapter in addition to a number of conference contributions. He is actively involved in research to unravel the mysteries of the nature. He worked as PI of many research projects funded by various government organizations. He served as the HoD of Atmospheric Science Department and the Director of School of Marine Sciences of Cochin University of Science and Technology.

Maria Brovelli has a degree with honors in Physics, PhD in Geodesy and Cartography. From 2001 to 2011 Head of the Geomatics Laboratory of the Politecnico di Milano. From 2006 to 2011 lecturer of GIS at the ETH – Zurich. Professor of GIS and Web Geoservices at Politecnico di Milano. Since 2011 Vice Rector for the Como Campus. Co-chair of ISPRS WG IV/5 "Web and Cloud Based Geospatial Services and Applications"; Charter Member of OSGeo; Member of the Advisory Board of ICA-OSGeo Labs Network (GeoForAll) and the Italian SIFET, Co-organizer of the NASA WW Europe Challenge; Conference chair of FOSS4G Europe 2015. In 2015, she was awarded with Sol Katz OSGeo Prize.

Blagoj Delipetrev is assistant professor at the Faculty of Computer Science, University Goce Delcev in Shtip, MAcedonia. He is currently hold the position of vice-dean.

Aida Hosseinibaghanam is a Ph.D. Candidate in the Department of water resources engineering at the University of Tabriz, where she has been a student since 2003. She completed her MSc. at University of Tabriz while her undergraduate studies was also at the same University. Her research interests lie in the area of hydro-environment. Aida has published 5 journal papers as well as 3 book chapters and over 10 conference papers.

Amira Idrees is an Associate Professor (staff member) in Information Systems Department - Faculty of Computers and Information - Fayoum University. My research area is Data mining, Opinion Mining, Business intelligence, E-Learning, and Cloud Computing.

Smart Kundassery is an Oceanographer pursuing Oceanograhic exploration of the outer Solar System. Nine among the retinue of natural satellites have the prospects of holding planetary Oceans. Among them, few exist as sub-surface waterbody sandwiched between an icy crust and solid core, as in the case of Jovian Europa. Few lie beneath an atmosphere, favoring air-sea interaction, like the Saturnian Titan. Along these lines, he pursues doctoral research at the Department of Atmospheric Sciences, Cochin University of Science and Technology. In the social front, he is an avid science communicator, felicitator and profounder of 'RiverineGiS', a geo-visual Web GiS framework (http://orgin.in) in the public domain.

Pradeep P. Lodha is an associate professor of civil engineering at GEC Valsad, India. He has been teaching the civil engineering subjects since 1995 both at under graduate and post graduate levels. He has widely published research papers in international journals and conferences and authored two books. He has been associated with R&D work in the area of hydrological modeling, water management, climate change impact assessment, geographic information system, and many allied areas.

Priya M. received her B.E degree in Computer Science and Engineering from Madras University, M.E degree from Anna University and going to pursue Ph.D. from Pondicherry University, India. She is currently working as an associate professor in Bharathiyar College of Engineering and Technology. She has presented and published 8 technical papers in International Journal and Conferences.

Masoud Mehrvand is studying as a Ph.D. student in department of Hydrology and Geohydrology at University of Stuttgart, Germany. He has earned his M.Sc. in the field of hydraulic engineering from University of Tabriz, Iran where he studied on applications of artificial intelligence-based models, Self-organizing maps (SOM) and wavelet transform incorporating remote sensing data into rainfall-runoff models.

Vahid Nourani was born in Tabriz, Iran in 1975 and received his B.Sc. and M.S. degrees in Civil Engineering from University of Tabriz, Iran in 1998 and 2000, respectively. He then continued his graduate study in Civil and Environmental Engineering in the field of Hydrology at Shiraz University, Iran and Tohoku University, Japan and was graduated in 2005. Nourani was with the Faculty of Civil Engineering, University of Tabriz as an Assistant Professor from 2005-2009; as Associate Professor from 2009-2011; as Full Professor from 2014 and with Dept. of Civil Eng., University of Minnesota, USA at 2011 and in this period, 55 Ph.D. and M.Sc. students were graduated under his technical supervision. His research interests include rainfall-runoff modeling, Artificial Intelligence applications to water resources engineering, Hydroinformatics and computational hydraulics. His researches outcomes have been published as 104 Journal articles, 2 books, 5 book chapters and more than 75 papers presented in international and national conferences. He is currently researching about incorporating remote sensing data into rainfall-runoff models.

Kalpana Ramanujam received her B.Tech degree in Computer Science and Engineering, M.Tech (CSE) degree and Ph.D. from Pondicherry University, India. She is currently working as a professor in Pondicherry Engineering College. She has presented and published 28 technical papers in various International Journals and in Elsevier publication.

Shahriar Shams is currently working as Senior Lecturer at the department of Civil Engineering, Institut Teknologi Brunei (ITB) since December 2012. Before joining ITB, he worked as an Assistant Professor in the Department of Civil and Environmental Engineering at Islamic University of Technology (IUT), a subsidiary organ of OIC. He obtained his PhD from The University of Manchester (UK) in 2008. He obtained his M.Sc. in Environmental Engineering and Sustainable Infrastructure from Royal Institute of Technology (KTH) Sweden sponsored by Swedish International Development Agency (SIDA). He has published over 40 refereed journals and conference papers. He has attended a number of short courses such as short course on Climate Change in Integrated Water Management from 4 to 22 July, 2011 at UNESCO-IHE, Delft, Netherlands funded by Netherlands Fellowship Programme (NFP), Advanced Training Programme on Integrated Water Resources Management (IWRM, Phase II) from 23 August to 14 September, 2010 in Stockholm, Sweden (IWRM, Phase IV) from 9 to 21 February, 2011 in Vientiane, Laos funded by Swedish International Development Agency (SIDA). He also participated in a Spring School on Fluid Mechanics and Geophysics of Environmental Hazards held at NUS, Singapore from 19 April to 2 May 2009. He has reviewed a number of papers under Scopus listed journals and currently an Editorial Board member in Brunei Darussalam Journal of Technology and Commerce (BDJTC). He has worked as a Consultant for European Commission Joint Research Centre (ECJRC) on two projects "Assessment of Open Source GIS Software for Water Resources Management in Developing Countries" and "Water Balance study: An application of WPS technologies". He also worked as an Environmental Specialist on a Project "Global Warming and Climate Change: A Rapid Assessment of South-Western Region of Bangladesh". His current research interest includes Integrated Water Resource Management, Hyroinformatics, Water pollution and Climate Change.

Derek Starkenburg received a Masters of Architecture from North Carolina State University, and was involved in design for ten years. I then chose to study meteorology, receiving an M.A. in earth sciences from the University of Northern Colorado, and finally my Ph.D. in atmospheric sciences from the University of Alaska Fairbanks. I enjoy using the graphic skills from design to illustrate the physical processes of our atmosphere, in order to teach and convey science to all ages.

Giorgio Zamboni obtained a degree in Computer Engineering and a PhD in Geodesy and Geomatics at Politecnico di Milano. He collaborates with the Politecnico di Milano on projects concerning geomatics topics.

Index

S

T

U

V

W

Y

www.ingramcontent.com/pod-product-compliance
Lightning Source LLC
LaVergne TN
LVHW061542290926
849162LV00004B/883